About Island Press

Island Press, a nonprofit organization, publishes, markets, and distributes the most advanced thinking on the conservation of our natural resources—books about soil, land, water, forests, wildlife, and hazardous and toxic wastes. These books are practical tools used by public officials, business and industry leaders, natural resource managers, and concerned citizens working to solve both local and global resource problems.

Founded in 1978, Island Press reorganized in 1984 to meet the increasing demand for substantive books on all resource-related issues. Island Press publishes and distributes under its own imprint and offers these services to other nonprofit organizations.

Support for Island Press is provided by The Geraldine R. Dodge Foundation, The Energy Foundation, The Charles Engelhard Foundation, The Ford Foundation, Glen Eagles Foundation, The George Gund Foundation, William and Flora Hewlett Foundation, The James Irvin Foundation, The John D. and Catherine T. MacArthur Foundation, The Andrew W. Mellon Foundation, The Joyce Mertz-Gilmore Foundation, The New-Land Foundation, The Pew Charitable Trusts, The Rockefeller Brothers Fund, The Tides Foundation, Turner Foundation, Inc., and individual donors.

The Wisdom
of the
Spotted Owl

The Wisdom
of the
Spotted Owl

Policy Lessons for a New Century

Steven Lewis Yaffee

ISLAND PRESS

Washington, D.C. • Covelo, California

ISLAND PRESS is a trademark of The Center for Resource Economics.

Library of Congress Cataloging-in-Publication Data
Yaffee, Steven Lewis.
    The wisdom of the spotted owl : policy lessons for a new century / Steven Lewis Yaffee.
       p.   cm.
    Includes bibliographical references (p.   ) and index.
    ISBN 1-55963-203-8 (cloth). — ISBN 1-55963-204-6 (paper)
    1. Spotted owl—Northwest, Pacific.   2. Forest management—Environmental aspects—Northwest, Pacific.   3. Forest policy—United States.   4. Wildlife conservation—Government policy—United States.   5. Habitat conservation—Government policy—United States.
    I. Title.
    QL696.S83Y34   1994
    333.95'16'0973—dc20   93-48897
                                                                  CIP

Printed on recycled, acid-free paper

Manufactured in the United States of America

10 9 8 7 6 5 4 3 2

Contents

Preface

One of the lessons of ecosystem science is that change across landscapes is more a matter of gradients than of clearly defined boundaries. It is difficult to determine exactly where one ecosystem ends and another begins. The same is true with dominant human values or ideas. It is often difficult to realize when important thresholds of change are crossed, particularly by those involved in traversing them. Historians may look back on events or social movements as key determinants of major change, but it is often difficult to view them as such while they are ongoing. The view of the forest is obscured by that of the trees.

I believe we will look back on the spotted owl controversy as a watershed event in American resource and environmental policy. It represents a transition point in biodiversity policy, shifting our understanding from a localized species perspective to one of larger landscapes. The case personifies the end of forest management based on exploiting presettlement forests, and the need for a shift to a set of policies grounded in concepts of long-term environmental and economic sustainability. It bears witness to the need to change organizations and decisionmaking processes, and indicates the problems and opportunities attending a change in national politics and a changing world economy. How we manage these transitions has considerable meaning for the quality of human and nonhuman life in the next century.

The controversy also represents a portal into a new century of conservation. One hundred years ago, the Progressive Conservation movement responded to a growing awareness that public natural resources were overexploited and degraded, and pushed for a major change in resource policy. While previous federal policies had aimed at disposing of federal lands into private hands, turn-of-the-century policies focused on reserving and managing these lands for public purposes. Agencies like the U. S. Forest Service adopted concepts of management and decisionmaking that were appropriate and functional for much of this century. Yet the history of the owl controversy and evidence from other areas of resource policy suggest that the realities of management today do not live up to the ideals of the reformers of the late nineteenth century, and some of the theories of the first hundred years of public resource management have become inappropriate as the world has changed. Indeed, resource managers are asking many of the same questions posed a hundred years ago: To what ends should public natural resources be put? What tools of management are appropriate? What skills and what kinds of leaders should agencies develop? What is the appropriate relationship of agency experts to the society around them?

Having recently celebrated the 100th anniversary of the establishment of the national forest system, the Forest Service (FS) is at a crossroads that is replicated in many other resource agencies. As a new generation of leaders take their place in these agencies, an unparalleled opportunity exists to craft new styles and directions. The wisdom provided by the spotted owl case can assist decisionmakers as they seek to redirect agencies and forge directions for the next century, and can help citizens and scientists understand the role they might play in this process of change.

I first got involved with the spotted owl case in 1988, when a chance meeting at a Washington, D. C. conference led to a conversation with Hal Salwasser, then Deputy Director of the FS's Fisheries and Wildlife staff. At the time, the FS was actively and painfully involved in trying to set direction for owl management in its Pacific Northwest region, and in Salwasser's view, the issue would only expand, involving federal and state agencies, policies and politicians, and interest groups of all shapes and sizes, in a major controversy. In his view, someone ought to look at the evolving situation and draw lessons out of it, so that hopefully the FS would be able to avoid or better manage such situations in the future. I was faced with an upcoming sabbatical, and was looking for a new situation to explore, and somewhat naively agreed to take a look at the case.

As I got into the case, I was amazed by its long history and complexity. Like the structure of an onion, it had many layers of meaning. Was it an endangered species case? Was it a forest management case? Was it a case involving problematic organizational behavior? Was it a controversy that had its origins in national environmental policy? Was it a case study of regional or national politics? Or was it an issue rooted in conflicts in values, and how values are shaped in our society? The case was always something more than owls versus jobs, but its meaning was not obvious. Five years later, the spotted owl controversy is still unresolved, though it has grown from a regional issue of the Pacific Northwest to a national battle over power, policy, and priorities.

The case lends itself to potent imagery: the logger wearing a "Spotted Owl Hunter" cap, with a bumper sticker proclaiming, "I like spotted owls—fried"; the aerial view of a managed forest, evidencing a patchwork of forest fragments in a quilt of clearcuts. On one memorable day of my research, I spent the late afternoon hooting for owls in the Olympic National Forest (a quest that was rewarded by an interview with a mother and juvenile owl) and the evening in Forks, Washington. The display windows of a vacant storefront in Forks contained artwork from local school kids describing their fears of the coming change in forest management policy: "My Dad's a logger. He cares about the forest. Leave us alone." Adjacent to the downtown stood a fifteen-foot-high cross with plastic owls roosting on the crossbar. In front lay a grave with a sign, "Here lies the hopes and dreams of our children."

The juxtaposition of the two sets of images—old growth and owls, community hopes and fears—was startling to me, and I wanted to assign blame somewhere, but it was not easy to do so in a simple, black and white manner. Prior to my visit to Forks, one Washington State environmentalist described Forks as "the town that evolution forgot," yet I found Forks not unusual, but reminiscent of other small towns and not that different in some ways from the suburban area in which I grew up. As a Baby-Boomer environmentalist, my underlying bias was that the conflict was about a generational change in values, and that as the older generation moved out of decisionmaking positions, the problem would go away. But I was surprised, when moving from interviews with environmental group staffers to timber industry officials to agency staff members, that the set of individuals in all three looked much the same, and indeed much like myself: same age, similar backgrounds, same socioeconomic class. None appeared incompetent. Most were sincere in what they argued and what they believed. It was hard to assign blame in any simple way.

As a symbol, the owl is ambiguous. In cartoons and children's books, we anthropomorphically associate owls with wisdom, in part because of their large-eyed studious appearance. Yet much Native American lore views owls as evil, moving silently through the night to carry off living creatures—a harbinger of death. Is the owl wise or evil, and does the owl controversy precede important social changes or the death of important social values? My own cut on this question is to view the controversy with optimism. While this book details numerous failings of individuals, organizations, and institutions, it is also clear that change does occur over time, and important change requires conflict and chaos to sort itself out. I believe that significant changes are needed in resource management policies and organizations, and that the trauma of this case can encourage such changes to take place.

While a number of books will be written about the spotted owl controversy, my primary interest lies in what it tells us about how groups of different people can come together to collectively make important social choices, and how such decisionmaking processes can be improved. I believe the owl case illustrates problematic tendencies in human behavior at all levels of social organization—individual, group, institution, and society. Indeed, the most frightening aspect of the owl case is not that agencies and individuals made some bad choices in dealing with the owl issue; rather it is that similar organizations and individuals dealing with comparable choices would have a good chance of making the same mistakes again. It is important to learn from the past to do better in the future. My hope is that the lessons of this book can help individuals recognize problematic behavior and deal with it, so that the owl controversies of the future will be dampened or better managed when they occur.

This effort was supported by many individuals and organizations and I want to thank them for their assistance. Hal Salwasser got me into this, and while his involvement in the case was described by different parties in various ways, all would attest to his intelligence and sense of perspective, and I owe him a debt of gratitude. Support for portions of the research underlying this book was provided by the USDA-Forest Service, the National Fish and Wildlife Foundation, the Conservation Foundation/World Wildlife Fund, and The University of Michigan, and I appreciate the opportunity that these institutions provided me.

A number of individuals reviewed draft materials, and provided comments that helped guide the editing process, and I very much appreciate the attention they gave to a very long manuscript: Elise Jones, Lenore Garcia, Steve Kellert, Hal Salwasser, and two anonymous reviewers. Barbara Dean, executive editor at Island Press, remained a consistent supporter of this book, and her ideas and encouragement were very helpful. Other individuals assisted in the gathering, organizing, and interpretation of materials: Todd Barker, Esther Bartfeld, Alan Clark, Bill Crown, Paul DeLong, Laurel Horne, Scott Mills, John Rosapepe, John Shenot, Heidi Sherk, Martha Tableman, and Jennifer Thomas. To all these individuals, I would like to express my appreciation for their assistance and encouragement. Numerous governmental and nongovernmental staff members were interviewed at length about various elements of the owl story and what it meant, and their perspectives and honesty were extremely helpful. While

I criticize agency decisionmaking, I remain impressed by the sincerity and competence of most of their staffs. I hope that this book is helpful to these individuals as they try to make sense out of their own experiences.

Just as the owl case reflects its sociopolitical context, I too have benefited from a set of individuals and forces that have influenced my thinking and provided encouragement and support. The faculty and students of the School of Natural Resources and Environment at Michigan are a special group, evidencing commitment to making the world a better place, and I appreciate the environment that they provide me. The strength and capabilities of the graduate students in resource policy and behavior in particular provide me with my greatest reason for optimism about the future.

To my daughters, Anna and Katie, who spent their early years drawing pictures of spotted owls and wondering why it took so long to write a book, and to Julia Wondolleck, my partner in all things, and from whom I have benefited in innumerable ways over the past fifteen years, I want to express my thanks for the time they gave to me, and the joy they put in my heart.

Finally, to my parents Shirley and Philip Yaffee, two retired federal employees, whose dinner table conversations of office and Washington, D. C. politics, no doubt planted the seeds of my interest in government and bureaucracies, I want to express my thanks. Neither is particularly environmental, yet the sense of caring, compassion, and humor that they embody is, I believe, at the heart of our relationship with the environment and with each other. This book is dedicated to them.

Introduction

Early each summer in Portland, Oregon, civic leaders hold a tourism and economic development event called the Rose Festival. One traditional component of the festival is a parade that features floats, marching bands, and the like. As a major influence in the region due to its extensive land-holdings, the United States Forest Service has traditionally participated in the Festival parade by sending individuals dressed as Smokey Bear and Woodsy the Owl. In the summer of 1989, however, the Bear and the Owl did not march. The Forest Service reportedly had received death threats against the pair and declined to send any of its employees into combat.

As anthropomorphic symbols of environmental concern, Smokey and Woodsy can hardly be viewed as threatening, but in the summer of 1989, as symbols of the Forest Service and visible reminders of the wildlife that occupy public forest lands, they personified an ongoing and extremely heated controversy over management of the national forests. On the surface at least, the controversy pitted the habitat needs of the northern spotted owl, one of three subspecies of spotted owl that ranges from northern California through Oregon and Washington, against the raw material supply needs of the timber industry, a significant component of the regional economy.

By the spring of 1989, court injunctions effectively had stopped the timber sale programs of the Forest Service and the Bureau of Land Management in Oregon and Washington. Environmental groups were forecasting total victory in their battle to save the old growth forest that provides the majority of habitat for the owl. Timber industry representatives were forecasting economic disaster of the magnitude of a quarter of a million lost jobs. And a further court ruling suggested that the U. S. Fish and Wildlife Service would most likely list the northern spotted owl as a federally recognized threatened or endangered species, a move that would tend to limit options for resolving the conflict. Polarization was extreme, basic values were threatened, and hostility and anxiety boiled out of the normal processes through which our society manages conflict.

Expressions of violence, such as those levied against poor Smokey and Woodsy, were not unusual. They were both heartfelt and frightening.

Evolution of the Spotted Owl Controversy

The spotted owl had first winged its way into public view more than fifteen years earlier. The 1970s were a time of increasing public interest in nongame wildlife values, and researchers at Oregon State University began to press for a change in public forest management practices to protect owl habitat. The issue grew and evolved through the 1980s. Forest Service (FS), Bureau of Land Management (BLM), and Fish and Wildlife Service (FWS) studies led to administrative decisions reluctantly made that were appealed by interest groups through administrative appeals processes and the federal courts, and often overturned. Few on either side of the issue were happy with the agencies' decisions. For example, only a handful of the more than 40,000 public comments received supported the Forest Service's draft decision on owl management in the mid-1980s.

In the year following the summer of 1989, timber and dependent community interests, horrified by the potential impacts of the court injunctions on timber supply and regional economies, "appealed" the judicial decisions to the U. S. Congress, an arena in which they had strong supporters. Agency officials, equally concerned about the uncertainty surrounding their resource management programs, and fearful of their loss of control over decisions determining the future of agency lands and personnel, also looked to Congress for a way out. But the most that the Congress could provide was a temporary, one-year solution in which timber was provided to mills in exchange for some level of protection of old growth stands. Proponents of more permanent solutions were stalemated and their proposals died a slow political death.

By the end of 1991, Congressional leadership, the White House and Executive Office, Cabinet officers, Oregon and Washington governors, and numerous staff members of the FS, BLM, FWS, affected state agencies, and interest groups of all sizes and shapes were heavily involved in the issue, often to the exclusion of other important business, yet no clear outcome was in sight. A plan framed by an interagency team of scientists was adopted by the Forest Service as a credible way out of the controversy, but it faced hostility from the Executive Office and Cabinet secretaries who tried unsuccessfully to find an alternate, more politically palatable solution.

After being caught with their political pants down, FWS leaders listed the owl under the provisions of the Endangered Species Act (ESA), turning up the level of anxiety and chaos several notches. Opponents of owl preservation called for a decision by the so-called God Squad, a high-level group of federal administrators which has the power to exempt projects from the provisions of the ESA, a decision that would for the first time since passage of the ESA elevate economic interests over endangered species concerns. Others called for dividing up the national forests into single purpose units, with some preserving old growth forests, and others being dedicated to tree farming. A mediagenic controversy, the owl/old growth issue received coverage by a variety of media outlets including a *Time* magazine cover story.[1]

An army of researchers combed the woods of the Pacific Northwest in search of owls, old growth, and redemption, and found some of the former but little of the latter. One researcher claimed that the spotted owl issue was sounding the death knell for endangered species elsewhere in the country, because public attention and expertise were being diverted from them. Another described the persistence of the issue as a "tar baby": "Once you come close to it, you're stuck and can never get away." Another described it as a black hole into which all energy and action disappears, never to reappear. More time was spent by more individuals representing more institutions certainly than any other endangered species issue, and perhaps more than any other contemporary resource or environmental issue, yet a definitive solution to the problem remained elusive.

The Spotted Owl Controversy as Historical Inevitability

In many ways, a controversy of this magnitude and tenacity was inevitable somewhere on federal public lands, but most likely on national forests in the Pacific Northwest. Historical inertia alone was enough to set the stage. Responding to public demands for lumber to build the American suburban dream, the Forest Service in the Fifties, Sixties, and Seventies evolved from custodian of national forest lands before World War II to a major timber producer, particularly as the timber base declined on state and private lands. This new role was in many ways reinforced by a traditional Forest Service self-image as a technically based, "Can Do" organization. In the Pacific Northwest, a region with forests of immense value, timber objectives outcompeted all other forest uses and agency officials tended to view nontimber resources as adjuncts of the timber management program.

The deification of timber in the Northwest was a self-reinforcing phe-
nomenon. As timber outputs grew, more agency staff were employed in
timber management, more local mills became dependent on the Forest
Service timber sale program, and local and regional economies, and hence
lifestyles, were grounded increasingly in public forest timber.[2] Political
forces generally follow social realities, and political leaders in Oregon,
Washington, and the Congress increasingly pushed for an expansive timber sale program. Through much of the period, Forest Service leadership
went along for the ride, for big timber meant big budgets, and timber
production was not inconsistent with the sustained yield–use orientation
that was well-established in agency traditions.

At the same time, post-War affluence and access created a rising public
interest in federal public lands for their recreational and other noncommodity values. Divorced from the locations where raw materials were
harvested to produce the goods that they were consuming, an increasingly
urban population viewed national forests as aesthetic, recreational, or
even spiritual resources. Their values shaped, and were in turn shaped by,
the growth and institutionalization of a large set of environmental interest
groups that by the late 1980s had become remarkably effective at advocating their interests. The Forest Service and other involved agencies were
not divorced from these shifts in values. As their workforces diversified
(both in terms of expertise and socioeconomic characteristics) and individuals who acquired their values in the 1960s and 1970s came of age in
the organizations, agency direction and methods were increasingly questioned from within.

The clash between these two sets of historical, economic, and political
forces in the spotted owl case was probably inevitable. It was a clash of
values only partly drawn along economic class or other sociological lines.
Other skirmishes had foreshadowed the battle. Wilderness disputes
clearly were grounded in these same value conflicts, but they had traditionally been easier to resolve by giving higher elevation, less productive
lands to wilderness proponents. But the old growth forests in the Northwest that were the best habitat for the owl were primarily low elevation,
very productive lands that were also the best habitat for loggers.

Other endangered species battles had been fought and resolved, but
they had involved largely site-specific, fairly endemic species facing specific development projects. The snail darter might require that a reservoir
not be completed in a small corner of east Tennessee, and the Mississippi
sandhill crane might necessitate moving one interchange of interstate
highway in southern Mississippi, but the spotted owl could require the

elimination of timber harvest activities over 11.6 million acres of land.[3] The character and magnitude of the issue, and the underlying clash of values, suggested at the start a battle of monumental proportions, one that would have precedent for and influence on a broader range of social actions and decisions.

The Spotted Owl Controversy as an Indicator of the Effectiveness of American Decisionmaking Processes

While the clash might have been inevitable, our public agencies and decisionmaking institutions contributed to, rather than diminished, the magnitude and polarization of the issue. Indeed, while the spotted owl controversy has been portrayed as an issue juxtaposing the biology of the owl against the economics of the region, the controversy raises an equally compelling set of concerns about how we make choices that are in our collective best interest. Just as the spotted owl has been used as an indicator of the status of the old growth forest ecosystem, the political controversy that has surrounded it can be seen as an indicator of our ability or inability to make tough public policy choices.

Clearly our ability to make an effective, binding decision on the issue did not improve over the fifteen years of controversy in direct proportion to the magnitude of effort and activity that went into it. By the mid-1970s, we had a pretty good idea that the northern spotted owl was more than likely rare, and probably dependent for most of its habitat on old growth forest in blocks of significant size. Certainly the experts were confident enough in their management prescriptions that, had this been a species that required less controversial action, such action would have been taken. By the mid-1970s, experts were also warning of a coming timber supply problem in the Northwest, as old growth was liquidated from public and private inventories and the region shifted to second growth production. But little was done to deal with the implications of these projections, and fifteen years later, the same questions were still being asked. Indeed, at the first meeting of the Oregon Endangered Species Task Force in June 1973, it was recognized that a map of old growth on FS and BLM lands was needed to make effective management recommendations. The Forest Service and BLM representatives at the meeting agreed to bring this mapped information to the next meeting of the group. Sixteen years later, a definitive map of old growth on federal lands was still not available.

While this dispute was intrinsically difficult to solve (and all public policy issues have a gestation period allowing issues to be defined, interests legitimized, and political support mobilized) most agencies and decisionmaking processes could have done better. The Forest Service and BLM were fairly myopic, bound by traditional objectives and longstanding operating practices, and hence slow to respond. The Fish and Wildlife Service was politicized and ineffective through much of the evolution of the issue. The interest groups created as much smoke as light through their lobbying activities. The Congress avoided with great effort many of the important questions and difficult choices. While difficult choices are inherently messy in a democracy, what happened certainly was not optimal. Could we have done better? Were we well served by our public institutions? Is there a way to deal more effectively with similar disputes in the future?

The Spotted Owl Controversy as a Symbol of the State of Natural Resource and Environmental Policy in the 1990s

The spotted owl controversy also provides insight into the underlying characteristics of many current environmental and resource policy issues, and what is needed to deal with them in an effective and enduring manner. The owl dispute is like most present-day environmental controversies, in that it involves: a multiplicity of interests and subissues in what often appear to be fairly simple issues, a significant amount of technical uncertainty and complexity, and a decisionmaking environment that is often portrayed as "zero sum," pitting those arguing for fairly intangible concerns against the advocates of more tangible ones.[4] The conflict also affects a large amount of geographic and intellectual landscape, requiring transboundary, integrative solutions from a political and administrative system that is inherently fragmented. It requires us to make choices that balance short-term costs against long-term benefits, and bind ourselves to a state of affairs that is probably rational in the long run, but may be hard to justify in the short term. Implicitly, the controversy also defines our society's obligations to future generations of humans and other life, and affects their quality of life many years down the line.

These characteristics are equally true of many other major environmental and resource issues. They repeat themselves in other endangered species–development conflicts, including the battle that pits the Pacific salmon against water users of the Pacific Northwest, and other land management controversies, such as the battle over oil development in the Arctic National Wildlife Refuge. Conflicts over the siting of new solid waste landfills and low-level radioactive waste sites express these characteristics, as do global-scale policy issues such as the battle over reducing CO_2 emissions to limit their impacts on global climate.

In a society that has significant slack resources, many of these competing demands could be satisfied with a minimum of conflict, but public policy choices increasingly are characterized by limited slack. A fixed resource base faced with rising demands for a variety of goods and services, a grim public sector fiscal situation underlain by the deficit spending of the past decade, and an economic future that is less likely to yield a state of affluence that characterized environmental and resource policy in the past four decades, all limit our ability to negotiate and/or spend our way out of pressing public problems. Indeed, much of the conflict that pervades current forest policy is a result of conflicting demands that are hard to reconcile given very little slack left to do so. While the owl dispute was not inherently a zero sum game in the early 1980s, by the end of the decade, few options existed to satisfy the expanded set of claims on public resources. Just like the progression in the owl case, environmental and resource policy decisions will be tougher to make in the next decade than they have been in the past forty years.

Policy conflicts are more manageable in a society that has a fairly stable coalition of interests determining overall direction, or strong leadership that is able to articulate a vision and build support for it. But as the owl conflict illustrates, neither a stable coalition nor strong leadership characterizes the current policy environment. The power to determine the outcome of American resource and environmental policy controversies is fragmented across a broad set of interests in society, and no single set of interests tends to dominate. While it was largely due to the efforts of environmental groups that the owl received protection at all, the rise of interest group politics and the decline of more holistic partisan politics has contributed to the fragmentation of public purpose and direction. For a variety of reasons, few elected officials have articulated a compelling vision for the future, nor have executive agencies like the Forest Service been effective at either building a political consensus or imaging a credible future. As a result, conflicts like those underlying the spotted owl case yield enduring impasses, rather than stable futures.

If public values are clear and consistent, policy decisions are more easily made, but present-day American values are anything but clear and consistent. In many ways, the owl dispute represents a conflict between a variety of substantive and procedural values that are central to many other policy disputes. What is the appropriate balance between economic growth and environmental protection objectives? How should societal value be measured and progress evaluated? To what extent should nonhuman lifeforms and future generations be considered in current decisionmaking? What should be the role of experts versus individual citizens in determining public choices? Should bounds be placed on the amount and character of participation by interest groups? What is the role of administrative agencies? What is the obligation of the government to losers in policy battles? These and other questions underlie the owl dispute and other environmental conflicts. How they are resolved in many ways determines the character of our society as expressed over time.

By and large, we resolve these kinds of fundamental choices iteratively, through small decisions on particular issues, and the way in which the owl dispute is ultimately resolved has a great deal to tell us about the state of American values in the 1990s. Climbing a mountain is partly a matter of putting one foot down after the other, and sooner or later a path is made, hopefully toward the summit. The many small choices made in the context of resolving the owl dispute have significant implications for a variety of societal ascents toward the future, including:

- *The role and character of the major resource management agencies in the United States, including the USFS, BLM, and the Fish and Wildlife Service.* Is multiple use technically and politically viable? What kinds of expertise are needed by these agencies? What types of leaders are required to deal with the challenges of the 1990s and beyond? What are appropriate organizational management styles?

- *The effectiveness and credibility of the USFS.* The owl controversy clearly is a watershed event for the Forest Service, hastening an ongoing process of change and disturbing a carefully cultivated, stable organizational existence. Throughout the later years of the controversy, agency leaders looked like pilots who were desperately seeking a landing field on higher ground, but unsure of where to look for one. Are there ways to rebuild the credibility and legitimacy of the agency? Can the traditional values of the agency be retooled for the challenges of the times? Are there new values and approaches to land management that must be adopted?

- *The character and significance of American endangered species policy.* There is an ongoing debate about the effectiveness and practicality of the approach to plant and animal protection established through the Endangered Species Act. Is the ESA flawed? How should such choices be made? Does a move toward concern with a broader notion of biological diversity supersede and overrule a single-species regulatory approach? What is the best way to promote species protection while allowing other important human activities to take place?

- *The methods of information collection and analysis used to inform important policy choices.* The ability to resolve the owl dispute was undermined by a lack of high quality information with which to make choices. Are there new approaches to collecting, organizing, and evaluating information that can assist decisionmakers in understanding problematic situations and making wise choices? What is the role of scientific versus political inputs into decisionmaking?

- *The mechanisms by which important societal decisions are made.* Can adversarial processes be reformed so that the benefits of collaborative problem-solving are realized? Can new decisionmaking institutions be created that are less problematic than those that gave us the owl drama?

- *The organization and balance of power between interests in society.* How extensive should be the rights of the public to challenge agency decisionmaking? Who will have a greater impact on public policy choices: Environmental or economic development interests? Political executives or career bureaucrats? Executive branch or legislative decisionmakers? Managers or scientists? State and local interests or national-scale concerns?

- *The underlying sets of values held by the American public as defined and pursued by governmental institutions.* What is our obligation to future generations of humans? How do our responsibilities as a member of the global community affect domestic policy issues? What is our responsibility to nonhuman life, present and future? Do we have the ability to implement democratic ideals and make difficult choices in an effective and just manner?

While small steps often determine overall direction, climbing a mountain is unlikely to be successful without a plan or an overall vision about how to reach the destination, and the owl case suggests the danger of simply muddling through. Putting one foot blindly in front of the other is as likely to lead an individual to the bottom of a crevasse as to the summit. In many ways, the choices that agency and elected leaders made in the owl case led us blindly into the tough choices of the early 1990s, rather than guiding us to the higher ground of effective and enduring decisions. Fortunately, the owl case also suggests some overall policy themes that must characterize resource and environmental decisionmaking if we are to avoid an endless repeat of the spotted owl drama and tragedy. In the dust stirred up by the owl case, there are patterns and themes that can move policymakers and managers toward an effective future. We are not doomed to relive the history of the spotted owl case like some kind of political and organizational purgatory.

What Have We Learned?

This study expands on the themes laid out above, and seeks to draw policy-relevant lessons from an examination of the historical evolution of the spotted owl issue. It seeks to provide insight into the following questions:

- How did we get to the situation of extreme polarization, ineffectiveness, and misinformation that existed in the late 1980s and early 1990s?

- What underlying themes help to explain the development of the issue and the inadequate response of agencies and decisionmaking processes?

- What do these themes imply about changes that are necessary in our resource policy institutions to enable them to better deal with the issues of the 1990s and beyond?

The study simultaneously raises some broader public policy questions: What are the appropriate bounds of political input into science-based decisions, and of science in policy choices? How should we deal with technical uncertainty when faced with the need to make important choices? What are the obligations of government agencies to dependent interests? How can we best bind ourselves to long-term objectives in the face of short-term incentives that lead us astray? While the study focuses in many places on the behavior of the U. S. Forest Service as it dealt with wildlife policy, the story of the FS is one repeated in many other local, state, and federal agencies as they implement a wide variety of resource and environmental policies. The lessons of the case are relevant to this broader set of agencies and policy areas.

While this study is critical of the past behavior of a variety of individuals, organizations, and decisionmaking processes, its purpose is not to cast blame for the failures of decisionmaking in the spotted owl case, for all of us who simultaneously use wood products and value wildlife have contributed to the controversy. Indeed, many of the behaviors and decisionmaking biases that are evident are rational in the short term or for a single individual or organization. Collectively or over time, however, they may be inappropriate for the greatest good of the greatest number for the longest period of time. The study's purpose is to draw lessons from the history of the owl case that will enable us to better understand the realities and idiosyncrasies of our public decisionmaking institutions, and improve them where possible.

It is important to draw out such lessons not only because of the immense amount of human and organizational resources that has been spent over the course of this controversy, but also because these types of disputes are likely to be more prevalent in the future. Disputes over large areas of habitat, brought by multiple and conflicting interests with comparable power to influence policy direction, mediated by agencies with imperfect information, where science has an important but imprecise role, will increasingly characterize environmental policy decisionmaking. It is important to do the best we can in dealing with and learning from these conflicts.

It is also important to learn from our past decisionmaking approaches because the nature of the choices in the future will be inherently tougher. The context of resource and environmental decisionmaking discussed in the previous section is not one that foreshadows easier choices down the line. Nor is the early experience from touted reforms, such as the Forest Service's New Perspectives program, and the federal government's efforts to solve its fiscal problems, particularly hopeful. In addition, an increasing global population with rising expectations about what constitutes basic needs will make unprecedented demands on natural resources. The fact that the political world of the 1990s is much less stable than the geopolitical landscape of the past 45 years makes hard choices more difficult to make, and even more difficult to enforce.

Finally, it is important to draw lessons from our experiences in the past because a window of opportunity is open today for the proponents of informed change. Out of conflict sometimes comes necessary and important change, and that is possible with our public land management institutions. Political alliances guiding public forest management that had been fairly stable over the forty years after World War II no longer have absolute control over policy direction, meaning that new arrangements and direction are possible. In addition, the science base for wise decisionmaking has improved, and the values base of the agencies is much more diverse and in flux. These and other factors suggest that there is a significant set of opportunities for needed change in agency styles and decisionmaking arenas. What exists is clearly not a blank slate. Its shape and texture are defined by the past as well as by present-day administrative realities. But there is, at minimum, some space to inscribe new themes and directions so that we can all better meet the challenges of the 1990s and beyond.

A Roadmap to the Book

Again, this book highlights lessons to be learned from the evolution of the spotted owl controversy. The research that is drawn on to frame these conclusions included extensive review of historical documents, observation of Congressional and administrative proceedings, and more than one hundred interviews (ranging from one to three or more hours) with federal and state agency officials in Washington, D. C. and the Pacific Northwest, and interest group representatives from both timber and environmental concerns.[5]

Part I describes the evolution of the spotted owl controversy. It purposely provides greater detail on the history of the case before it achieved national prominence because the birth and adolescence of the controversy are not well known or understood, while its adulthood was front-page national news. In many ways, it was the developmental stages of the dispute that resulted in the intractable, seemingly unsolvable controversy so prominent in the early 1990s. It is important to understand why such issues develop and endure in our political system, in order to offset the likelihood that other disputes will attain similar prominence and notoriety.

Chapter 1 describes the origins of the controversy, from its conception in the values and styles of the Forest Service and the Bureau of Land Management in the 1940s, 1950s, and 1960s, to its birth in the scientific research and administrative choices of the 1970s. Chapters 2 and 3 discuss the efforts of the agencies in the early to mid-1980s to muddle their way through the issues underlying the owl controversy, at the same time that the scientific base for decisionmaking was expanding, and environmental interest groups were becoming more effective at pursuing their interests through the courts. Chapter 4 describes the Forest Service's last chance to resolve the issue administratively through the preparation of an impact statement and decision on owl management, published as a final decision in December 1988. It also chronicles the expanding involvement of the Fish and Wildlife Service as the owl became an endangered species issue as much as a forest management issue. Neither the FS, the BLM, or the FWS were able to control the unraveling issue, and Chapter 5 describes the chaos that ensued in the early 1990s as Congress debated spotted owl, timber supply, endangered species, and ancient forest legislation; several groups of interagency experts were convened to find a way out of the conflict; the Bush administration sought a minimally credible political solution to the controversy; the Supreme Court heard arguments related to the case; the so-called "God Squad" met to decide the significance of the owl; and the Clinton administration tried to find a way out.

Part II identifies and discusses underlying themes that help explain the historical evolution of the owl case and suggest the state of resource policy and management in the early 1990s.

Under any circumstances, the spotted owl dispute would have been particularly difficult to resolve and would tend to persist. Chapter 6 explores the characteristics of the owl issue that makes this so. Chapters 7 and 8 discuss the effectiveness of American decisionmaking processes at dealing with these kinds of issues. Chapter 7 examines the underlying characteristics of American policy processes and the problems they cause for resolving complex, science-based issues like the spotted owl case. Chapter 8 describes the incentives that a variety of actors in the policy process face, and how they affect the behavior of these individuals and groups. Many of these behaviors exacerbate the underlying problems in our policy processes. Forest policy of the 1970s and 1980s helped fuel the controversy and did little to assist its resolution. Chapter 9 evaluates the efficacy of public lands and endangered species policy as they guided decisionmaking on the owl issue, and examines the particular effect that the Reagan and Bush administrations had on the implementation of natural resource and environmental policy. Since the responsibility for public land management resides with the federal agencies, their ability to do their job with authority and sensitivity to public values is critical to avoiding future owl-like battles. Chapter 10 focuses on the organizational values, traditions, and culture within the FS that tended to downplay the significance of the owl issue and deal with it in a way that was counterproductive in the long term.

Part III moves toward a prescription for action, so that American resource policies and agencies are more effective in dealing with the issues of the 1990s and beyond.

In outlining reforms, it is necessary to have a robust understanding of the environment in which proposed changes will take place, and Chapter 11 describes the organizational and political context for change in the 1990s. While the context for change is problematic, there are reasons to be optimistic that necessary change can occur, and the chapter concludes by reviewing these reasons. Our public agencies and choice processes can do better, and Chapters 12 and 13 identify a set of specific action items for agencies, decisionmakers, and citizens, so that we are better positioned to deal with the challenges of the future. Chapter 12 focuses on ways to improve organizational management and decisionmaking processes, while Chapter 13 takes a broader cut and defines a set of policy reforms needed to update and adapt natural resource policy for the challenges and opportunities of the future.

ONE

The Evolution of the
Spotted Owl Controversy

1

The Birth of a Controversy: 1945–1977

While the spotted owl controversy matured as a major public policy dispute in the mid-1980s, its origins lie in the context of the styles and objectives of forestry and forest policy as they developed in the 1950s and 1960s, and in the change in values and national politics in the 1970s. Just as our individual behavior is guided and constrained by the genes of our ancestors, public policy choices are influenced and often defined by historical trends. In the spotted owl case, battles in the courts and the Congress in the 1980s and early 1990s reflected in part decisionmaking styles and patterns of behavior that had been established many years before. These traditional behaviors included:

- An overwhelming adherence to timber production as the primary organizational objective of the Forest Service and the Bureau of Land Management in western Oregon and Washington;

- A tendency to view resources other than timber, such as wildlife or recreation, as either secondary or adjuncts of the timber management program;

- A proclivity to try to find solutions to difficult choices through elaborate technical analyses and planning processes whether they were warranted or not; and

- The development of a FS organizational image and style as a tightly controlled, "Can Do" agency that ironically made change in direction more difficult.

For the Forest Service, many of these styles were set down as fundamental operating principles by Gifford Pinchot, its creator and first Chief at the turn of the century,[1] but they were expanded upon and locked into public policy in the period following the second World War. While the basic reason for establishing the national forest system was to ensure future timber supplies,[2] not a lot of timber was cut on national forest land before the 1940s. The national forests were by and large sleepy backwaters under little pressure to produce timber because there was an ample supply coming from private lands. Indeed, the Forest Service was under pressure to keep national forest timber off the market so as not to undercut the prices private companies could get for their timber. Less than 2 billion boardfeet (bbf) of timber was cut per year on all national forests before 1940, even though total domestic softwood lumber production exceeded 30 bbf annually in the 1920s.[3]

After the second World War, rapidly rising demand for new homes coupled with declining supplies of timber from private lands led to pressures to open up the vast timber storehouses that the national forests represented, and the Forest Service responded. From the 1940s through the mid 1960s, timber sold from the national forests rose from about 2 bbf annually to approximately 12 bbf.[4] The boom generated revenues for the federal treasury and helped to build state and local economies because a portion of timber receipts is paid to states and localities to support public services such as roads and schools.

The post-War timber boom signified a rebirth for the Forest Service, an agency that had largely served a custodial function in the previous half century. Agency budgets and staff increased, and a strong national constituency emerged to lobby for agency programs, at least as long as they centered on an active timber sale program. Local, state, and federal politicians increasingly recognized the political value of public timber, since it supported local jobs and schools, and helped to satisfy the American dream of lumber-hungry, single family suburban homes. They developed relationships with the forest products industry and the Forest Service, which reinforced the economic forces in support of an enhanced timber program.

For the Forest Service, these trends tended to exalt timber from its position as king to that of a deity. The organization's leadership was increasingly timber-oriented, its measure of success became how well a line officer could "get the cut out," and its new recruits were primarily foresters or forest engineers trained in programs that tended toward more specialization with less understanding of the broader social objectives that the public lands serve and that founding father Gifford Pinchot sought. An enhanced emphasis on timber production was rational and productive for the Forest Service no matter what model of bureaucratic behavior you subscribe to—the agency as budget-maximizer, turf-maximizer, or political power-seeker.[5] For the Forest Service in the 1950s and 1960s, the objectives converged, making timber management the overriding organizational guiding light.

On the surface at least, the Forest Service remained committed to the hazy concept of multiple use as developed by Pinchot at the turn of the century, but it tended to promote other resource objectives as adjuncts of the timber management program. Multiple use came to mean maximum production of resource outputs over the long term: through government's increased use of various forms of cost–benefit analysis, maximum production generally came to mean maximum numbers of things, particularly dollars and user-days.

Agency objectives for wildlife resources on public lands at this time were defined almost exclusively as producing adequate numbers of game animals, activities that fortuitously were consistent with the enhanced focus on cutting timber and measuring agency benefits by counting user-days. For example, to grow a lot of deer, wildlife managers sought to create openings in the forest that would provide browse. Openings provided via timber cutting hence also generated a lot of huntable animals, and the foresters and wildlife managers marched hand in hand in facilitating the multiple uses of the national forests, uses for which there were both markets and political support.

Demand for all forms of outdoor recreation boomed after the second World War, and the Forest Service was not unaware of this national trend. While forest recreation had been encouraged since the early days of the agency, the Forest Service expanded its efforts in response to a perceived national need, and an accurate sense of a new and potentially large political constituency for forest management oriented toward providing recreation opportunities—a constituency that was being sought aggressively by the Forest Service's old nemesis, the National Park Service. Timber management and outdoor recreation were seen by FS leaders as potentially complementary activities.[6] Access provided through forest roads was seen to benefit outdoor recreationists of all types, particularly in a time when "driving for pleasure" was the top-ranked outdoor recreation activity nationwide,[7] but enhanced access also facilitated an expanded timber sale program. An enhanced forest fire protection program was seen to benefit all of the above resource values.

While a national movement was developing in support of setting aside good-sized chunks of public lands as wilderness areas, and the Forest

Service had administratively set aside a small amount of its holdings as primitive areas primarily in response to the interest of a specific district ranger or forest supervisor, wilderness preservation was an odd and uncomfortable notion to agency leadership. Wilderness got in the way of maximizing use, particularly in a time when timber was an increasingly valuable component of national forest lands, and it represented a single purpose set-aside in violation of the principle of multiple use. Further it limited agency discretion and hence violated the leadership's consistent interest in maintaining control. Indeed, the agency opposed the concept of statutory wilderness throughout the 1950s, a concept that eventually became federal law in the Wilderness Act of 1964.[8]

The dank, inaccessible old growth forests of the Pacific Northwest were seen as a valuable short-term wood products storehouse that was simultaneously the antithesis of good long-term forest management. The trees in these forests were generally 300 to 1000 years old, and, classified by the Forest Service as "overmature," they had long passed the point where they were efficiently adding new wood fiber to their biomass. The large amount of dead, dying, or down material in the forest was seen as a perfect entry point for insect infestation or forest fires. The old growth forest was seen largely as a biological desert, providing a home for few significant animals, none of which were in much demand. In addition, the inaccessibility of these forests made them of limited use to forest recreationists.

The agency's goal for the old growth forests was to change all this, by substituting even-aged stands of mostly Douglas fir that would be managed scientifically and efficiently on a sixty-to-eighty-year rotation to maximize the long-term sustainable yield of wood fiber, and along the way produce deer and recreation user-days. The problematic old growth forests were a blank slate on which the ideal multiple use, maximum-sustained-yield forest could be created. But over time, the old trees would have to go to enable agency leaders to realize the dream.

As a concept, the notion of protecting old growth forests also conflicted with the Forest Service's image of itself, as laid out by Pinchot and built on and magnified by subsequent leaders. Fundamentally, the Forest Service is about control. To plan the development trajectory of a huge portfolio of landscapes scattered across the country over the next 50 or 100 or 150 years, you have to believe in your understanding of forest science and your ability to manage the landscape effectively. To control the operations of an organization that is similarly diverse and scattered geographically, it is necessary to institute management mechanisms to ensure compliance with organizational goals. To maintain control over the long-term direction of an agency, you have to buffer it from the ebb and flow of political winds that blow from Washington in two and four year cycles.

To accomplish both biological and organizational management and to protect the agency from the vagaries of local and national politics, the Forest Service developed and implemented a set of operating styles and principles that were well established by the 1960s. Direction was to be set based on scientific and technical analyses that generated "right answers" to the allocation questions implicit in forest management.[9] Policy interventions were of the form of landscape manipulations such that growing trees or wildlife was made similar to an industrial problem where manipulating a set of factors of production would yield an appropriate product. The objective was once again maximum sustained yield where waste was either over- or under-production.

When conflicts emerged over direction of the national forests, the Forest Service worked hard to keep the controversies within the framework its leaders had established. By and large that meant that when faced with conflict, agency leaders tended to try to keep resolution within the ostensibly technical choice processes they had established or to create new ones that would yield appropriate answers to questions in debate. Planning processes were created early in the history of the agency both to provide consistent management direction for the long periods of time involved in forest management, and to deal with competing images of the future in a way that was well controlled by the agency.

To maintain control over the organization, the Forest Service evolved a remarkable set of management control devices that subtly but effectively controlled its staff. In his classic study entitled *The Forest Ranger,* Herb Kaufman identified information, budget, and personnel systems that tended to enhance the compliance of the Forest Service workforce with the overall direction of the organization.[10] By the 1960s, the Forest Service was a fairly militaristic, "Can Do" agency that promoted and rewarded individuals that mirrored the values and objectives of the agency's leadership, and tended to select against individuals who disagreed. While the Forest Service's district rangers and forest supervisors had remarkable amounts of discretion at the forest level, they exercised it primarily within the overall themes defined by the organization. Leaders worked hard at building their employees' identification with the agency, transferring staff members regularly, building on the symbolism of Smokey the Bear, the Forest Service uniform, and the like, and developing an image of the agency as a professional, science-based unit of government that was above the chaos of politics that characterized most other federal land management agencies.

By the 1960s, the combination of organizational styles and behaviors described above had succeeded remarkably well for the Forest Service. It had an expanding budget, a set of supporters in the federal budget process, and an esprit de corps that was the envy of Washington. In addition, the Forest Service's objectives and styles fit well within a government and a society that was heavily oriented toward post-War economic development, and the agency's contributions to local economic development matched a federal administration that was increasingly involved in local-level social programs. Finally, the agency was contributing directly and obviously to the material well-being of the American populace.

Overall, the 1940s, 1950s, and 1960s were a great time to be in the Forest Service. The agency's mission was growing, clear, and valued by society, and its methods of land management and organizational control were well-tested. While the agency had been challenged occasionally over site-specific controversies, by and large it had won those challenges and was in control of its destiny. There were concerns in the agency that timber and other interests were beginning to ask too much from national forest lands in a way that would preclude good multiple use management,[11] but overall, agency leaders were looking forward to a future that, in their view, would continue with the themes of the near-term past.

The development of these organizational styles and themes predated the development of the spotted owl controversy yet, in many ways, contributed directly to it. The controversy reflects a clash of values and styles as much as anything else. It illustrates a conflict between the 1960s Forest Service and the communities, industries, and politicians that came to depend on national forest resources, and a society in evolution to an expanded set of values in the 1970s and beyond. While the agency tried hard to fit the changing realities of the 1970s and 1980s into its time-tested standard operating procedures, it was trapped by its traditions and clients, and ultimately failed. In doing so, America was bequeathed a conflict of epic proportions—one that was probably inevitable even before the spotted owl winged its way into public view.

The Rise of Environmentalism

In the late 1960s and early 1970s, the winning combination of objectives and styles that the nation's public land management agencies had refined in the post-War period began to break down. An increasingly urban and affluent population began to change what they asked of the public lands and the agencies established to administer them. The population of urban areas grew dramatically following the post-War baby boom and continued a migration of individuals from rural to urban areas that had been ongoing for more than one hundred years. By 1960, more than two-thirds of the American population lived in urban areas; by 1970, almost three-quarters of the population was classified as urban.[12] At the same time, post-War affluence gave people the time and money to satisfy their material needs and seek leisure time activities. The development of an extensive national highway system and the automobilization of American culture facilitated an outmigration of city dwellers seeking fun and relaxation in the nation's playing grounds.

Hitching its (station) wagon to an Airstream, the new American family increasingly sought its recreation destiny on the federal public lands, the bulk of which were national forests that had previously been inaccessible. The traditional forest recreationist was an individual from a nearby rural community who hunted or fished on national forest lands. The new recreationist was more likely to be seeking opportunities to camp, hike, watch birds, or drive for pleasure.

The new recreationists brought with them different attitudes about the purposes and values of the public lands. Whereas traditional forest users viewed the public lands as a source of hunting and fishing opportunities, subsistence, and job-creating industrial wealth, the new users tended to view the public lands as objects to view, experience, and use nonconsumptively. National forests were aesthetic and recreational resources. Wildlife on national forests had scientific, recreational, aesthetic, and other values that went beyond its worth as game. The visual and environmental impacts of timber management activities were questioned increasingly as this new set of public lands constituents began to express their concerns about the obvious direction that the Forest Service and the Bureau of Land Management had set for their lands.

It is interesting that the development of attitudes supportive of protecting noncommodity values in public lands came at a time of rapidly rising consumption of commodities produced from raw materials from the public lands, and that the obvious conflict between the two was largely unseen by the new America. Timber cut off national forest lands was being used to build the woodframe houses of Levittown and countless other suburbs that were left behind in the summer migration to the cool forests and quiet lakes of public forests. Minerals and oil and gas mined from public lands fed an industrial machine that produced the automobiles, roads, and hula hoops that American consumers loved.

While consumption continued unabated, changes in what users valued in public lands were reinforced by a broader awareness of environmental values and impacts. Books like Rachel Carson's 1962 study, *Silent Spring*, and incidents like the 1969 Santa Barbara oil spill suggested that there were serious side-effects to the lifestyle to which America was becoming accustomed. Battles, such as those over a proposed pumped storage hydroelectric project on Storm King mountain in the Hudson River Valley, and an intensively developed recreation project proposed by Walt Disney for Mineral King Valley in the Sequoia National Forest, were protracted and highly visible. The media increasingly portrayed such conflicts as pitting "good guy" environmentalists against "bad guy" developers, a message that was simple to convey and one that appealed to longstanding American sensibilities. Live television coverage of the 1969 *Apollo XI* moon landing, with the attendant pictures of the earth as a lovely blue ball floating in a sea of blackness, powerfully conveyed a sense of environmental fragility and human dependence on the earth that had not been possible before.

By 1970, environmental quality concerns were squarely near the top of the American public's list of social priorities. For example, while only 35% of a nationwide sample considered water pollution to be a serious problem in 1965, 74% were concerned by it in 1970.[13] A similar survey regarding the seriousness of air pollution showed an increase in concern from 28% in 1965 to 69% in 1970.[14] The first Earth Day, held in the spring of 1970, prompted interest and activity in primary and secondary schools, colleges, and communities. At a time when the Vietnam War continued to drain the national psyche, environmental quality was something everyone could be in favor of.

Environmental interest groups proliferated and matured. New groups were created in response to local and regional controversies. Old groups were bolstered by heightened interest at the local level. While a number of groups had been active politically for many years, by and large their concerns had mirrored the public interest in conservation, which primarily meant management of public lands to produce animals to hunt and fish. Yet nongovernmental organizations are only effective if they continue to attract members and contributions for financial, staff, and political support. Changing national attitudes opened up new groups of the population to be tapped. Marketing strategies had to relate to the concerns of an increasingly urban, middle-class population; and advocating the protection and preservation of public lands and associated resources fit this agenda well.

Both the new and reborn environmental interest groups became skilled at manipulating the political arena. Their skills at lobbying, while still adolescent, were increasingly effective and organized. Coalitions of groups formed around specific policy issues. Their ability to mobilize significant numbers of supporters and their developing expertise became both a potent weapon and a resource valued by elected officials.

Politics is the art of the possible, and environmental politics was very possible in the late 1960s. Elected officials generally act only when the political momentum is in their favor, taking public stands on issues only when their constituents have made their interests known. Much of what characterizes successful politicians is their ability to sense constituent support and/or find ways to build it prior to the time when they are forced to take action on an issue. The public mood of the late 1960s and early 1970s made it clear that politicians could get elected on environmental platforms. A stand against pollution or for better natural resource management was a stand for God, motherhood, and all good things, and many elected officials scrambled to get on top of the environmental wave. Congressional leaders such as Senators Ed Muskie (D-ME) and Henry Jackson (D-WA) fought each other for the title of Mr. Environment. Even as unlikely an environmentalist as President Richard Nixon worked hard to take visible leadership on the issues.

The competition for political leadership on the environmental issue, the swelling and maturing environmental constituency, and an increased belief in federal regulation as appropriate public policy led to numerous legislative victories for the environmentalists in the early 1970s. Laws aimed at improving air and water quality, protecting marine mammals, reducing noise, limiting the use of pesticides, and improving land use planning in the coastal zone were all fairly expansive elements of federal policy fought and won by an increasingly effective and entrenched environmental constituency.

Two other laws, passed as additional products of this period of legislative environmentalism, would have significant effects on the ability of nongovernmental interests to challenge forest management decisions in years to come: the National Environmental Policy Act (NEPA)[15] and the Endangered Species Act (ESA).[16] NEPA was a short law that simply required federal agencies to consider the environmental effects of their major day-to-day activities: a seemingly rational, fairly innocuous requirement that was not expected to have major effects on the way government worked. Signed into law on January 1, 1970, NEPA required that any "major Federal action significantly affecting the quality of the human environment" would have to be evaluated for its environmental effects. Since NEPA applied equally to projects carried out or permitted by federal agencies, projects ranging from a subsidized housing project underwritten by the Department of Housing and Urban Development to a timber sale offered by the Forest Service would require completion of an environmental impact statement. Impact statements were to be done in a systematic, interdisciplinary manner that considered alternatives to the proposed action, and its long-term and irreversible consequences.

While NEPA has been implemented primarily as a procedural law (that is, it mandates that federal agencies consider but not necessarily act on a project's adverse environmental effects), it has had a major long-term substantive effect by opening agency decisionmaking processes to view in ways not thought of in the 1960s. In many ways, NEPA became an environmental full disclosure act which provided information to interested parties outside of government. And because full disclosure is never entirely possible, it became new grounds for legal challenge and delay as judges decided that they could recognize inadequate consideration of impacts. NEPA also completed the standardization of government project analysis begun on water resources projects in the 1950s and 1960s, and as a result, offered intervenors greater certainty as to what the decisionmaking process was supposed to look like and hence how to affect it.

As the last piece of environmental bandwagon legislation passed in response to the wave of environmentalism in the late 1960s and early 1970s and set in motion prior to the oil embargo and resultant energy crisis, the 1973 Endangered Species Act was expansive, substantive federal policy. Previous legislation passed in 1966 and 1969 had generated a process whereby wildlife species that were in danger of becoming extinct either domestically or worldwide could be granted federal endangered status and would be protected on federal lands and banned from international commerce. The 1966 law also required the Interior, Agriculture, and Defense departments to preserve the habitat of endangered species, "insofar as is practicable and consistent with their primary purposes," a requirement that had little practicable effect on the behavior of agencies like the Forest Service and the Bureau of Land Management.

The 1973 law went far beyond earlier efforts at endangered species protection. While previous laws recognized educational, historic, recreational, and scientific values in endangered species, the 1973 act added aesthetic and ecological values, and listed ecosystem conservation as the first of several purposes of the statute. It offered protection to endangered species, subspecies, or isolated populations of animals or plants, and provided two categories of protection: endangered, for plants or animals in imminent danger of extinction; and threatened, for organisms likely to be endangered in the future. A listing as "threatened" was much less restrictive and generally allowed habitat management to take place. Decisions as to what species warranted being stamped federally "endangered" were to be based solely on biological grounds, and species could be added to the federal list if they were threatened by almost anything, including "the present or threatened destruction, modification, or curtailment" of necessary habitat.[17]

Most important for our purposes is that federal agencies were now required to do what was necessary to protect listed species, insuring "that actions authorized, funded, or carried out by [the agencies] do not jeopardize the continued existence [of a listed species] . . . or result in the destruction or modification of [critical] habitat."[18] What was necessary would be defined by the Fish and Wildlife Service via an interagency consultation process through which FWS would issue a biological opinion on the impacts of agency projects. If the FWS foresaw an unmitigatible impact on a listed species, it would issue a jeopardy opinion following which the agency could proceed at its own legal peril. Finally, the law gave anyone the right to sue anyone else (including both the FWS and the development agencies) for violations of the act. As we shall see, this legal lever became potent by the end of the 1970s and was one of the engines driving the spotted owl controversy onward.

Passage of the federal endangered species laws in the late 1960s and early 1970s symbolized the shift in public values in natural and environmental resources. Interest in nongame and endangered wildlife species proliferated, reinforced by emotion-charged media coverage. Rising to become a major force in modern America, television understood the visual qualities of wildlife and endangered species issues. The image of furry animals or majestic birds drawing their last breath is extremely powerful. Television shows such as "The Wild Kingdom" were very popular and helped to spread the word. School children sent their lunch money to the Department of the Interior to help preservation activities.[19]

While many of the legislative changes of the early 1970s were largely symbolic acts on the part of Congressional representatives who were seeking the political benefits of ridership on the environmental bandwagon,[20] their overall substantive and procedural direction was to create a force in law and politics that would require the public lands management agencies to change their traditional ways of doing things. NEPA would require the agencies to open up their decisionmaking processes and explain their traditionally unquestioned professional judgments. The ESA and an expanding public interest in protecting noncommodity resources would mean setting aside areas from multiple use management. With a minimum of biological information, these two laws could also be used as legal devices to delay or change land management direction if management would harm a listed species' critical habitat. Both laws and the shift in environmental values would run head on into the direction of the Forest Service and the Bureau of Land Management in the public forests of the Pacific Northwest.

The Spotted Owl: Focal Point for Scientific Research and Advocacy

In the Pacific Northwest, a region that was somewhat schizophrenically proud of both its environmental amenities and its resource commodity industries, shifting public values generated a heightened interest in nongame wildlife, and legitimized the interests of scientists in the region in studying animals like the northern spotted owl. While the spotted owl had first been recorded in the Pacific Northwest in 1893, very little was known about the abundance or biology of the bird by the late 1960s, and that attracted the interest of Eric Forsman, an individual who became the leading authority on spotted owl biology in the 1970s. When Forsman started graduate work on the biology and ecology of the spotted owl at Oregon State University in March 1972, he rapidly became concerned about the future of the owl:

> From the very beginning [of my master's research work] it was obvious. In the first year of the study we must have found around thirty or forty pairs of spotted owls, and it was immediately obvious that there were conflicts all over the place because most of the pairs we found were in timber sales. Everywhere we went we'd find spotted owls and find boundary markers for a timber sale. They were right smack in the same area. We started letting the Forest Service know right off the bat that we were counting those sorts of things.[21]

In the 1960s and 1970s, decisions about the status of endangered or threatened species were best guesses made by biologists often operating with very little information. Advocating the policy and management implications of their expert judgments was simply an extension of this role. Besides, few individuals or groups outside the old boy network of scientific experts in government and universities had the knowledge or credibility to play this role. Forsman and his advisor, Howard Wight, were convinced that their early research indicated that the owl was in trouble, and they began to lobby for changes in policy and on-the-ground management to protect the owl while additional research was carried out. As leader of the Oregon Cooperative Wildlife Research Unit, a division of the U. S. Fish and Wildlife Service housed at OSU's Department of Fisheries and Wildlife, Wight was in a position to work through official channels. In July 1972, only five months after Forsman began his studies, Wight sent identical letters to his boss in Washington, D. C. and regional officials in the Forest Service and the BLM, informing them of the early results of Forsman's research:

Forsman has located 37 sites where pairs of spotted owls can regularly be seen. All of these pairs are inhabiting old-growth timber. No spotted owls have been located in any other type of habitat. These birds seem to require climax forests and are very sensitive to any alteration of these habitats. Preliminary observations indicate that if a forest habitat is altered the birds are forced to move, and usually do not reproduce the following year. They move to the nearest old-growth they can find, but with the current timber management practices in Oregon this is becoming more and more difficult. . .[22]

Wight noted that it was too early to provide a definitive recommendation on the size of the habitat areas needed by the owls, but recommended that the agencies proceed cautiously in harvesting over-mature Douglas fir not only in Oregon, but in Washington and northern California as well: "It is extremely discouraging to us to know that we are dealing with a species that has a very specific requirement for old-growth timber habitats and to know that these very habitats are a high priority for timber harvest by forest management agencies."[23] Wight also noted that under current law, the owl lacked official status as a federally protected species, and that made it extremely vulnerable:

At the present time, the spotted owl is under consideration by the Committee on Rare and Endangered Species as a bird fitting the 'rare' classification. Under the new Bill pending in Congress dealing with endangered species, I understand this rare category will be eliminated and a listing of species 'threatened with endangerment' will be substituted. This is in conformity with the philosophy that it is as important to manage our nation's wildlife to keep a species from the endangered list, as it is to restore a species on the endangered list to a more secure status. In our view, the spotted owl certainly fits this 'threatened with endangerment' category because of the very rapid disappearance of its vital habitats under current timber cutting practices. Because no official action has been taken on the spotted owl by the Committee, the bird is under no official designation as needing help. Thus, there is no reason for the BLM, the Forest Service, or private timber companies to provide any management to preserve the habitats of this rare and little known forest bird. . . . [While both the FS and the BLM have expressed an interest in the owl problem,] it must be recognized that at present time the spotted owl has no official designation as a rare or endangered species and is, thus, highly vulnerable.[24]

The FWS Washington office responded to Wight's memo by revising it and sending it under the signature of the FWS Director, Spencer H. Smith, to the Chief of the Forest Service and the Director of the BLM.[25] The Chief responded in September, 1972 by at least giving lip service to the needs of the owl. He forwarded information to the FS regional offices, and suggested that interim guidelines be prepared to protect habitat until more complete information was available.[26] Whether the Chief's memo had any impact on the ground is at best questionable. For example, Eric Forsman sent a letter to the District Ranger on the Klamath District of the Winema National Forest in southern Oregon outlining the coincidence of owls and timber sales. Out of six nesting pairs located on the district:

five are on habitats marked for cutting, proposed for cutting, or already cut. . . . Five out of six, Mr. McQuown, and I do not consider this an atypical sample for the Klamath Falls–Ashland area. The Forest Service and the BLM are systematically destroying the Spotted Owl population in your area, and yet I would bet that not a single timber sale impact statement produced on the entire Winema National Forest makes any mention of the destruction of Spotted Owl habitat.[27]

The District Ranger responded to Forsman a week later by pointing to the potential impacts that habitat set-asides for the owl could have:

Your [letter] suggests that nesting habitat for each pair ranges from about 200 to 600+ acres and that pairs are found a minimum of about 1 mile apart. I think we can both understand the magnitude of the set-aside areas if there are many nests involved. Owls have nesting demands, but other resources have demands too. Hopefully, additional research will clarify the limits of tolerance on the demands of all resources on a dwindling land base. So long as man retains a sense of perspective that allows him to recognize that we are dealing with a problem rather than a disaster, I shall remain confident that sound judgments . . . will prevail.[28]

In many ways the owl was in a Catch 22 situation: Under current law it was not considered a protected species, hence there was no reason for agencies to protect it, particularly since it was not in their organizational interests to do so. Yet their actions could in the long term result in the owl becoming endangered, at which point available habitat would be limited and the likelihood that it would recover would be low. What agency officials did not recognize due to limited information and a fundamental desire not to know more, because knowledge could only conflict with current management directions, was that as the options for protecting the owl got slimmer, the problem for the agency would not go away, rather it would get worse.

As Eric Forsman continued to uncover additional owl–timber sale conflicts, he increasingly became a visible advocate for owl protection. According to Leon Murphy, head of the fish and wildlife staff in the Forest Service's Portland regional office at the time, Forsman "began an almost one-man campaign to alert forest managers to the fact that [the owl] and his relatives must have this special habitat if they are going to survive."[29] Forsman describes a set of these conflicts on Corvallis, Oregon city watershed lands as the first time the owl issue escaped the confined discussions among agency biologists and managers:

When we first started the study, one of the first areas we went to was the Corvallis City watershed, which is a thin area that is jointly owned and administered by the Forest Service and the City of Corvallis. They both have land inside the watershed. One of the first three pairs that we found on the watershed was right in the middle of the city's planned sale for that year. The City cuts a certain amount of timber each year on the watershed to generate revenue for the city water program and the streets and everything else. I went to the city and informed the guy who administered the city watershed that "Hey, we've found this pair of owls one hundred feet from the edge of your timber sale this year. It's a rare species and I think you ought to re-consider that unit that you're going to cut there." In response, he basically said, "Sorry, this is not something that we do out here." We were rebuffed.[30]

Floyd Collins, Supervisor of the Water & Waste Plant of the Corvallis Utilities Division, laid out his reasoning in a letter to Forsman:

It is vital for the protection of the water supply that all resources are managed as a total program. We can not exploit one at the expense of another. Our current timber harvesting program bears heavily upon this overall operational plan. If, for example, all the old growth timber was to be left standing and allowed to continue decaying, it would increase the amount of windfall timber, thus causing addi-tional fire hazard and soil damage. The soil damage in turn would affect the water quality. If a fire were to develop in the area of old growth timber, it could possibly destroy the entire vegetative cover of the watershed. If this were to happen the watershed would be of little value to the citizens of Corvallis. Also, the dead and decaying timber, either standing or down, can increase the potential for addi-tional diseases which could affect the remaining healthy timber stands; thus jeopardizing the entire watershed and all the available resources. . . .[31]

While Collins did not go the remaining step and link the protection of old growth in the watershed to the downfall of Western society, he did note the economics of the matter:

You have suggested that approximately 300 acres per nesting site be set aside to remain as old growth timber. In viewing the economics alone on a 300 acre parcel, you have placed a value of 2 million dol-lars per nest on these areas. This is based on the price of timber at today's values. I'm sure that you can appreciate the responsibility that the City of Corvallis has to the citizens of this city to manage the watershed and all of its resources in a manner that provides maxi-mum service to the citizens. We attempt to follow that goal in all matters applying to the watershed area.[32]

Forsman responded to Collins noting that "You say in your letter that 'We cannot exploit one [resource] at the expense of another,' and yet your current management policy is completely at the expense of the old-growth community and the Spotted Owl,"[33] and decided he wanted to pursue the matter through different channels. At a City Council meeting, he warned community leaders about the effects of management on the city's lands:

> The media picked up on it, and it started getting written up in the Corvallis papers. It got to the point where an old time forester, T. J. Starber, who at one time was a professor at OSU, started a letter writing campaign about this silly ass spotted owl stuff, and he'd been working in the woods forever and had never even seen one of these damn things. So I wrote a letter back, and it escalated into sort of a letter writing thing in the local paper. It kept building, and once the media picked up on it, then the Forest Service became aware very quickly of the fact that they had a possible conflict. By the way, the city never did drop that timber sale.[34]

The Oregon Endangered Species Task Force Takes the Lead

The increased visibility of the spotted owl as a public controversy, the heightened general public interest in nongame wildlife, and the soon-to-be-enacted ESA led to the first official response to the conflict: the creation of an expert committee to study the issue. Creating committees to study budding policy issues is a time-honored practice that is both rational and strategic. Gathering information and soliciting expert advice is a reasonable way to determine the dimensions of a problem and to frame appropriate policy and management interventions. But designating a task force or a blue ribbon commission also is a great way of giving the semblance of action without necessarily changing course. The extent that such groups have an effect on the policy process is largely in the way that they end up defining the dimensions of an issue and the legitimacy they give to political interests supportive of the groups' ideas. Even if an agency has no real interest in changing course in response to expert advice, creating a study committee generates information and political currency that at some point may force the agency to make a choice they never intended to face. For the spotted owl controversy, the Oregon Endangered Species Task Force (OESTF) began a slow process of problem definition and legitimation that forced the agencies to respond more than a decade later.

In May 1973, John McKean, Director of the Oregon State Game Commission, established an interagency scientific task force to begin inventory, research, and management work on endangered Oregon wildlife.[35] Eric Forsman believes that the OESTF was formed largely in response to the developing public controversy over the spotted owl:

> The attention on the species kept building and began to appear in the media, the FWS became aware of the problem, Chuck [Meslow, Assistant Unit Leader of the FWS Coop Research Unit] and I were going and talking to anybody that would listen about this animal that was out there and it was being ignored. And that the habitat was being cut right and left. So, in response to that, this interagency committee was formed. . . The impetus for this thing was the owl, basically. . . It was obviously an interagency problem. The animal occurred throughout western Oregon and needed to be managed. The ownerships were all mixed together and it needed to be managed on a broad basis rather than area by area.[36]

Charlie Bruce, an Oregon Department of Fish and Wildlife (ODFW) non-game biologist who was a member of the Task Force from its inception, believes that the soon-to-be-passed federal Endangered Species Act was a source of pressure for action at the state level:

> Basically, the Endangered Species Act came along and everybody and their brother started going off in their own direction, saying we've got to put together a list. And our director said, hey wait a minute. Let's all get together on this. At the time, John McKean was the director. [His action] was actually something of a miracle because John was one of the original graduates of the OSU Fish and Game School. He grew up killing eagles and hawks. But times had changed. . . . I was the token nongame biologist for our department at the time, which was still the Game Commission, so it was still, you know, there was lots of, "Oh, you're the guy who's in charge of the dickie birds" type of stuff going on.[37]

Chuck Meslow also believes that the interaction between federal objectives and state interests were forces present at the birth of the task force, but in a different way:

I think it was for all the wrong reasons. . . . I don't think John or the agency had any great love for the owl. That certainly wasn't it. Or endangered species. John was a states' rights-er. And he was pissed because he had just lost control of the marine mammals through the Marine Mammal Protection Act, and he had lost control of the Columbian Whitetail deer by the Endangered Species Act. He would rise to that bait, that those were his animals, and that he or the agency was to manage any damn well way they thought and he didn't want the feds mucking around with it.[38]

While the OESTF was created with a mission larger than just dealing with the spotted owl, it rapidly focused on the owl's situation, in part because the scientists on the committee felt the problem would be easy to deal with. According to Chuck Meslow:

[The Task Force] started off as an endangered species subcommittee and it rapidly devolved to where the principal emphasis was put on spotted owls. . . . They looked at this, and they said, "Hey, here's one that by God seemed to be a simple one." It looked like we could identify the habitat that was involved and all we had to do was address reserving enough of that habitat and we ought to be able to solve this one.[39]

At the first Task Force meeting held on June 29, 1973, representatives from the Oregon State Game Commission, the BLM, the FS, the FWS, and OSU agreed to place their "highest priority on establishing guidelines to preserve representative habitat for [the spotted owl]."[40] Reporting on preliminary research results, Howard Wight indicated that the habitat of the species was restricted to "old growth forests (e.g., a forest untouched by man)",[41] with a territorial requirement of a minimum of 300 acres per pair with nest sites approximately one mile apart for contiguous territories. He also noted that over half of the seventy known pairs of owls were located in timber sale areas, and recommended that the Task Force "talk in terms of reserving representative areas of the old growth community for the spotted owl and other species." The BLM and FS representatives reiterated Wight's concerns indicating that "all old growth forest on BLM lands will be harvested in less than 30 years," and "a similar situation exists on National Forest lands."[42]

At its first meeting, the Task Force took three sets of actions that defined their direction for the next few years. First, they adopted a resolution that recognized the "unique characteristics of old growth for-ests and their rapid disappearance from Oregon, and the importance of these communities to a variety of wildlife species," and proposed research into the habitat needs of old growth-dependent species.[43] Second, they adopted interim guidelines for spotted owl management that were to guide agency action until the Task Force's studies were completed. The guidelines included 300-acre set-asides around each known owl location, on which timber planning would cease. Third, to identify areas where old growth should be reserved to protect owl habitat, the chairman of the Task Force suggested that a map be developed that charted remaining old growth forests, on which wilderness areas, wilderness study areas, re-search natural areas, and spotted owl nest sites could be overlaid. FS rep-resentative Leon Murphy and BLM representative Dave Luman com-mented that "a map of remaining old growth forest would be developed by their respective agencies before the next meeting."[44]

In many ways, the OESTF was off to a reasonable start. It represented the major habitat-managing agencies in Oregon (though Washington was not included at this time). Ostensibly, it was basing its decisions on science-based expert judgment. And it was providing advice on manage-ment direction, while pursuing needed information.

At the same time, approaches established by the Task Force early on affected the course of the controversy in ways the scientists on the Task Force did not intend. For example, basing their recommendations for in-terim (and later final) management direction on *minimum* acreages needed to protect the owl tended to understate real habitat needs. Their solutions tended toward protecting isolated islands of reserved habitat, such that a protection strategy would look like old growth cookies cut from and scattered across the regional landscape. In addition, they as-sumed that information necessary for constructing recommendations was available and could be obtained from the agencies. All of these were un-derstandable assumptions at the time that were well grounded in scien-tific conservatism, but nevertheless evolved into a management program that was inadequate to meet the biological needs of the owl population.

As it turned out, perhaps the most impacting bias of the Task Force's recommendations was its use of minimum acreages necessary to protect the owl. The 300-acre figure was Eric Forsman's best guess as to the mini-mum area of old growth that owls were using. According to Forsman:

There had been no telemetry work up to that point. We were sitting in there and saying, "Well, what do you think we ought to do? How much habitat do these birds need?" And they look at Chuck and I, and we said, "Well, all we know is we have yet to find a pair of them in an area where there is less than about 300 acres of old growth." That's how scientific it was. Then they said, "Okay, 300 acres, we're going to go with it." . . . It was the biggest mistake we could have made, because it turned out in the late Seventies, after we started looking at some telemetry data, it was obvious that 300 acres wasn't even close to being enough in most areas.[45]

Chuck Meslow points to the counterintuitive nature of concluding that a small animal could need a lot of reserved acreage:

I think where Eric was coming from at that point in time was that he had looked at spacing between owls and looked at the situation, and then you look at a little brown bird this big, and you look at all those big trees, and my God, I can't imagine that the animal can require more than 300 acres. . . . It just is pretty inconceivable that an animal that big is going to have demands or requirements for something greater than that.[46]

The Task Force's actions were also affected by the bureaucratic environment in which habitat management was to take place. The agencies' response to the Task Force's recommendations were uniformly negative. Neither Forest Service nor BLM managers took the owl issue seriously, particularly since remedial actions would conflict with their well-established direction of timber-directed multiple use management. The agencies' response to the OESTF's recommendations for interim management is illustrative. In August 1973, the Forest Service and the BLM rejected the Task Force's recommendations because:

(1) Rigidly-cast prescriptions tend to become the accepted practice, even though they are intended to be of temporary nature only; (2) it appears reasonable to assume that the present old-growth (200 plus years) stands located within acceptable elevation limits, contain spotted owl populations, even though specific sightings of the birds or nests have not been made; (3) we are confident that further analysis will verify that sufficient old-growth timber stands exist to provide interim, one-year State-wide protection of habitat without the 300-acre restriction around all the *known* spotted owl nesting and observation sites; (4) determination of the desired State-wide

production level of these birds should be resolved before total protection of all sighting areas in prescribed-size blocks is undertaken; and (5) the "management by individual animal location" philosophy, when applied to all species which may be identified as requiring old-growth stand habitat, presents a land management spectre of considerable magnitude.[47]

While last in their list of reasons for rejecting temporary protection, the fifth reason was probably closest to the mark. Remember that at this time, only seventy nest sites had been identified. At 300 acres per site, protecting known sites would involve some 21,000 acres of public land scattered across a BLM and FS land base of 31 million acres in Oregon (10 million acres in western Oregon)—a trivial amount of land given the broader scheme of things. But if you view temporary set-asides as the opening of a door that you strongly prefer to leave closed, the temporary measures represented a threat to the agencies, one that there was little reason to take on voluntarily.

The agencies were giving lip service to the needs of nongame wildlife species at the time. For example, the BLM's Wildlife Program Activity Policy Statement issued in February 1973 noted that "increasing public interest in non-game wildlife species and concern for species threatened with extinction will shift management efforts," and set a long-term objective of maintaining "a maximum diversity of wildlife species in sufficient numbers to meet public demands. This will be accomplished by means of habitat management."[48] Management is the operative word here. Reserving areas for owl habitat infringes on this norm, and opens up a Pandora's box of unknown proportions.

At the field level, there was even less awareness of the owl or other old-growth related species, let alone actions to protect them. According to Eric Forsman:

> At that time, you could walk into a Forest Service office in a district and say, "I just found a pair of spotted owls on your district" and they'd look at you and say "What?" They literally didn't know what they were, didn't know that they were out there, period. And that didn't just apply to Spotted Owls, it applied to a lot of other things.[49]

A lack of basic information about "what was out there" hampered the development of management recommendations in 1973, and amazingly continued to limit productive discussions for the next sixteen years. In

1973, land managers had little idea of what nongame animals resided on their lands, and more fundamentally, what their land base consisted of. Information about forest cover was generally collected in a spotty fashion at the local level for the specific purpose of timber management. There was little interest in collecting detailed, geographic-based information on nonproductive lands given obvious agency direction and hence the day-to-day information needs of land managers. Even where it existed, aggregating such information into a common form on one map was at least seen as unimportant, and might provide the basis for outsiders to challenge the decisions of line officers—forest supervisors and district rangers—lords of their domain.

Getting a single map of old growth habitat suitable for spotted owls that was acknowledged to be accurate by agency and nongovernmental groups, a seemingly straightforward task, was elusive in the mid-1970s and remained so through 1989. FS representative Leon Murphy and BLM staff member Dave Luman were supposed to bring such a map to the second meeting of the OESTF, but were unable to do so. According to minutes taken at the meeting: "Such information is probably available at the National Forest or BLM district level but would be very difficult and time consuming to compile."[50] Charlie Bruce described the events:

Leon Murphy and Dave Luman said, "We'll go back and start pulling together some information." They came back to the next meeting and said, "Well it's not there. We can't tell you how much is there. We can get bits and pieces from the forests and that sort of thing, but nobody's got any sort of comprehensive inventory." And of course it's still the same debate today. What is old growth? How much is there? I guess from an ecological standpoint to me it's even more disturbing because we're obviously cutting some trees out there and if we don't know what's out there to begin with, how can you realistically develop any sort of land management program that means anything in the long run?[51]

Chuck Meslow identified some of the reasons it was so difficult to get the needed information:

They didn't map forests in that fashion. They don't map at that scale. . . . Then you get back to how a bureaucracy operates. The forest supervisor is pretty much God on his chunk of turf, and the way things are mapped and typed isn't exactly the same when you abut boundaries. You could say you wanted a map of all D4-3bar Doug fir type in the state, and you still might get quite disparate results out of it. . . . I don't know, but one can become suspicious that they didn't want that information assembled in one place. And I suspect that anyone that was looking down the line could see how that could be used and misused by folks on the other side of things. . . . One can wonder if it was intentional to prevent people from easily grasping the big picture. I mean, here's a way you can look at where the spots are on the map in one year, and you can look at where they are three years later and you can see the change. And it's hard to hide what's going on.[52]

Meslow did suggest that getting a definitive definition of old growth that could then be mapped with confidence was not entirely straightforward:

Let's give the foresters and everyone that's worked with this some credit in that it is a very difficult thing to define. You're dealing with a gradient, a continuum, and it isn't like defining what is a pole-stage stand. How many overstory trees does it require before you have enough of an element of old growth present so that it is really old growth? . . . [Nevertheless,] it's a matter of settling on a definition, and no one wanted to push hard enough to settle on a definition.[53]

In spite of this inherent difficulty, a rough proxy of old growth characteristics could have been used by the agencies to plan management action. For example, at the August 1973 OESTF meeting, following Dave Luman and Leon Murphy's pronouncements that neither the Forest Service nor the BLM had an appropriate map, Bill Nietro, a BLM district wildlife biologist, demonstrated how he could identify potential spotted owl sites using aerial photographs. Glenn Juday, a doctoral candidate in the OSU Botany Department, volunteered to map remaining old-growth stands using infrared Earth Resources Technology Satellite (ERTS) photographs. The Task Force agreed to cover the cost of materials for the mapping project.[54]

The mapping project was carried out over the next six months. Juday used ERTS photographs, cross-checking areas with U-2 aerial photos and field observations. Old growth areas were generally characterized as 200+ years old, with a diameter (dbh) of 30 inches.[55] By the Task Force's September 1974 meeting, copies of the map had been reproduced, and a companion map was developed by the Oregon Wildlife Commission identifying existing Forest Service roadless areas, wilderness and wilderness study areas, spotted owl locations, and land ownerships. The total cost of the mapping effort was $827.

Reviewing the maps, the OESTF concluded that "Of the 103 spotted owl pairs that have been located to date, only six known pair locations appear to be reasonably secure."[56] Based on his mapping work, Glenn Juday identified thirteen clusters of old growth–spotted owl areas that in his view needed some type of immediate protection. The Task Force decided to push for a complete census on the thirteen areas to identify all spotted owl pairs, and to work toward management recommendations for preserving old growth forests and associated wildlife species to be completed by December 1975. OWC staff member Bob Mace, chairman of the task force, suggested that field work be continued "to inventory climax vegetation of all types for the entire state to head off conflicts that are sure to come."[57]

In the next two years, considerable effort was put into identifying owl locations, particularly by the BLM, but very little additional work was done to map old growth habitat. Getting the two federal agencies to produce or concur on a statewide map of suitable habitat remained elusive. Inventory work focused on the thirteen "Critical Old Growth Forest Areas" that Glenn Juday had identified. By November 1976, the BLM had inventoried 63% of its western Oregon lands for spotted owls, and had identified 193 nesting pairs or sightings. By the end of 1976, the Forest Service had located 79 owls or nest sites.[58]

Research into the habitat needs and population biology of the spotted owl continued and began to produce results. Two efforts were underway. A project funded cooperatively by the California Department of Fish and Game (CDF&G), the Forest Service's Region 5 (San Francisco), and the National Park Service was being carried out by Gordon Gould, a CDF&G wildlife biologist. And Eric Forsman's graduate thesis work on the status of the owl in Oregon was being funded by the OSU Cooperative Wildlife Unit and the Forest Service's Pacific Northwest Research Station out of LaGrande, Oregon, with Jack Ward Thomas as project leader and patron.

Presented as his master's thesis in October 1975, Eric Forsman's work largely confirmed his prior ideas about the size and status of the owl in Oregon. He identified 123 owl sites in western Oregon, including 116 pairs.[59] Ninety-five percent of the sites were characterized by unharvested old-growth conifer forests. Only three pairs occupied second growth forests with minor components of old growth. As reported in his thesis:

Timber harvest occurred or was scheduled in 52 percent of the owl habitats located during the study. In most cases, timber harvest within an occupied habitat did not drive the owls completely out of the area, because only small portions of extensive forest areas were harvested. When portions of small forest areas (less than about [200 acres]) were harvested, however, owls often disappeared from these areas. Two pairs located in old-growth forests which had been subjected to very light overstory removal indicated that, under some circumstances, owls could tolerate this type of harvest activity. Clearcut harvest, however, eliminated roosting, nesting and most foraging in the affected areas.[60]

Forsman's long-term management suggestions emphasized two strategies: (1) establishing long rotation stands that would be managed on a 300–400 year rotation; and (2) creating preserves of low elevation old-growth vegetation.[61] New preserves were needed because areas that had been set aside by the federal agencies as wilderness or roadless areas had limited potential to provide good quality owl habitat. They were largely high elevation areas characterized by subalpine or timberline vegetation that was no good for growing owls.[62] Forsman concluded his thesis emphasizing that the spotted owl is an indicator for the broader old growth ecosystem, and that protective actions for the owl would benefit other species as well:

Areas managed or preserved for spotted owls will also provide habitat for other species of plants and animals which comprise the old growth community. Such communities are becoming increasingly rare, and are, therefore, becoming important as research and recreation areas, as well as sanctuaries. Without old-growth coniferous forests, numerous species can be expected to decline in numbers, perhaps to the point of acquiring rare or endangered status.[63]

As the research work in Oregon and California began to generate results that confirmed the owl's preference for old-growth acreage, individuals critical of the studies' conclusions argued that spotted owls were found mostly in old growth, because that is where researchers went looking for them. Forsman and Meslow attempted to answer this concern by conducting parallel inventories in even-aged second growth stands in the Oregon Coast Range during the 1976 field season. Produced late in 1976, their report identified owl densities in second growth of 0.05 pairs/mile, while old-growth areas yielded an average density of 0.58 pairs/mile. In their view, "the low index in 35- to 60-year old second-growth indicated that such forests provided marginal spotted owl habitat at best. . . . If patches of old-growth were present, owls were often located near them; this reinforced our belief that such areas are important as nest and roost sites."[64]

While research appeared to be confirming that the owls needed some change in management direction, little progress was made on coming up with a management plan for the bird, and very limited protection was being provided on the ground. The December 1975 deadline that the OESTF had set for itself to establish management recommendations came and went. The biologists were busy doing owl inventories, and the managers were happy to avoid having to deal with the issue. According to Eric Forsman, part of the Task Force's problem was having to make recommendations that would so obviously impact current agency direction:

What happened was that the committee would come up with a plan, we'd send it around to the regional forester and to the state director of the BLM and they'd say, "We can't live with this." We went through that for years. . . . The Forest Service and the BLM attitude from the beginning has been stall. Whenever somebody makes a management recommendation that you don't like, you say "You don't have adequate data. We are not willing to make an important management decision that's going to affect people's lives to a great extent based on this complete lack of data." That was the game being played with this committee from the very beginning. The committee would come up with something and we'd see this. Any recommendations had to be approved by the respective directors and chiefs in each of the agencies. So everybody had to agree. We'd come up with something and send it off and somebody would come back—usually the regional forester or state director—and say we don't like this. So it'd go back to committee, and they'd mull it over. That went on and on and on.[65]

By the late fall of 1976, pressure was building for the OESTF to take some action regarding management guidelines, and surprisingly, it was the BLM biologists who seemed to push the process along. According to Charlie Bruce: "BLM was curiously more receptive to protection than the Forest Service, and it had always been the other way around. The Forest Service had always seemed to be a little more up front on fish and wildlife sorts of things [until the owl issue came along]."[66]

It is clear that the personalities of key individuals in policy and management controversies have as much to do with actions that are taken as anything else, and the BLM district-level biologists were becoming more vocal about the needs of the owl. The process of inventorying BLM owl habitat gave them a common problem to work on; their meetings became a forum for sharing ideas and concerns. At this point in time, Forest Service and BLM biologists were generally isolated and unempowered, situated in scattered districts across large areas of ground, and lacking in the critical mass that gives legitimacy to criticism and concerns. The work on the owl inventories helped the BLM district biologists overcome their low status in their agency and push for changes in management direction. In many ways, also, the BLM's dominant use management of the Oregon & California (O&C) lands in Oregon for timber production freed up agency biologists from spending their time growing deer.

At the same time, the arguments of the BLM biologists were bolstered by statutory changes that occurred in the mid-1970s. Up until 1974, the BLM had no obligation to worry about nongame wildlife on its forest lands in Oregon unless they contained a federally listed endangered species, and the spotted owl was not on the federal list. In October 1974, Congress passed the Sikes Act Extension which required federal land management agencies to help protect state-listed sensitive species.[67] To implement the Sikes Act in Oregon, the BLM signed a memorandum of understanding with the Oregon Wildlife Commission in May 1975.[68] Since earlier that year, the Oregon Wildlife Commission had adopted a list of state threatened and endangered wildlife that included the northern spotted owl as a threatened species, the BLM for the first time had a somewhat-legal obligation to consider the owl in its timber management planning. According to BLM staffer Bill Nietro, "The owl was a state-listed species, and that's the only reason we dealt with it."[69]

The passage of the Federal Land Policy and Management Act (FLPMA) in October 1976, which established a multiple use mandate for the BLM and included wildlife as a major use of BLM lands, reinforced the agency's interests in at least giving lip service to noncommodity values on the lands it managed. Emulating the Forest Service, its more powerful and professional relative, BLM had sought FLPMA as a way to upgrade its longstanding second-rate status as a public lands agency. While the O&C lands are covered by the provisions of FLPMA, "in the event of conflict or inconsistency" between FLPMA and the O&C Act regarding timber management,[70] the O&C Act takes precedence—a concession given to timber interests in Oregon during the negotiations over FLPMA. Nevertheless, FLPMA's passage increased the power and legitimacy of the BLM biologists from zero to a small, but positive, number. Agency officials began to consider how they could insert nontimber objectives into ongoing timber management planning.

In 1976, BLM staff members began defining the planning process that would generate updated timber management plans in the early 1980s. BLM timber management plans were revised every ten years, and the plans due in the early 1980s had to conform to the provisions of NEPA, the Sikes Act, FLPMA, and the Endangered Species Act. While O&C lands were still primarily timber production lands, the latitude with which BLM timber planners were used to operating was at least constrained by the necessity to prepare judicially defensible environmental impact statements. Having lost several lawsuits on NEPA grounds, BLM planners needed guidelines that would allow them to avoid obvious conflicts with sensitive species, such as the state-listed northern spotted owl.

The BLM's need for management guidance led to the reactivation of the OESTF, after two years without a meeting. On November 17, 1976, the BLM staff outlined five possible alternatives for planning guidelines, ranging from no owl-related action to 200–300 acre old growth set-asides at each owl nest site.[71] The BLM biologists recognized the biological desirability of the nest-site preserves but noted that "while desirable, it would probably be economically unjustifiable with our present knowledge of the bird's requirements."[72] They recommended that a less-impacting alternative be selected. This alternative would "establish protected areas for enclaves or known concentrations where no timber harvesting would be permitted but certain other forest management practices could be accomplished (spraying, etc.)."[73]

To the BLM biologists, this approach was the "most acceptable method for preserving 'reasonable' populations of spotted owls in Oregon." Identifying chunks of habitat that would include the full latitudinal range of the owl in Oregon would provide the best chance for protecting genetic variability in the population.[74] It was also viewed as the more politically feasible alternative: "While tentative areas have been outlined for habitat protection, this does not necessarily mean that *large* acreages of old growth forest are involved. Included in these areas are acreages of private lands, as well as areas that have already been cut over." Nevertheless, the tone of the BLM biologists involved in the 1976 decision was clearly one of frustration and advocacy for owl protection:

Untenable decisions should not be reached which will further place BLM resource managers in an unfavorable limelight, resulting in public pressures that ultimately force us to accomplish the job that was our decreed responsibility in the first place. Under the Endangered Species Act of 1973 (and others) it is our full responsibility to be sure that our land management actions do not bring wildlife species to the brink of questionable survival. The Bureau establishes an "allowable cut" for the timber industry, but it is of equal importance that we establish an "allowable survival" for all forms of wildlife. Some administrators become excellent in administering and allocating various resources for selected industries with which they deal, while failing to develop an understanding of the esthetic appreciation which many millions hold for the multitude of non-game wildlife forms.[75]

If the BLM biologists expected action at the November OESTF meeting, they were disappointed. The FS reported that 79 owls or nest sites had been located on FS lands, with approximately 1.5 million acres surveyed. According to BLM biologist Mayo Call, while the meeting was held to review new information obtained by the agencies during the preceding two years, very little new information was presented: "Unfortunately, BLM was the only agency with any substantial new data to present. . . . It was disappointing that the Forest Service had not taken the initiative to collect further data on the occurrence of the spotted owl, comparable to BLM's data, during the past two years so that a more complete picture of occurrence could have been presented at the meeting."[76]

Little new information was available from the FS because the agency remained uninterested in finding out the answers to the owl questions. Eric Forsman's work was pretty much the only owl-related work ongoing on national forest lands. Guidance was sent to Oregon national forest supervisors in October 1976, identifying the owl as a nongame habitat diversity indicator species that would be used in preparing Biological Unit Management Plans (BUMPs) in fiscal year 1979.[77] In addition, interim guidance was to protect known nesting sites until these plans were prepared in which "actual habitat needs for the maintenance of a viable population" would be specified.[78] In reality, though, the "present status of management for spotted owls on the National Forests rests with each Supervisor to deal with known sighting or nesting sites individually,"[79] and few supervisors were inclined to worry very much about spotted owls. Prompted by the upcoming November OESTF meeting, Leon Murphy had also requested information on owl sightings in Oregon national forests in mid-October 1976.[80] Otherwise, the only owl work ongoing on the national forest system was the result of the personal interest of the stray wildlife biologist.

Pressure from the BLM staff for a Task Force recommendation pushed the OESTF to schedule another meeting a few weeks later. Leon Murphy sent out an urgent request for help in compiling a map identifying known spotted owl sightings, so that FS data could be combined with BLM data. Murphy also noted what to concerned FS managers was a ray of hope that the owl might not need large blocks of old growth habitat:

> As you are aware, several years ago researchers at Oregon State University who had initiated spotted owl research were pressed by members of the Committee for minimum habitat requirements to protect spotted owls. With great reluctance, the researchers identified 300 acres in an "old growth" condition as the minimum. Since that time, numerous observations by BLM biologists indicate that spotted owls are living and/or nesting in old growth areas of less than 300 contiguous acres. If any Forest Service biologist has gathered sufficient site-specific information which can be utilized to more clearly identify habitat requirements for a nesting pair of owls, please send me the details.[81]

Meeting on December 13, 1976, the OESTF was able to formulate interim owl habitat management guidelines. Leon Murphy's efforts had resulted in a total of 131 known owl sightings on FS lands, while the BLM reported 193, yielding a total of 324 individual known habitat areas in Oregon. The consensus of the OESTF members, and associated owl experts, was that "if a minimum of 400 pairs of owls were maintained on the federal lands throughout western Oregon, the species will survive and could be delisted as a state threatened species."[82] Responding to Eric Forsman's fears about habitat fragmentation, the Task Force recommended that the 400 pairs be maintained in clusters of habitat, so that forty 10-pair concentrations would be scattered up and down the west slope of the Cascades and the Coast Ranges, generating enough interchange for long-term species viability.

The Task Force further recommended that the federal agencies be asked to follow a one-year interim policy that protected all existing and new owl sightings and nest sites, while a management plan was produced that specified how the 400 pairs would be accommodated. The FWS representative to the Task Force was ecstatic. In his view, the approach taken by the OESTF was unique and forward-thinking. "The pre-recovery team is the only one in the United States" that he knew of.[83] The group set a January 1, 1978 deadline for themselves to finalize a management plan.

After about four months went by, the BLM and the FS agreed to follow the interim direction until January 1978. A subcommittee of the OESTF worked on preparing a draft plan, and both the FS and the BLM worked on mapping old growth lands. By October 1977, the agencies had completed their mapping work. All Oregon national forests had mapped old growth (forest type D5) lands, and a few forests and ranger districts had undertaken intensive survey work, but most had not. In total, approximately 297 locations with owls were identified on FS lands. Based on habitat work, the FS staff estimated that 270 pairs could be accommodated on national forest lands.[84] The BLM had gotten responses from owls at 217 locations on their lands, and had located habitat areas that could support 86 pairs.

Not surprisingly, the allocation of habitat areas by agency in the draft management plan reflected these numbers. The 400 pairs were split, with the BLM taking responsibility for 100 pairs, and the FS for the remaining 300. Their guidance on habitat requirements was taken from Eric Forsman's master's thesis work, including a minimum of 1200 contiguous acres per owl pair, with a core area of at least 300 acres of old growth. The remainder of the territory was to be managed so that at least half of the acreage was in stands of 30+ year old forests. Old growth was to be at least 200 years old, and contain an average of eight to ten overstory trees, with a developed understory greater than 30 years of age. Each enclave was to include the home ranges of a minimum of three owl pairs, with six being ideal. Core areas for each pair were to be separated by approximately one mile, and enclaves (of 3 or more pairs) were to be 8 miles apart (not to exceed 12 miles).[85]

The draft plan was finalized and ratified by the OESTF at a meeting on October 17, 1977. A few small but significant changes were made in it. The federal agency representatives felt that the state agencies should also bear some responsibility for owl protection. While the representative from the Oregon Department of Forestry said that his agency was unlikely to be supportive, the OESTF changed the draft guidelines so that 20 owl pairs would be maintained on "Other" ownerships.[86] This left BLM holding the bag for 90 owl pairs, and the FS obligated to 290 pairs. The other major change was a change in wording that is important for what it represents: The word "Enclave" was deleted from the document, replaced by "Management Area." This change in wording made the plan more consistent with agency traditions, and potentially less demanding when the agencies began implementing it. Symbolically, the change made the plan less threatening, because while management was done all the time, enclave sounded too much like preservation.

While it had taken more than four years, the OESTF had given birth to an owl management plan. Their recommendations reflected their expert judgments based on four years of research and modified by their sense of political and administrative realities. They were the product of a truly interagency group of scientists and managers. A cursory inventory of old growth habitat had been carried out, and owls located, so that a clustered management scheme could be implemented. If nothing else, the task force members had hung on through a contrary administrative environment and proposed a small shift in federal land management when a window of opportunity appeared. In the best of worlds, the plan might actually protect the owl. According to Bob Stein, OESTF Chairman, "The implementation of this plan should assure the future existence of the spotted owl in Oregon and prevent it from becoming an endangered species."

2

Muddling Through: 1978–1981

The Oregon Endangered Species Task Force's plan pushed the spotted owl onto the agendas of the FS and the BLM, and their response over the next seven years was to try to muddle through: do the minimum possible, and maintain control by keeping decisionmaking within normal agency operating procedures. While this approach was rational from the perspective of agency leaders, it was doomed over the long haul by changes in the external environment of the agencies. Throughout the late 1970s and 1980s, major changes were occurring in law and legal interpretation, the power and skill of interest groups, public values, and science and information—all of which would have the long-term effect of wresting control over issues like the spotted owl from traditionalists in the agencies. Neither BLM nor FS leadership viewed the owl issue as particularly threatening through the first half of the 1980s. Supported by political executives from the White House on down, agency leaders tried to do more of the same, when the same would not do. Ironically, by not taking steps to provide minimal protection for owl habitat at a time when neither the scientific information nor the environmental groups were particularly demanding, agency leaders missed their best chance for resolving the controversy without major impacts on their missions.

Early in 1978, both the BLM and the FS agreed to follow the Task Force's recommendations, subject to ongoing agency planning processes.[1] According to the memorandum sent from the BLM State Director to his district managers, "The spotted owl management recommendations represent a rare example of a coordinated interagency approach to a common problem."[2] The response from the director of the Oregon state forestry department was less enthusiastic. The Oregon State Forester indicated that he would consider maintaining habitat when and if owls were located on state lands.[3]

It would have been surprising, of course, if the federal agencies had not supported the recommendations, since they had been cleared by the staff prior to the Task Force's decision. Besides, even though they included some commercially valuable old growth timber, the land areas involved were fairly insignificant. Eighty-seven thousand acres out of a Region 6 national forest base of 24 million acres is a pretty small amount. Even comparing it with an estimated 4+ million acres of old growth forest in Oregon at the time, the total amount of old growth set-asides represented about 2% of remaining old growth, and that assumed that the owl habitat areas would be new reservations, an assumption that was not true.

The fact that the FS and the BLM responded at all positively to the issue of owl protection was important, given that they had rejected weaker recommendations just four years earlier. It is true that the Task Force had better information in 1977 than in 1973, and the issue had been in play for four years which gave it a historical inertia and institutional credibility that spawned advocates and promoted action. Nevertheless, the actions of the federal agencies should be taken as more important than they actually were for on-the-ground owl protection, for they signaled a shift in national and regional politics, administrative realities, and the law that governed public lands decisionmaking. The actions also established a pattern of administrative response to the evolving owl issue that was organizationally understandable, but inadequate to deal with the political and scientific realities of the spotted owl case.

The legal environment of administrative decisionmaking had clearly changed by 1978. NEPA, and its attendant requirements for disclosing environmental effects of agency actions, increasingly were perceived by agencies as something real, rather than simply a Congressional symbol. Early actions by federal agencies that flaunted the intent of NEPA had gotten the agencies into trouble, and given nongovernmental intervenors a new tool with which to muck around in agency decisionmaking. The BLM had received this lesson on the knee of several federal court judges. According to one BLM staff member, when NEPA was passed, "it was just unbelievable. People in our district management thought it never applied to the BLM. It was just too far out. We're here to do things and cut trees. Agencies I guess tend to drag their feet as long as they can rather than getting something and trying to read it and implement. They wait till it's forced down their throat."

BLM had gotten into trouble earlier in the 1970s by preparing a single environmental impact statement on their entire 170 million acre grazing program,[4] and one on the timber management program in western

Oregon.[5] Lawsuits by environmental groups resulted in court orders to prepare plans and associated EISs on an area-by-area basis. BLM staffer Bill Nietro felt that this policy environment was the primary reason that the BLM state director bought into the spotted owl plan: "I think our state director at that time saw that he wasn't going to be able to get by without doing something for the owls. But he told us he demanded a quick fix. 'Give me a plan and let's go with it.'"[6]

By 1978, the Endangered Species Act was also being proven a significant piece of federal law, with the potential to affect agency choices whether they wanted it to or not. The celebrated Tellico Dam case demonstrated the significance of both NEPA and the ESA to federal agency decisionmakers. The Supreme Court's decision in 1978 that elevated the ESA Section 7 mandate to absolute status, taking precedence over all other agency mandates, was seen for part of what it was: an additional lever into internal agency decisionmaking that was real, potent, and capable of causing much organizational distress. Whether you believe that the FS and the BLM sought to consider ESA objectives in agency choices because (1) they believed in the endangered species mandates, (2) they were good soldiers carrying out Congressional intent, or (3) they feared the potential for chaos that could come from not considering endangered species in their decisions, all have the same effect on decisionmaking: Agency officials who could in earlier times disregard concerns about sensitive non-game wildlife species like the spotted owl, could no longer afford to do so.

The passage of FLPMA and the National Forest Management Act (NFMA) in 1976 were also important influences on the direction of the BLM and the FS. Both raised the significance of wildlife as a component in multiple use land management both directly and by establishing planning processes that allowed nongovernmental organizations a larger role in agency decisions. NFMA also contained a specific, yet vague, requirement that land management plans for the national forests "provide for diversity of plant and animal communities based on the suitability and capability of the specific land area in order to meet overall multiple-use objectives, and . . . provide, where appropriate, to the degree practicable, for steps to be taken to preserve the diversity of tree species similar to that existing in the region controlled by the plan."[7] While its meaning was still unclear in 1977 and 1978, the NFMA "diversity" language was at least a small influence in the FS's decision to support actions to protect owl habitat.[8]

The old growth issue itself was beginning to get some play. For example, the Siuslaw National Forest had developed an interim position statement on "Older Forest Communities," which argued for preserving options for the management of older forest communities while the land management planning process was underway.[9] In addition, Glenn Juday, the OSU doctoral student who had provided the first map of Oregon old growth to the OESTF, finished his dissertation work on old growth in the Oregon Coast Range late in 1976 and published an article on old growth management in *Environmental Law* in 1978.[10] The article pointed to the values of and threats to old growth in the Oregon Coast range, and noted the necessity for the FS to protect and manage a portion of the remaining old growth forest under the agency's control. Juday proposed that national forest plans have an explicit old growth management plan within them, and that Congress enact an explicit statement that the ecological integrity of the forest ecosystem must be maintained in national forest management. His work was one of the first statements of the need for comprehensive FS old growth protection that got much play outside of the fairly narrow set of administrative and university experts that had been involved in the issue up to that time.

The broader political environment of public forest decisionmaking was clearly in flux through the 1970s, and was an additional factor underlying agency support of a minimal level of spotted owl protection in 1978. By 1978, it was clear that neither the FS nor the BLM could avoid dealing with demands of the 1970s environmentalists. Battles over clearcutting and wilderness designation on the national forests, and oil and gas leasing, grazing, and desert and roadless area protection on BLM lands were fought in administrative and judicial appeals processes, with the policy direction of both agencies increasingly getting mired down in protracted, visible, public controversies. The enactment of both NFMA and FLPMA were partly in response to the new politics of public land management as it emerged in the early 1970s, with prime supporter Senator Hubert H. Humphrey noting that the purpose of NFMA was to "get the practice of forestry out of the courts and back to the forests."[11] While the FS hoped that NFMA would let it regain control over the direction of forest management, the battles preceding the act demonstrated that the agency's freedom of action could be constrained by outside parties.

The way that the NFMA and much of the other 1970s environmental legislation was written also represented a level of Congressional involvement in day-to-day administrative management that was unseen earlier, and was seen as a threat to agency decisionmakers. The NFMA, the ESA, and the Clean Air Act all prescribed specific behaviors that in an earlier day would have been left to agency discretion. For example, NFMA mandated that a group of scientists be formed to frame regulations that would identify maximum size limitations on clearcuts of different tree species, choices that hitherto had been totally within the discretion of FS line officers. (Ironically, the specification of clearcut size also increased habitat fragmentation, and exacerbated some of the sensitive species habitat problems that would develop in the 1980s.) Congressional attitudes were that the agencies had screwed up, and that they could not be trusted to make choices in the public interest. In addition, the post-Watergate and post-Earth Day generation of Congressional representatives was more independent of Congressional leadership or traditional agency objectives. The politics on the Hill were changing, and the federal land managing agencies had to adapt or risk more Congressional incursions into their administrative discretion, an action they viewed as organizational rape.

While changes in the legal and political environment help to explain why the agencies agreed to support owl protection measures in 1978, they do not fully explain the nature of the agencies' response. If denial is the first wave of response of an individual or an institution to an unwanted change in their lives, coopting the change is the second. After an agency is forced to respond to a policy shift, its response generally is to implement it, but to do so in a way that minimally changes day-to-day operating procedures. The old dog may learn new tricks, but will perform them in a way that looks remarkably similar to old behaviors.

The nature of the response by the FS and the BLM in dealing with the 1978 spotted owl management plan reinforced a pattern of behavior that remained consistent throughout the next ten years of issue history. By and large, the agencies tried to put owl and old growth protection into their existing standard operating procedures. Four elements of response can be seen: First, establishing small set-asides for owl protection as a matter of national or regional policy, with actual implementation dependent on the decisions of forest-level managers; second, making all choices on an interim basis, with the "real decision" made in ongoing planning processes; third, doing enough to avoid losing control by having the owl listed as a federally threatened or endangered species; and fourth, maintaining an aggressive timber program that increasingly precluded options for owl management. Each of these elements of agency response was present in the 1978 decisions of the FS and the BLM, and was equally a part of agency direction ten years later.

By the late 1970s, the FS was used to responding to new demands with small, isolated set-asides that had minimal impact on the timber program, even though they appeared to violate the concept of multiple use. Wilderness designations had been made that, up to this time, were largely higher elevation lands with limited value for commercial timber operations. Research natural areas had been created, as had national recreation areas. As a matter of regional policy, it was not a big leap to set aside owl habitat areas, particularly if they were not a threat to the timber program, and generally, FS officials planned on accommodating their designated share of owl pairs in existing reserved areas. For example, in the Willamette National Forest, the forest with the largest owl pair allocation, and the largest timber producer in Region 6, FS officials planned on locating owl habitat areas within existing wilderness, roadless recreation, and research natural areas.[12] On the adjacent Mt. Hood National Forest, Forest Supervisor (and later Chief) Dale Robertson responded to a regional office request for information about the "resource values that will be traded off to provide spotted owl habitat on a sustained basis"[13] as follows:

> The folks on the Mt. Hood compliment your staff and the Oregon Endangered Species Task Force for a job well done. . . . A preliminary review suggests that the Mt. Hood can provide spotted owl habitat on a sustained basis for a minimum of 35–43 pairs. To assume this responsibility, it will not involve so much a question of resource values to be traded off, but one of judicious planning and coordination of transportation corridors, timber harvest, recreation facilities and other activities. A reduction of timber yield is not anticipated. Management for the above pairs can be accomplished without entering programmed commercial forest lands proposed in the Timber Management Plan for the Mt. Hood National Forest.[14]

The second element of agency response was to make owl habitat decisions on a temporary basis, with permanent decisions to be made within established forest and district level planning processes. According to BLM Oregon State Director Murl Storms, "BLM will begin to implement the management plan immediately, consistent with ongoing planning efforts. . . . Final draft EISs on all western Oregon master units are scheduled for completion within the next five years. . . . At such time as the planning system is finalized by district, we will then proceed to initiate development of specific wildlife habitat management plans (HMP) for spotted owls."[15]

The FS, preparing to embark on NFMA-required planning, responded to the OESTF's spotted owl plan with the same themes: "Based on our present analysis, we can meet the [OESTF's 290 owl pair] target, and will take necessary action to ensure that allocation of lands to this resource are incorporated into our land management planning process. . . . The final decision on the number of owl pairs, their location, and distribution will be made through public involvement in developing land management plans for each planning area.[16]

To some observers, the nature of the FS and BLM response was simply a matter of the agencies trying to maintain control over the issue by putting it into their standard operating procedures. To others, this represented decisionmaking cloaked in a lot of smoke and mirrors. For interested nongovernmental parties, it was at least difficult to figure out what on-the-ground policy was at any point in time, and required a lot of effort to participate in forest- and site-level decisionmaking that affected the future of owl habitat. Tying owl protection to forest- and district-level planning also delayed the ultimate decision as the planning processes themselves generated a huge amount of conflict over a variety of issues that took a long time to resolve.

Besides temporary habitat set-asides, the third portion of the agency response was to carry out enough research and participate in interagency planning activities to demonstrate a good faith effort at owl protection, and in the process, avoid a decision to list the owl as an endangered or threatened species. The fear of having the owl listed was a consistent force creating incentives for agency action throughout the history of the owl controversy. The correspondence within the agencies throughout this period describes their concern at the idea of having the owl listed as a threatened or endangered species.[17] Eric Forsman described the impact of having the Endangered Species Act looming on the horizon: "I think the biggest threat even back then was the concern that the species would be listed. That was always the hammer that was being held over their heads. That if we don't do something, the Fish and Wildlife Service . . . was going to have to do their job, and list the owl. Back in those years, that was the biggest threat. The environmental groups were not nearly as well organized as they are now in terms of the political pressure that they were bringing to bear. They were there, and they were interested, but they weren't in court suing people . . ."[18]

While a variety of motives could underlie a fear of listing, the major concern of agency leaders most likely was that having the owl as a listed species would limit their ability to manage their lands as they thought appropriate, and would involve the FWS in FS and BLM decisions. For example, one of the rationales for the Siuslaw's position on old growth was the fear of having old growth-dependent species listed. "The restrictions associated with endangered or threatened status would deny management options to land administrators. . . . The objective to retain our management alternatives is an important Forest policy."[19]

In taking these actions, the agencies were not ignoring the issue completely; rather they were placing it within SOPs in a way that diminished its organizational impact and threat. Small owl protection areas, set aside temporarily while research and planning processes worked things out, might have been okay if other agency direction insured that there would be adequate options down the line, so that when decisions were made, they could be implemented. But the other clear dimension of agency behavior was to continue full steam ahead on timber production in a way that raised questions about options for owl protection in the future.

By 1978, there was a clear understanding on the part of governmental and nongovernmental timber interests that there would be timber supply problems in the not-too-distant future. OSU forest economist John Beuter described a coming timber supply shortage in a January 1976 report. His report forecast that given that "current policies and actions will persist, then declines in harvests are forecast within the next 30 years for western Oregon as a whole, and for all timbersheds in western Oregon except the North Coast."[20] Timber harvest was expected to decline by about 22%, compared with 1975 harvest levels, by the year 2000, with a corresponding reduction in direct employment of 27%. According to Beuter, "None of the projections . . . was meant to be a prescription. The gap between the current situation and reasonably possible capability is merely an area for policy consideration and negotiation. Evidently there is a considerable amount of leeway." The leeway, however, was all in federal timber, and the obvious solution to getting through the coming timber famine, the so-called "Beuter gap," was to expand production from national forest and BLM lands.

Expanded production from national forest lands was what Pacific Northwest timber interests expected, and the FS planned for, in spite of evolving public values toward other uses of the national forest system. The late 1970s could have been a time for the FS to expand and retool its constituencies in line with the shifting public values, but instead agency leaders continued doing what they knew best, and politically what they probably had to do. The 1977 appointment of Dick Worthington as Regional Forester in Region 6 is a good indication of agency direction at the time. In a newspaper interview, Worthington, who was returning to the Northwest after heading timber management in the Chief's office in Washington, D. C., highlighted his plans for increased timber yields:

We won't approach the high-yield investments of Weyerhaeuser and some of the timber companies, but we will manage some of our land as intensively as they do. Nationwide, we could more than double harvests, but that's not to say we will if land-use planning or other decisions reduce the potential. . . . We'll lose our shirt if we don't harvest more of the old growth that is dying and going to waste. . . . I can take you into the Mt. Hood or Pinchot forest and show you dead tree after dead tree. We can't afford to continue doing some things the way we have been. . . . No one knows how much wilderness we need. [I hope that the new roadless area review ordered by Assistant Agriculture Secretary Rupert Cutler] resolves any constraints we have on land that most folks think does not constitute wilderness so we can get on with multiple-use management. . . . As a football coach who stresses blocking and tackling before fancy passing, I want to get our people more versed in the basics of good forestry. You have to know how to use a shovel to put out a fire before you drop retardants from the air.[21]

In the face of an agency that was gung ho on expanding the cut off its lands at the same time its harvest practices were being constrained by outside pressures and forces, what actually happened on the ground in response to the OESTF's first plan depended on the good graces of the individual district ranger or forest supervisor. Those places where the line officers were supportive, and wildlife biologists were vocal, like on the Siuslaw, actually protected owls and owl habitat. In many other places, however, line officers barely went through the motions, and owl habitat areas were more policy than reality, and BLM districts echoed this pattern.[22] As it turned out, the response of the agencies to the OESTF's plan satisfied no one for very long. While the FS and the BLM hoped that they would get by the owl roadblock, neither timber nor preservation interests were happy with agency actions, and their voices were getting louder and more potent with time.

Interest Group Action Begins: Appeals and Political Pressure

While the agencies' responses to the spotted owl plan were tentative and of little impact, they were also the first actionable decisions of the agencies on spotted owl management. Barring no specific statutory deadline that agencies must meet, as long as no decisions have been made, they cannot be challenged. No decisions means no appeals of decisions, which lets the agencies maintain control and minimize political discomfort. It also means that the agencies can send mixed messages to different constituent groups. To timber, "Don't worry. This owl business will blow over with little effect on your interests. Trust us." To environmental groups, "Don't worry. We'll protect the owls. Trust us." Until the agencies take a stand by buying into a specific direction, it is hard for nongovernmental entities to figure out what is real agency intent, and it precludes their appealing any choices. Hence, the agencies consistently made choices by appearing to not make choices.

The FS and BLM decisions to follow the OESTF's spotted owl plan were the first time interest groups had something to shoot at, and both environmental and timber interests responded as best as they could given the resources and modes of action available to them. Environmental groups had been active in Oregon for many years, though most were fairly mainstream birding, hunting, and fishing clubs. A budding wilderness movement began organizing in the state as early as the 1950s. For example, the Sierra Club established its first chapter outside of California in Oregon in 1954, and the Oregon Cascades Conservation Council was formed in 1960 to focus exclusively on the issue of wilderness, but the political strength of the state's timber industry limited their effectiveness.[23]

In the 1970s, the situation began to change as Oregon developed an image as an ecologically conscious place. Governor Tom McCall signed the first state-level disposable beverage can law and a landmark land use law, and discouraged growth by asking potential immigrants to stay where they were. The social activism of the 1960s, the attitudes created about government during the Vietnam War and Watergate crises, and the development of a national environmental movement with considerable legislative success in Washington, D. C. during the early 1970s generated the conditions for a new style of environmental activism in the state, pressing for action on wilderness and other environmental issues.

Increasingly, battles over wilderness were fought over lands with value as timberland. In the 1930s, the FS had administratively set aside much of the higher elevation roadless areas, which were later designated by Congress as wilderness in the 1964 Wilderness Act. Environmentalists began to move down the mountains from "rock and ice" roadless areas to more valuable middle-elevation lands, just as commercial timber interests had finished logging lower elevation lands and were moving up. The major wilderness-oriented environmental groups, the Sierra Club and The Wilderness Society, saw the need to move the politics along in their direction as fast as possible, yet knew that their image was seen as too radical by many Oregonians, particularly those living east of the Cascades where many of Oregon's roadless areas lay. They decided to create a new group that would not carry the Sierra Club stigma, and in February 1974, helped to midwife the Oregon Wilderness Coalition (OWC) into existence.

As OWC developed, it moved the wilderness debate away from recreational and aesthetic arguments to a focus on wildlife and ecological values. According to FS historian Dennis Roth, Jim Montieth, OWC executive director for almost all of its existence, "was instrumental in forming a statewide scientific network in support of wilderness."[24] By 1976, Congressional testimony on Oregon wilderness focused more strongly on "the needs of certain wildlife species for undisturbed habitat; with concern for critical soils; and above all, with an understanding of scientific, historical and ecological values."[25]

OWC also created a new style of environmental activism in Oregon, as the staff aggressively pursued its issues.[26] By 1977, Montieth had given up on getting relief from administrative planning processes, and felt that: "the only way we can insure that the majority of de facto wilderness will have even a remote chance of remaining wild is to meet the problem nationally (or at least regionally) in the courts. . . . Without a nationwide suit which legally challenges the process being used to destroy wilderness, conservationists stand to lose more roadless areas to development. Given our limited resources, if bad decisions are not challenged in the courts regionally or nationally, we have no method to stop the destruction of perhaps 90% of the wild lands. National court action . . . will give us the de facto wilderness to work with in five years."[27] Montieth's decision led to a policy of aggressively seeking delay in administrative decisions over roadless areas, while simultaneously nationalizing the issue so that national environmental politics would overcome regional political biases.

The OWC style also set up the organization for a conflict between its Sierra Club patrons and its new directions. According to Roth, "The Sierra Club leadership in Oregon thought that the OWC was politically naive, while OWC prided itself on its Indian and sportsmen constituency and felt that the Sierra Club was an 'elitist western Oregon recreation group' populated by the 'wine and brie set.' . . . OWC was convinced that they were charting a new course for Oregon wilderness, one that avoided 'left wing urban anti-hunting groups' and sought out new constituencies to bring the wilderness movement more in line with basic western lifestyles."[28] The conflict between the two organizations started in the late 1970s and continued through the life of the spotted owl controversy, as the OWC became the Oregon Natural Resources Council. ONRC aggressively moved the wilderness and owl issues onto the national agenda, in spite of the concerns of the national environmental groups that the potential costs of a backlash from timber-dominated members of Congress were not worth the benefits claimed by ONRC.

Besides OWC, other seeds of environmental activism were growing in the Pacific Northwest. National Wildlife Federation, generally an old-line, middle-of-the-road organization, developed a legal program in the 1970s, attaching activist-oriented legal clinics to law schools around the nation. One of the NWF clinics was located at the University of Oregon in Eugene, and became an important tool for environmental activists seeking to change agency behavior through court action.

OSPIRG, the Oregon Student Public Interest Research Group, a student-funded environmental activist group located in Portland, was also created in the 1970s, and one of its part-time staff members, Cameron LaFollette, got involved in the old growth and spotted owl issues. In the summer of 1978, LaFollette started tracking the old growth issue and wrote a report for OSPIRG entitled, "Saving All the Pieces: Old Growth Forest in Oregon."[29] Her writing was the first visible comment on the spotted owl plan, outside of those made by the internal agency experts. In the June 1978 copy of OSPIRG *Impact*, she wrote, "The decisions of this plan do not have a solid grounding in biology; they are entirely political in nature. Old growth forests are highly valued by the public land management agencies for their timber production. The spotted owl plan looks very much like an effort of these agencies to have their cake and eat it too."[30]

LaFollette's assessment claimed that protecting 400 owl pairs would mean that 1100–1600 existing pairs would be lost, and noted that no one really knew what constituted a minimum viable population level for the species. "Since we do not know this, it is wiser to leave ourselves a margin of error by providing for as many owls as possible. . . ." In commenting on the decision to leave final choices of owl habitat areas to the land planning process, she noted, "Historically, neither the Forest Service nor the BLM have placed high priority on wildlife, especially nongame species. There is every reason to believe that the spotted owls will receive the least productive and healthy lands for their habitat—that which is least valuable for timber production."

LaFollette also pointed out that the owl is not the "crux of the problem: it is the old growth forest community which is highly complex both in flora and fauna. There is no biological sense in using one species as an indicator for the welfare of all the other old-growth dependent species, each of whose needs are vastly different from those of the spotted owl."[31] According to LaFollette, "The spotted owl management plan, ostensibly designed to protect this diminishing species, may actually endanger the birds and the old growth forests they require for survival. By rescuing the spotted owls from the federal threatened and endangered species list, federal agencies like the Forest Service and the Bureau of Land Management have neatly circumvented any positive action in regard to old growth management. In fact, the spotted owl management plan is fast becoming the only old growth management scheme. This plan has become more than a recovery plan for a species that was sliding toward extinction; it is substituting for a real old growth preservation plan."[32]

After attending a January 1979 meeting of the Spotted Owl Subcommittee of the newly formed Oregon–Washington Interagency Wildlife Committee (OWIWC), LaFollette realized that the spotted owl plan had been adopted with no explicit environmental analysis or public review, and recommended that an environmental impact statement with full public comment be prepared on the owl plan.[33] Her analysis, written as a study for the Cascade Holistic Economic Consultants (CHEC), an "advocacy research" nonprofit organization based in Eugene,[34] became the basis for the first administrative appeals on spotted owl management. It also established the direction of interest group legal strategies for a number of years. The spotted owl was not a federally listed endangered or threatened species, and NFMA was too new to try to figure out what its diversity language meant for agency behavior. That left NEPA as a tool for delay and documentation, one that had been proven effective through environmental lawsuits through the 1970s.

In an appeal dated February 11, 1980 and filed on behalf of OWC, CHEC, the NWF Law Clinic, and the Lane County (Eugene) Audubon Society, with NWF, the Oregon Wildlife Federation, and the Portland Audubon Society as Intervenors, University of Oregon/NWF Law Clinic attorney Terry Thatcher and Legal Intern William Devine argued that a region-wide EIS should be prepared on the spotted owl plan, and that timber sales in owl habitat be suspended until an acceptable impact statement was produced.[35] An almost identical appeal was filed before the Interior Board of Land Appeals, the administrative review body with jurisdiction over BLM decisions, on February 25, 1980.[36]

NEPA requires the preparation of an environmental impact statement for any proposal for major federal action significantly affecting the quality of the environment, and the appellants argued that by adopting a plan that protected 400 owl pairs, the FS was implicitly deciding the fate of other owls and their associated old growth habitat and hence had proposed a major federal action. The Regional Forester responded by stating that the owl pairs protected under the management plan represented minimum numbers only, and the decision was not a decision: real choices would be made in the forest planning process.[37] As to public review and comment on the plan, the Regional Forester noted that while "the many working meetings of the Oregon Endangered Species Task Force were not widely publicized . . . these work sessions were certainly not private and members of the public did occasionally attend."[38]

In response to the Regional Forester's argument that supporting the owl plan was not a proposal of a major federal action, the appellants quoted the NEPA regulations that defined when an action constituted a proposal: "[A] proposal exists at that stage in the development of an action when an agency . . . has a goal and is actively preparing to make a decision on one or more alternative means of accomplishing that goal and the effects can be meaningfully evaluated."[39] They also noted that under the judicial interpretations of NEPA that existed in the late 1970s, impact statements on many pieces of an agency action, such as would be the case if owls were dealt with in thirteen forest plans, were not sufficient to satisfy NEPA requirements if they missed cumulative effects.[40] The Regional Forester was not convinced, viewing owls as simply one of the many uses of national forest lands: "The spotted owl is an important resource on National Forest lands—but there are many other important resources. Writing separate Regionwide management plans for all these important individual resources would be extremely time-consuming and expensive. It is precisely this type of fragmented functional planning that NFMA sought to eliminate by requiring one plan which would meld the individual use demands into a single allocation document."[41]

In his decision on the appeal dated August 11, 1980, the Chief of the FS supported his Regional Forester, though he set out a different argument for why NEPA had not been violated: "No NEPA violation has occurred because the plan is an affirmative protective measure to implement interim controls to ensure protection of the spotted owl resource, pending consideration of spotted owl management in NFMA Section 6 plans. . . . Implementation of the SOMP is not an action, but a restraint on action. The measure allows no more adverse impact to spotted owl habitat than was previously allowed before 1977; it reduces adverse environmental impacts pending preparation and adoption of Regional and Forest plans."[42]

The environmentalists had lost this first skirmish in what would become a decade-long administrative, judicial, and political battle. Not surprisingly, the decision on the BLM appeal reiterated many of the same
themes, with the Interior Board of Land Appeals ruling on February 25,
1982 that the BLM's use of the interagency owl guidelines did not constitute a major federal action requiring an EIS. The appellants decided
not to try their luck in the federal courts at the time. But time was on
their side, for in deciding the appeal in favor of the Regional Forester, the
Chief did provide owl preservation advocates a tool for future use: the
Regional Plan was to include "a proper biological analysis to determine
the number and distribution of spotted owls which constitute a viable
population . . .,"[43] an analysis that could provide another lever into
agency decisionmaking.

The environmental groups were working other angles on the old
growth and owl protection issues. Seattle Audubon had corresponded
with the FWS about the agency's assessment of the status of the owl population, and had been told that the FWS did not intend to list the species
as endangered or threatened. A January 1982 status review indicated that
the FWS considered the northern spotted owl a "vulnerable species," and
they included it in a list of "Sensitive Species" published in August 1982,
but they did not consider it threatened with imminent extinction. Nevertheless, FWS managers understood that the future of the owl was dependent on effective habitat planning and protection on the part of the FS
and the BLM. According to FWS Olympia Area Manager Joseph Blum,
"The success of ongoing planning and associated management actions
will dictate future recommendations."[44]

The environmental groups also continued to press on the wilderness
issue, though proponents of wilderness designation had their good days
and their bad. Most were very disappointed by the outcome of the FS's
second roadless area review analysis, dubbed RARE II. The outcome of
the agency's analysis was that of the 3.4 million acres of land studied in
Oregon, the FS recommended 370,000 acres for wilderness designation.
Senator Mark Hatfield (R-OR) introduced a 600,000-acre wilderness bill
in 1979 that was opposed by both sides. Powerful House Ways and Means
Committee Chairman Al Ullman (D-OR) wanted less wilderness than
Hatfield's bill proposed, while Jim Weaver (D-OR), the freshman Congressman who was the patron of the environmental groups, wanted more.
Sierra Club and The Wilderness Society opposed the so-called permanent
release language of the bill, which would permanently open up lands that

were not designated wilderness to other purposes, including logging. The controversy deadlocked and resulted in a pouting Mark Hatfield closing the window of opportunity for Oregon wilderness designation: "Mr. Weaver and company . . . are not going to have an Oregon wilderness bill in this Congress," said Hatfield.[45]

The environmental groups, particularly those organized at the local or regional levels, also continued to be involved in unit-level planning where, in theory at least, real choices would be made. For example, the Umpqua National Forest, a westside forest in south-central Oregon, published a final environmental impact statement on its land management plan in 1978, which was quickly appealed by the Umpqua Wilderness Defenders on the grounds that the EIS gave inadequate consideration to old growth, spotted owl, and other wildlife habitat issues.[46] Similarly, proposed timber sales on the Willamette National Forest were appealed by the University of Oregon Law Clinic in 1979 due to their impact on old growth and spotted owl habitat.[47]

Where much is at stake, strategic behavior on the part of one set of interest groups almost always generates comparable behavior on the part of opposing interest groups, and timber began to be a much more visible force countering the efforts of the new activist environmental groups. While it had been an overwhelmingly influential force in the Northwest for many years, its mode of influence traditionally had been quiet yet compelling pressure in state and federal political arenas, and on the public land management agencies. As the activities of environmentalists began to be mildly threatening to their interests, the timber industry stepped up efforts to counter owl protection measures.

Up through the 1970s, the environmentalists had not been a real threat to industry control of the direction of public land management. Gordon Gould's early work in northern California generated a public relations response from the National Forest Products Association (NFPA) and the Western Timber Association as early as 1975, with both interpreting Gould's results in a very selective way to conclude that the owl was neither endangered nor threatened in California. The industry also began to play the research game, by supporting field research on owls and old growth management in the late 1970s. The results of most of these efforts did not contradict the information being acquired by the agency experts, but did tend to stress different conclusions about management implications. For example, an NFPA-funded owl survey in northwest Washington in 1976 concluded that there were real differences between owl occurrence in old growth and second growth, but that "the continued harvesting of old-growth forests will probably result in a decrease but not extirpation of the spotted owl population in Washington."[48]

Other quasi-scientific responses of the industry were more clearly advocate science. Most of the owl biologists felt that the activities of Dr. Robert Vincent, a consulting biologist for the industry, fell into this camp. His comments on the Umpqua Wilderness Defenders (UWD) appeal of the Umpqua National Forest Plan are illustrative. Concluding that the owl was neither endangered nor threatened and would survive well in a managed forest, Vincent argued that:

> The northern spotted owl is an adaptable bird. It is found from southern British Columbia to San Francisco, from sea-level to 6000 feet. It lives in rain forests, and it lives in arid, mixed-conifer forests of southern Oregon and on the east slope of the Cascades. It nests in fir, in pine, in cavities, in abandoned nests of other animals, in platform-like areas on large limbs, etc. It lives in mixed coniferous forests with almost no understory or ground cover and lives in Douglas-fir forests with a dense canopy and several thick understories.
>
> This owl has a remarkable ability to fly long distances in search of suitable habitat and an untapped potential to adapt. Spotted owls live along major highways and along logging roads. In fact, one lived for several months on the window sills of the upper stories of a retirement home in Medford; others have been accurately reported to me from downtown Springfield, Gresham, and San Diego.[49]

New Information, New Demands, Limited Agency Response

The early responses by the timber industry were made in response to a fairly minimal threat, but new information generated by the research work of Eric Forsman and others started to point to an expanded set of habitat requirements for the owl. Forsman's radiotelemetry work, completed in 1980, suggested that owl home ranges were much larger and contained more old growth than had previously been thought. In studies of fourteen adult birds conducted in 1975–76 on the H. J. Andrews Experimental Forest in the Willamette National Forest, and on BLM lands in the Coast Range in 1980, mean home range sizes were 3859 acres in the Cascades and 4728 in the Coast Range. The minimum amount of old growth found within a home range was 740 acres, and the minimum amount of old growth found within the combined home ranges of an owl pair was 1009 acres. All of these numbers suggested considerably larger habitat requirements per pair than had previously been estimated.

The new information obviously had management implications given that the agencies were protecting minimum land areas for owl pairs, and the members of the Oregon Endangered Species Task Force knew the implications of the updated information. Formally, the OESTF had been incorporated into a broader Oregon-Washington Interagency Wildlife Committee, with ODFW's Bob Stein as its chairman. Based on Forsman's telemetry results and studies in the Klamath National Forest in California, the OWIWC agreed to forward a revised plan to the FS, BLM, and Oregon Department of Forestry in March 1981. The proposed revision stated that: "On each site managed for a pair of owls, maintain an old growth core area of at least 300 acres around the nest and an additional 700 acres of old growth within a 1.5 mile radius of the nest. If 1,000 acres of old growth does not exist within a 1.5 mile radius of the nest, then substitute the oldest stands available and manage them in the same manner as old growth."[50]

Upping owl protection set-asides from 300 to 1000 acres of old growth per pair was destined to be controversial, and the Committee noted that they were "fully aware that timber harvest will be influenced by the recommended increase" in habitat areas, and that additional studies were needed to verify the habitat requirements for the owl. To respond to the agencies' concerns, the Committee changed the recommendation so that forests be managed "so that the option to provide 1,000 acres of old growth per pair within a radius of 1.5 miles of nest sites is maintained." Their recommendation was to guide management for five years, at which time new information could be used to revise the plan.[51]

The response of the agencies to the revised owl plan was not overwhelmingly supportive. According to BLM's Bill Nietro, "Formal acknowledgment of the acceptability of these revisions was not forthcoming from the Bureau of Land Management or the Oregon Department of Forestry. The U. S. Forest Service initially objected but subsequently agreed to incorporate the revised standard into its planning activities."[52] According to Eric Forsman and Chuck Meslow, "Both agencies balked at accepting the 1,000-acre guidelines, opting instead to manage for only 300 acres of old growth. However, the Forest Service did agree to 'retain the option' to manage for an additional 700 acres of old growth in each spotted owl management area if it eventually became apparent that 300 acres were not enough."[53] In fact, the FS never formally adopted the

Committee's 1000-acre recommendation and used 300 acres as its guidance for land management for the next several years.

The Coos Bay Plan and Changing Political Realities

The BLM did not have the luxury of waiting out the preservationists as they moved through an extensive and confusing planning process, and perhaps nothing set the tone of what was to come in the evolution of the spotted owl controversy as the response to the BLM's proposed Coos Bay plan. The Coos Bay district plan was the first ten-year timber management plan issued under the 1970s court order and the first plan issued during the Reagan administration. A final EIS was issued in March 1981 for public comment, and it immediately got a response.

The EIS adhered fairly closely to the 1970s recommendations of the spotted owl committee, including protection of 16 pairs of owls, based on 300 acres of old growth per pair. "The plan proposed to place 15% of the mid-aged and old-growth stands in the commercial forest under constrained management for wildlife."[54] Once again, the biologists were forced to play the minimum game: "BLM biologists had said that preservation of 25% of these age classes in the commercial forest overall would provide optimum conditions for wildlife, but that 15% would maintain most species at minimum populations."[55] Nevertheless, the older growth set-asides resulted in a 7% reduction in the allowable cut on the Coos Bay district.[56]

The Coos Bay plan was the first time owl management translated directly into lost timber, and the industry responded with media attacks and political pressure in the Northwest and in Washington, D. C. For example, a June 17, 1981 article in the Coos Bay newspaper quoted a number of timber managers about the impact of the owl plan. "By locking up large chunks of land for spotted owls, the federal agencies are reducing the amount of acreage available for timber production. They are decreasing the allowable cut, and making the remaining timber more expensive. That cost ends up coming right back to your table and mine," said John Rollin, logging manager for Champion Building Products in Gold Beach.[57] Jim Izett, timber manager for South Coast Lumber Company, Curry County's largest private employer and a firm that was 95% dependent on federal timber, argued that more information was needed before a choice was made: "I don't think we should be panicked into locking up all that timber before we find out some more answers. Animals cope. Animals can learn to survive under different conditions. If they can't they die out—

like the dinosaurs. I've gotten along just fine without the dinosaurs, and I think I could live without spotted owls, too."[58]

The industry developed numbers describing the potential impact of owl protection. In a "white paper" subtitled "What You Always Needed to Know About the Northern Spotted Owl—But Didn't Know to Ask," and prepared by Dennis Hayward of the North West Timber Association, a timber lobbying group very active on this issue in the early 1980s, impacts of owl management (at 300 acres of old growth per pair) on FS and BLM lands in Oregon would be a loss of 300 million board feet, a loss in timber receipts of approximately $120 million, and a loss of more than 2000 jobs in the state.[59] On lost timber receipts alone, the plan would cost approximately $300,000 per owl pair.

The Coos Bay plan could not have hit at a worse time in terms of the economic health of the industry, as the recession of the early 1980s started to take hold. According to Jim Izett of South Coast Lumber, "The timber industry is in bad shape. The withdrawals of land for wilderness areas have hurt. The high interest rates have hurt. It's almost getting so that people can't afford to buy a home anymore, everything is so expensive. People don't understand the situation we're in. I've always thought maybe if we could somehow manage to take toilet paper off the shelves for six months, all over the country, people would realize how important the timber industry really is."[60] Senator Mark Hatfield summarized the political response to the Coos Bay plan: "If I'm reduced down to a question, and I don't think it's that simple, where I'd have to choose between a spotted owl and the welfare of humanity and human beings, I'm going to choose my fellow human beings."[61]

The reaction brought intense pressure on BLM's leadership, which began a process of flip-flopping on the issue, trying to find a way to respond to the pressure and a decision that would endure. In May, the Oregon State Office clarified its direction for district-level planning, so that spotted owl allocations had to be considered in the EIS, but did not have to be contained in the preferred alternative.[62] An urgent call for help to the Interior Department lawyers got a clarification of the lawyers' opinion about the obligations of the agency under conflicting direction from FLPMA and the O&C Act. In their view, O&C-mandated forest production took priority over FLPMA-mandated wildlife protection.[63] BLM Director Robert Burford went beyond the lawyers' opinion, and sent a clear message as to the D. C. office's (and the Reagan administration's) priorities: "State-listed species will be considered, but habitat . . . will be provided only to the extent it comes from noncommercial timber lands, or whenever a land use allocation . . . coincidentally can provide a secondary benefit for these species."[64]

The Oregon State Director took Director Burford's advice to heart, and issued a draft decision on the South Coast-Curry Timber Management Plan in the Coos Bay District that proposed a 14% increase in the level of timber production from the previous plan.[65] To achieve this level of production, the plan proposed to abandon most of the protective measures provided in the earlier EIS for spotted owls and other old-growth-dependent wildlife. The proposed decision would reduce the numbers of owls on the district to three or four pairs.

It was the environmentalists' turn to respond, and ONRC, the National Audubon Society, and The Wildlife Society began to consider petitioning the FWS to list the spotted owl as an endangered species, or trying to get Congress to amend the O&C Act. In October, 1982, the Oregon Fish and Wildlife Commission found the BLM's proposed decision to be in violation of state wildlife policy, the Sikes Act, FLPMA, the ESA, and the BLM's own planning regulations. The Oregon Department of Land Conservation and Development (ODLCD) found that the proposed decision would be inconsistent with the Oregon Coastal Zone Management Program. Indeed, even though the BLM backed off its draft decision in a proposed final decision, the ODLCD served the BLM with an order prohibiting implementation of the final decision, an action authorized through provisions of the federal Coastal Zone Management Act that mandates that if a state has an approved coastal zone management plan, all federal actions have to be consistent with it.

In response to the controversy, a solution was negotiated between BLM and ODFW in September 1983 that indicated that for the next five years the BLM would manage habitat so as to maintain a population of 90 spotted owl pairs.[66] They would also work together on research and monitoring, and revisit the issue no later than October 1, 1988. While BLM leadership remained fairly unconcerned about owl protection, they were concerned about the impacts of a decision by the FWS to list the owl as a threatened species, because the ESA would clearly override the O&C Act, and land managers would be stuck with a much more restrictive situation than if they adopted the spotted owl plan's recommendations. In addition, if Congress began looking at the O&C Act, they might also start to revise the revenue formula by which the eighteen western Oregon counties receive 50% of timber revenues, an amount double that received from FS lands, and about five times what the counties would get in taxes if the lands were privately owned.[67] In addition, the Forest Service became increasingly concerned about the implications of the BLM decision on management on FS lands. If the BLM protected no owls, it would presumably increase the number of owls to be protected on FS lands, an outcome FS leadership clearly disdained.

According to Forsman and Meslow, "The Coos Bay experience apparently convinced BLM that it was not going to be possible to depart significantly from the interim spotted owl agreement without serious political and legal repercussions," and the other district-level timber management plans followed the interim guidelines closely.[68] The battle over the Coos Bay plan also demonstrated the rising strength of interest groups in favor of owl protection, and the commitment of some elements of Oregon state government to owl protection measures. The political landscape of federal forest management was changing, and the Coos Bay experience suggested that nongame wildlife issues would have to be addressed in a serious way by federal land management agencies in spite of the direction emanating from Washington.

3

New Science, New Directives, More Muddling: 1981–1984

The Forest Service had made it clear that their real response to any need to protect spotted owls would come in the NFMA-mandated planning process, and new scientific approaches to analyzing sensitive species issues were developing just as the FS was attempting to finalize guidance on how the forest plans should be constructed. Some FS staff members supported and embraced many of these new approaches, and tried to employ them in spotted owl planning with mixed results. In some areas, the new scientific approaches provided a method of analysis that gave insight into the level and character of owl protection measures; in others, they became a tool for nongovernmental intervenors to use in challenges; and in others, they became a technical veneer that made agency actions even more difficult for outsiders to penetrate and understand. Regardless, developing scientific knowledge in the early 1980s influenced the direction of the owl controversy, and transformed it into an issue of the 1980s and 1990s.

The concept of biological diversity and what it meant for land management underwent an evolution in the 1980s. In the early days of wildlife management, wildlife management meant growing game animals, and that was a goal consistent with timber management. According to Eric Forsman, "The early wildlifers, mostly, were thinking about managing hunting species, things like deer and elk, and their perception was that clear cut harvest was good for most of those species and therefore that type of management was good for wildlife. And they had been saying that to the forest managers, and forest managers had no reason to quibble with that. They thought that was great. So everybody was going along on the assumption that things were hunky-dory."[1]

In the 1970s, more attention was put on nongame species and biological diversity, and diversity was defined largely as species-level diversity, measured by the number of species in an area. Studies of song birds and other species led wildlife researchers to emphasize the creation of "edge" as a means of promoting diversity. Larger numbers of species were generally associated with edges, the boundaries between types of land cover, and the way to maximize diversity was to create edge. Again, this wildlife objective was consistent with timber management practices that employed a lot of patchy clearcuts, because in the process, they created a lot of edge.

The problem with this concept of diversity is that edge effects promote early successional species at the expense of species associated with later successional stages, including the spotted owl. In the 1980s, the scientific understanding of diversity started to expand to incorporate this reality.[2] By the mid-1980s, it was generally considered that the term *biological diversity* encompassed genetic, species, and ecosystem levels.[3] Unfortunately for timber managers, under this more inclusive definition, biological diversity was not necessarily maximized by creating a lot of edge through forest disturbances, and for the first time, diversity objectives were in opposition to timber objectives.

As interest grew in ecosystem-level processes and diversity, an enhanced understanding developed of old growth as an ecosystem. FS researchers such as Jerry Franklin and Jack Ward Thomas undertook a lot of the pioneering work on the old growth system, and as their understanding grew, they became advocates for increased attention to old growth as the object of more research and protection.[4] The late 1970s and early 1980s were also a boom time for research and writing in the area of conservation biology, a literature that contributed to our understanding of the science of biological diversity and its management implications. Research that tested and applied island biogeographic theory, explored the genetics of diversity, and examined the effects of forest fragmentation on species and ecosystems generated publications and understanding with implications for the management of national forests and their associated wildlife.[5]

By the early 1980s, the regulations had been defined and put into place for implementing the NFMA, and the interpretations of the group of scientists involved in writing the regulations defined and expanded on the concepts of diversity contained in NFMA in accordance with the developing science base of diversity and conservation biology. The NFMA regulations defined diversity as "the distribution and abundance of different plant and animal communities and species within the area covered by a land and resource management plan."[6] They set as a goal the management of fish and wildlife habitats so as "to maintain viable populations of all existing native vertebrate species in the planning area and to maintain and improve habitat of management indicator species."[7] The regulations did not define what constituted a viable population or a management indicator species though they did require that "population trends of the management indicator species will be monitored and relationships to habitat changes determined."[8]

What constituted a viable population became very important, as concepts of population viability began to substitute (on the surface at least) for the seat of the pants expert judgment that characterized endangered species decisionmaking in the 1970s. No one really knew what insured a viable population for management purposes, and the scientific literature on the subject was in its infancy. An influential 1981 paper by Mark Shaffer, a FWS scientist, established a set of operative concepts of minimum viable populations. In Shaffer's view, "A minimum viable population for any given species in any given habitat is the smallest isolated population having a 99% chance of remaining extant for 1000 years despite the foreseeable effects of demographic, environmental, and genetic stochasticity, and natural catastrophes."[9]

One of the first FS staff members to begin to interpret the concept of population viability for forest planners was Dr. Hal Salwasser, a regional wildlife ecologist in Region 5, and his interpretation of a viable population was as follows: "A viable population is capable of persisting on the Forest in the face of anticipated environmental changes, both natural and mancaused. It is not a maximum, nor even a desired population level. Rather it is the minimum level, below which the population is considered to be subject to local elimination due to habitat loss, disease, harassment or some other factor. Threatened or endangered species are considered to be below viable levels."[10]

Salwasser attached a graphic that clearly indicated that the legal threshold level for population viability was at the "minimum" level. While he no doubt intended his definition to be consistent with Mark Shaffer's notion of a minimum that is truly a viable minimum, meaning that the species can weather a variety of stochastic events, unfortunately, the subtleties of this definition of viability were lost on an agency used to providing the minimum to wildlife interests. Increasingly, minimum viable populations meant choosing the lowest number that was contained in the data, or that researchers would define. Minimum viable populations became minimal viable populations as agency officials were forced to make planning and management choices.

Drafting the Pacific Northwest Regional Guide

The FS's response to the owl issue is demonstrated well in the development of the Pacific Northwest Regional Guide (first called the Regional Plan), and the ambiguous and confusing direction established in it. NFMA and its cousin, the Resources Planning Act, established a step-down planning process, in which national assessments are translated into regional guidance which in theory directs forest-level planning. Region 6 was in the process of developing its Regional Plan when the owl controversy heated up, and the Chief, in denying the environmental groups' appeal of the nondecision over interim spotted owl management in his August 1980 decision, promised that a biological analysis of owl management alternatives would be forthcoming in the Plan.

A draft owl management analysis and proposal was prepared in the fall of 1980 and grappled with the notion of population viability. The draft analyzed the effectiveness of three alternative owl densities: A-Subsistence, B-Midrange, and C-Saturation, each with 300 acres of old growth established per owl pair, and varying on how far apart spotted owl management areas were located. At the subsistence level, 282 owl pairs would be supported on national forest lands; at the midrange, 693 pairs; and at the saturation level, 1298 pairs. The analysis located fragments of habitat in a network of appropriate spacing, and the resulting total number of habitat areas represented the number of owl pairs that could be supported. Alternative A used a maximum distance between spotted owl management areas of 12 miles, an estimate of the maximum distance owls were likely to disperse, while Alternative C used 2-mile spacing.

An interdisciplinary team (ID team) was formed to analyze the three alternatives, through a process that compared the alternatives to four "must do" objectives, and four "want to do" objectives. They also compared the three alternatives to a set of eleven concerns gleaned from the planning process's public involvement process. Based on this analysis, the saturation-level alternative scored the best and the subsistence-level alternative was determined not viable. Nevertheless, the planning team was concerned about Alternative C's feasibility in other realms: "While all ID team members agreed that management for spotted owls at this upper population density range would best meet spotted owl management objectives considered, it provided the least opportunity to manage for other resources, especially timber. It was also recognized, that attainment of the maximum population density portrayed on the Oregon and Washington base maps, is probably not attainable. Biological factors such as territorial needs of the birds, plus legal, economic and political issues, do not make this high range alternative desirable, and perhaps not even possible."[11]

Since the team's first set of analysis did not come up with the right answer, they tried again. At an interagency coordination meeting on spotted owl management held on November 5, 1980, they asked biologists from FWS, BLM, ODFW, and the FS for help: "With apprehension, the biologists agreed that any population density below 40% of the existing maximum population presented (750 pair for Oregon and 541 for Washington) would certainly not represent a viable population,"[12] and that "management of the spotted owl at or below the 40% level should be regarded as a subsistence or disaster density and would not represent a viable population over time."[13]

Based on the advice of the owl experts, the ID team recommended that spotted owls be managed on national forest lands at the 56% population level of the potential maximum population for Oregon and the 50% level for Washington. Since these numbers translated to 425 pairs of owls to be protected on Oregon lands, a 47% increase in owl protection over the level recommended in the OESTF's spotted owl plan, and 270 pairs on Washington national forest lands, they were not acceptable to regional leadership, and the ID team went back to work on a different way to provide a less impacting yet mildly legitimate FS owl plan.

The ID team was encouraged by FS researchers Drs. Jack Ward Thomas and Jerry Verner to incorporate some of the new understanding of viability resulting from the application of population genetics, and they invited Dr. Michael Soulé, a California geneticist, to consult with them on the numbers of owl pairs that would constitute a minimum viable population. Soulé was the author of one of the first texts that claimed the label of conservation biology and is an expert on the effects of genetic variables on population viability. He had published pathbreaking work in the late 1970s on the effects of genetic inbreeding on survival and, based on experimental work with fruitflies and theoretical modeling, suggested a rule of thumb for minimizing the effects of genetic inbreeding: maintain a breeding population of 500 individuals.[14] In a December 1980 meeting, Soulé offered his advice to the ID team that if the northern subspecies of spotted owls could be maintained at or above a population of 500 breeding pairs, genetic stability would not be a problem. He noted, however, that "this number does not represent the pairs required to distribute the owl throughout its natural range."[15]

While it was not intended to be definitive, the magic 500 number became very important. The next draft of the spotted owl portion of the Draft Regional Plan contained the same three alternatives, and an identical analysis of objectives and public concerns, but changed its overall assessment such that its recommended alternative would protect 510 owl pairs. This assessment dropped any discussion of the 40% of maximum population that had been so influential in the previous draft, and instead worked off the subsistence level Alternative A. Alternative A had been judged not viable largely because it relied on a number of spotted owl habitat areas with only a single pair of owls in them. The experts felt that many of these pairs would fail over time, and the new draft computed a minimum viable population based on the subsistence density with an extra set of owl pairs thrown in to replace those that would fail over time. This analysis produced a minimum viable population of 350 owl pairs. Finally, they compared the 350 pairs against Soulé's magic 500 number, and came up with a recommended alternative of 510 pairs (296 in Oregon national forests; 214 in Washington national forests). The team noted that, "the recommended alternative was not developed through calculations entirely based on scientific data. In some cases, it represents the best expert opinion from biologists and other managers in relation to considering the above mentioned criteria plus issues and scientific data discussed in the many meetings and materials presented."[16]

The magnitude of owl protection called for in this draft plan was a concern to both owl experts and timber proponents. FWS experts continued to push for 425 pairs of owls to be protected on national forest lands in Oregon, noting in a letter to the Regional Forester that "it is our position that the Forest Service should not eliminate potential future options, and should make certain that population and habitat determinations have enough buffer to allow for all uncertainties."[17] To those concerned about the impact of the owl guidelines on timber objectives, the proposed guidelines were moving in the right direction (from 695 to 510 pairs) but had not gone far enough. In particular, regional FS leadership was still not satisfied with the ID team's analysis and proposed decision.

The final Draft Regional Plan put out for public comment in May, 1981 dialed protection back one more step, though did it in a way that attempted to finesse the underlying issues. Remember two items were in dispute: the number of owl pairs that constituted a minimum viable population, and the amount of old growth acreage required by each pair. The Draft Regional Plan walked a confusing line between the different opinions. Included as appendices were both versions of the spotted owl task force's management plan (their May 1979 version recommending 300 acres of old growth per owl pair, and their February 1981 update recommending 1000 acres of old growth). At the same time, while the Plan stated that current research indicated that the owls needed 1000 acres, it simultaneously maintained interim management at protecting 300 acres per owl pair.

The Draft Regional Plan also clearly identified the legally mandated minimum viable population (MVP) as the subsistence density modified by the 75% of single pairs procedure, which with updated calculations produced a protection goal for national forest lands of 375 pairs of owls (263 in Oregon and 112 in Washington). But the plan recognized the uncertainties involved by requiring forest planners to analyze the effects of seven alternatives: (1) their share of the MVP protecting 300 acres per owl pair; (2) their share of the MVP plus 30% more at 300 acres per owl pair; (3) their share of the MVP plus 60% more at 300 acres per owl pair, a perverse interpretation of the number of pairs recommended as a minimum by the interagency scientists (they had recommended 60% of the maximum biological potential); (4–6) each of the above three alternatives at 1000 acres per owl pair; and (7) minimal owl protection with no impact on commercial timber land.

In you think the guidance for owl management provided in the Draft Regional Plan is confusing to the casual reader, it was even more so to the forest-level staffers who had to use it in preparing their forest plans. Numerous questions about what the Plan required resulted in the regional office sending out a memo clarifying direction in November 1981,[18] which raised the level of confusion at least one notch. A response to questions posed by the forest supervisor on the Gifford Pinchot National Forest brought the following Regional Office response in March 1982:

As documented in the Draft Pacific Northwest Regional Plan, habitat characteristics presently considered necessary to support one pair of northern spotted owls includes an old-growth core area of at least 300 acres around the nest and an additional 700 acres of old growth within a 1.5-mile radius of the nest. The Draft Regional Plan does not, however, establish 1000 acres as a "minimum standard." . . . The ultimate management objective is to maintain a viable population of northern spotted owls. At this time, maintaining 1,000 (300/700) acres of old growth within a 1.5 mile radius of the nest is considered a habitat requirement to accomplish that objective. The October 28, 1980, letter giving interim management direction still holds. This only requires 300-acre protection core areas. You should, however, maintain your option to provide 1,000 (300/700) acres per pair of owls for all pairs protected . . . if it will not impact other programs to do so.[19]

The memo from the Regional Office also noted that other than protecting each forest's owl pair allocation of the MVP at 300 acres per pair, and requiring that the EIS attached to a forest plan show analysis of several alternatives, the Regional Plan made "no final decisions on spotted owls; these are made as Forest Plans are completed and the impacts made more visible."[20] You can read this as "Hey, don't yell at us. We're not making decisions. That's what forest supervisors will do through the forest planning process."

While the publication of the Draft Regional Plan did not mean much in terms of day-to-day management of the national forests in the near term, its inclusion of the 1000 acre figure of old growth set-asides, even in an innocuous way, was sure to trigger a response on the part of timber interests. In addition, the designation of targets for forests in Washington, the first time FS direction on owl management in forests in Washington was set out somewhat clearly, also expanded the reality of owl protection into a new set of potentially aggrieved parties, including timber interests in Washington, and their political allies in Washington, D. C. Timber groups stepped up their media efforts, calling the owl "our billion dollar bird,"[21] and wrote their Congressional representatives who wrote the Forest Service asking what was going on. The FS wrote back explaining the situation, and generally pointing out that no final decisions had been made. Rather, as a letter to Senator Bob Packwood noted, "The final decision on the number of spotted owl pairs will be made through public involvement and the land management planning process."[22]

Congressional inquiries are always viewed as important items for agency response, but their impact on decisionmaking can be blunted. Inquiries from Executive-level political bosses are much more serious, and Reagan administration appointees were mobilized to pressure the FS on its owl direction. For example, John Crowell, Assistant Secretary of the Department of Agriculture in charge of the Forest Service, and former Louisiana-Pacific executive in Oregon, wrote a memo to FS Chief Max Peterson in July 1981 that can only be interpreted as political pressure:

I am interested in preservation of the spotted owl, but I am also increasingly concerned about the costs of doing as much as is proposed. . . . We are proposing devoting between 112,500 acres supporting $1,125,000 worth of timber and 958,000 acres support-ing timber worth $9,584,000,000. . . . This works out to from $300,000 to $1,600,000 per year per pair.

The magnitude of those figures suggests that the guidelines need to be very carefully reviewed. Reduction in the number of pairs of spotted owls to be protected on each forest should be considered, and that the numbers assigned to each forest be located in areas adja-cent to similar areas in adjoining forests. There also seems to be enough indication from the increasing knowledge about Spotted Owls that they may be able to live successfully in territories which include second growth timber rather than in pure standards of old growth, although they evidently prefer the latter. Thus careful re-cordation of observations of owls and additional research about this species is obviously called for in view of the high timber yields in-volved.

Perhaps we should talk some about this with Dick Worthington at the Regional Foresters meeting.[23]

Besides providing some minimal direction for owl management in na-tional forest planning, the Draft Regional Plan also contained a statement of research needs related to wildlife and old growth ecosystems,[24] and everybody was interested in more research. Preservationists and their sup-porters in the agencies were interested in research that would establish owl dependency on large areas of old growth forest and provide needed information into the management process. Researchers wanted owl-related work funded because it let them pursue interesting research ques-tions, while putting food on their tables and getting them out of the bind of having to give advice based on scientifically challengable information. Timber interests and their allies in the FS were interested in research to prove that the owl could survive in managed second growth stands and to delay taking action on owl set-asides. Finally, agency leadership sought research to get them out of the hot seat, by either providing credible an-swers of a politically appropriate type, or by letting them maintain control over the timing and character of owl decisionmaking. Regardless of mo-tive, all interests sought more information about owls and their associated habitat, and it was easier for the policy process to provide research funds than it was for it to provide decisions about tradeoffs between owl and timber interests.

Clarifying Direction on Minimum Viable Populations

While the researchers were off gathering information, the national forests in Region 6 had the very real task of carrying out NFMA-required planning, and that meant figuring out what NFMA required. One forest supervisor—on the Wallowa–Whitman National Forest in eastern Oregon—tried out the notion of putting all his old growth set-asides in wilderness areas.[25] The response from Associate Deputy Chief J. B. Hilmon, drafted by Hal Salwasser, noted the importance of a well-distributed population: "The alternative which you have described may or may not be legal. A viable population, for planning purposes, is one which has the estimated numbers and distribution of reproductive individuals to insure its continued existence throughout its existing range in the planning area."[26] According to Salwasser:

> We had come to the conclusion [early in 1982] that we needed to clarify what the planning regulations said about population viability. We had a forester out in the Northwest that had said, well we're going to put all the spotted owls in the wilderness areas. And we said, no, we don't think that's going to work, because if you've got all these populations in the wilderness areas—twenty, thirty miles apart— chances are that the birds aren't going to be able to get from one to the other. Eventually, genetic, demographic, or catastrophic things are going to wipe them out, one at a time. The thing will just kind of unravel on you over a long period of time. . .
>
> We were going to get a lot of inconsistency. We didn't know at the time what we were going to get as far as the full range of approaches, but we were pretty sure that if forests came back and said that they were going to manage for all their old growth type wildlife species in the wilderness areas, that we probably would not be able to sustain population viability for a lot of species . . .[27]

The response to the Wallowa–Whitman became FS policy via a memo sent from Hilmon to the Regional Foresters.[28] Drafted by Salwasser, the memo was an important extension of NFMA policy for it not only required the maintenance of a viable population of a species, but that the species must be maintained across its historic range. Hence, forests that were planning to maintain a species on isolated forest fragments could no longer do so, nor would fragments connected by transportation corridors necessarily satisfy the viability requirements of NFMA, as interpreted by the agency. Complying with the guidance would lead planners to maintain a grid of wildlife habitat, with chunks of habitat separated by dispersal distances, looking like so many measles evenly distributed on the body of the westside forests. In undertaking their analysis of alternatives and impacts, forest planners were only to consider alternatives that maintained adequate populations in an appropriate distribution over the landscape. Alternatives could result in smaller populations than currently existed, as long as they maintained minimum numbers in an appropriate distribution.

These changes were also incorporated in the regulations that implemented NFMA when they were opened up for revisions in the fall of 1982. One of the reasons that the regulations were revised was because Reagan administration officials felt that the 1979 regulations did not insure that forest planners adequately considered economic impacts. According to Hal Salwasser, "While they opened the regulations up to fix that economic stuff, we used that opportunity to fix the biological stuff at the same time. Of course there is a lot of difference of opinion about how well it got fixed." [29]

The revision of the NFMA regulations was issued on September 30, 1982, and contained the new thinking of FS staff on the issue of population viability. A viable population was "one which has the estimated numbers and distribution of reproductive individuals to insure its continued existence is well distributed in the planning area." The regulations also indicated that "in order to insure that viable populations will be maintained, habitat must be provided to support, at least, a minimum number of reproductive individuals. . . ." [30] Hence, the forest plans were required to do two things: maintain a sufficient number of individuals to insure population viability, and insure that habitat for a species was distributed throughout a national forest.

Developing Minimum Legal Requirements for Wildlife Protection

The FS leadership's concern with consistency in Region 6 forest planning generated administrative guidance that locked in the concept of a well-distributed population and the figure of 1000 acres of old growth required per owl pair. Minimum management requirements (MMRs) were specified for national forest resources (timber, fish and wildlife, etc.) to act as a bottom line on what forests had to do to comply with their legal obligations as they went about forest planning, and more time was spent developing MMRs for wildlife than for any other resource. Regional Office staff understood that "The impact of meeting the wildlife requirements on the ability to produce other resources is potentially significant."[31]

The first set of regional standards for wildlife MMRs produced in June 1982 did not provide much insight into NFMA requirements, and simply advised the forests to specify the assumptions that they used in planning. The forest supervisors were "very uncomfortable with the potential consistency that such an approach warranted"[32] and called for additional direction. A revised draft set of MMRs was sent to forest supervisors in September 1982. This draft incorporated the new viability definition including the requirement to ensure adequate distribution across a species' range.[33] In an appendix, it included the northern spotted owl as an indicator species for the mature and old growth habitat in the Coast and Cascade ranges, and specified the following habitat requirements for the owl: 1000 acres of old growth, including a 300 acre core area with an additional 700 acres within 1.5 miles of the core area. Home ranges were considered to consist of 4500 acres and habitat areas were to be no more than 12 miles apart, measured from the centers of the core areas rather than the edges of the core as was included in the earlier plans, in effect making the habitat areas closer to one another.

Agency staff understood the implications of these guidelines on timber resources, but their primary concern was ensuring legally defensible consistency.[34] Nevertheless, when the final set of MMRs was sent to forest supervisors in February 1983, they noted, "When incorporating Minimum Management Requirements in the analysis, every effort should also be made to utilize lands not suitable for timber production and lands managed for less than full timber yields in meeting requirements that constrain timber production. . . ."[35]

Guidance provided by the MMRs was controversial and confusing at the forest level. Forest planners wondered just what well-distributed meant in the pattern of habitat needed to support indicator species. In response, the regional office prepared a memo in March 1984 that analyzed several alternative patterns of habitat distribution and concluded that a grid of wildlife habitat with chunks of habitat separated by dispersal distances was most likely to maintain viable populations. The Washington Office reviewed the proposal and concurred in its interpretation, effectively locking in the requirement that a species be well distributed across its range. The Regional Office sent their clarification to forest planners on April 16, 1984, and provided specific direction on the distribution requirement. Included in this memo was an answer to another lingering question: what happens to points in the grid where there is no habitat that can currently support owls? The answer was that planned habitat patterns should "utilize suitable habitat when available, and utilize capable habitat when no suitable habitat is available."[36]

This elaboration had the potential effect of significantly increasing areas managed for wildlife, since capable habitat is "habitat that is capable of providing for a species' needs, but where the habitat is not currently suitable, and thus the species is not currently present." Hence, owl set-asides might include areas that were not yet old-growth in a condition capable of supporting a population of owls. Again, the folks at the forest level asked, are you sure that is what you want us to do? "To some planners, this resulted in more than the 'minimum,' since more habitat than was needed to maintain the current species distribution was, therefore, assigned to that species. They perceived this to be an increase, and they felt that their wildlife biologists were viewing the new habitat as an optimum rather than a minimum."[37] After considering the forest planners' concerns, the Regional Office ratified the expanded direction, though they agreed to monitor its impacts as they appeared over time.

The Timber Industry Responds

From timber's perspective, the movement in the state of planning for spotted owl protection from 1979 to 1984 was all in the wrong direction, and the NFMA revisions and the MMRs pushed things even further. Instead of 300-acre old growth set-asides that could probably be clustered in existing wilderness areas or in areas planned for scenic corridors or the like, the MMRs required 1000-acre reservations spaced grid-like across the national forests, with not just existing old growth set aside, but areas capable of being owl habitat as well. Industrial watchdogs cried foul: "The efforts of Region 6 to establish minimum legal requirements for planning and management of the national forests has revealed the severity of the 'minimum viable wildlife populations syndrome.' The potential for drastic timber/wildlife trade off first became obvious with the Spotted Owl Management Plan (SOMP), but we are now seeing that this is only a portion of the iceberg. Unless immediate action is taken, hundreds of millions of board feet of timber supply may be lost in land allocations of mature and old-growth timber."[38]

The objections of the industry fell along several lines. First, that by establishing a minimum for a single forest resource, the MMRs violated the FS's multiple use mandate. Second, that the NFMA regulations and the MMRs went far beyond the intent of NFMA and hence were not legally valid. Third, that the requirement of a well-distributed population was neither mandated by NFMA nor supported on technical grounds.[39] In industry's view, the FS had taken an unnecessarily narrow view of options to protect species viability:

> While timber, to a great extent, and other resources such as recreation, to a lesser extent, are expected to increase outputs through intensive and creative management, wildlife management—especially for non-game old-growth related species—has looked at only one approach; i. e., habitat preservation. . . . Biologists within the Forest Service admit that other alternatives do exist. It appears a philosophical decision, that the "natural" way is the only way, was made early in the total process and all efforts from that point on have been made to defend the decision. . . . [An alternative approach would consist of] a human-manipulated exchange of members of the population [which could] be considered for each alternative and compared to the alternative costs of removing lands from the timber base to provide habitats to assure natural exchange.[40]

Industry had complained consistently that decisions were being made that would set aside wildlife habitat with no analysis of their economic ramifications. Ever since John Crowell's letter to the Chief in July 1981, there had been pressure on the agency to produce a region-scale analysis of economic impacts of wildlife habitat protection. The consistent response from the agency was that no aggregate analysis was possible since the economic effects could only be determined when the forest plans were completed, and the regional memos and guide were not really final decisions but simply advice to the national forests. While there was truth in this response, at the same time, not addressing the issue of economic impacts insulated the FS from having to deal with the implications of their actions. The MMRs and other regional guidance might not have represented legal decisions, but for all practical purposes, they did define choices that would emerge fully in the forest plans. The impact of the minimums could have been analyzed, but the agency strategically chose not to do so.

The fact that the MMRs and associated guidance had been developed by FS staff without NEPA review did not escape industry's attention, and they argued that the guidelines should be fully assessed in an impact statement: "Region 6 has set new 'standards' for the forests which have not been thoroughly analyzed in an EIS process. By dictating narrow standards to resolve viable population and habitat issues the Region has allocated a tremedous (*sic*) amount of resources, which has not been quantified, and which has not gone through the NEPA process."[41] . . ."There is considerable feeling in industry that there should be direction to the Regional Offices that any minimum standards be included as part of the Regional Guide and fully exposed in the associated EIS."[42]

The Final Regional Guide and Environmental Group Response

The environmental groups were also aware of the fact that wildlife policy was being created without NEPA review, and the publication of the Final Region 6 Guide gave them an administrative action that could be acted on. The Final Guide was produced in May 1984, and while it had taken three years to move from draft to final stage, very little changed in its guidance on wildlife habitat planning, a strange outcome given the fairly substantial development of policy in the intervening three years. While part of the reason that the Guide took so long was because the NFMA regulations changed in midstream, most of the delay was because of the controversial nature of the decisions being made and guided in it.

The Final Guide treated owl management almost exactly as it did in the 1981 draft. Forest planners were to consider a range of alternatives in their analyses that included a share of the minimum viable population, which was defined as 375 owl pairs (263 in Oregon; 112 in Washington), and a level 30% above MVP. The 1981 SOMP guidance on habitat requirements was taken to be the best direction for planning purposes, consisting of 1000 acres of old growth, 300 acres of which were in a central core, with owl management areas separated by a maximum of 12 miles. The Final Regional Guide clearly indicated that NFMA required maintenance of an MVP that was well distributed across a species' range. Interim management guidelines, however, were based on the 1979 OESTF's plan, including a 300-acre core of old growth. The Guide looked like it had not fully incorporated information generated in the previous four years, and that it was walking a line between what was seen as biologically (minimally) legitimate, and politically and economically correct. Forest plans, where real decisions would be made should they ever be finalized, would include a range of alternatives grounded in minimum biological requirements, while day-to-day management was unaffected. The level of owl protection at the forest level remained fairly minimal and dependent on the good graces of individual forest managers.

In resolving the 1980 Oregon Wilderness Coalition appeal, the Chief had said that the Regional Guide would take care of everything: It would provide a biological analysis that would justify planning and management action. Environmental groups were waiting to see how the Guide came out, and were surprised and disappointed by the outcome. In many ways, the timing of the Regional Guide could not have been better for the regional environmental groups. The early 1980s had been a period of dramatic growth for environmental organizations nationally and regionally. The appointment of antienvironmentalist James Watt as Secretary of the Interior, and other visible Reagan administration directions, resulted in memberships in and contributions to environmental groups doubling and tripling in a fairly short period of time. The groups had a wonderful symbol to attack.

Even the more conservative of environmental groups got on board. For example, the National Wildlife Federation had always had the benefit of an extremely diverse constituency, ranging from hunters to preservationists, but its ability to act was stymied by its conservative wing. The Reagan administration gave NWF a tool to rally much of its membership around. According to Andy Stahl, an NWF staff member in Oregon in the mid-1980s and one of the prime movers on the owl issue, "James Watt turned the National Wildlife Federation into an activist organization. It took someone of Watt's outrageous behavior to awaken the slumbering giant of these sportsmen. It gave the activist staff of the Federation a handle: 'It's time we acted. If not, here's what James Watt has in mind. . . .'"[43]

In the early 1980s, it also became clear to the environmental groups that the only way that environmental quality goals were likely to be advanced, given the political forces at play, was through the action of nongovernmental groups. To pursue their goals, environmental organizations had to act more aggressively through the courts while becoming surrogate agencies, providing information and expertise that traditionally had come from the federal and state agencies, agencies that were increasingly politicized in opposition to environmental goals, and decimated by budget cuts. New offices of national groups were established around the country, and the Pacific Northwest was one of the areas to witness expansion of environmental group activity.

NWF created a new office in Portland in 1982, with Terry Thatcher, formerly at the NWF-funded law clinic at the University of Oregon, as staff counsel, and Andy Stahl as staff forester. The Portland NWF office was created partly in response to pressure from the timber industry on the University of Oregon. The industry and many state politicians had always viewed the University as suspect, and here was a state-supported law school that was providing the primary source of legal expertise to challenge the state's largest industry. An enormous amount of pressure was brought to bear on University administrators, and one of the responses was to disassociate the clinic from NWF. In some respects, just like the Watt appointment had the perverse effect (from the perspective of his proponents) of mobilizing environmental supporters, so did industry's move on the law clinic. NWF's Portland office became the major player pushing the owl case along.

In 1984, the battle for Oregon and Washington wilderness designation came to an end, with passage of bills that set aside about a million acres in each of the two states, and regional groups started looking to new strategies to protect additional acreage. The Oregon Wilderness Coalition had mutated into the Oregon Natural Resources Council in 1983, representing a broader set of interests and approaches than wilderness alone. ONRC was particularly pained by the outcome of the wilderness battles. According to FS historian Dennis Roth, "For years they had proclaimed their goal of protecting all of the 3.4 million acres of RARE II roadless land. They had recruited local groups to support almost all of these areas and thus it was impossible for them to engage in political negotiations that appeared to be 'selling them out.'"[44] The other prime player on the Oregon wilderness bill was the Sierra Club, who was much more willing to engage in the inevitable political negotiations. The result of the difference in interests and attitudes was a split in the environmental community between hardline regional groups and more moderate national groups, a distinction that continued to influence interest group activities throughout the remainder of the owl controversy. Nevertheless, with the wilderness battles pretty much over, regional environmental groups started looking around for other handles on old growth protection, and the FS Regional Guide appeared.

The old growth issue was also building as an issue separate from wilderness, and one with scientific and political credibility. A study by a Society of American Foresters task force published in 1983 helped push the old growth issue along, by recognizing old growth as an end in itself, and legitimizing a preservation policy.[45] The group urged the land management agencies to refine and adopt a formal definition of old growth, and inventory their lands to identify old growth, and concluded that: "At least until substantial research can be completed, the best way to manage for old-growth is to conserve an adequate supply of present stands and leave them alone."[46]

By late 1984, it was also becoming clear that the environmental groups in Oregon could win in court. The Mapleton case was a major turning point in perceptions of the groups about the power that they held, and hence their ability to overcome timber's historic political dominance in the region. Andy Stahl explains: "The first forestry issue that NWF got into in the Northwest was the issue of landslides in the Oregon Coast Range. We started an initiative to amend the state forest practices rules. We were able to get the court to enjoin all timber sales on the Mapleton Ranger District [of the Siuslaw National Forest] because they had never done an EIS. They were caught in the transition from unit plans to forest plans, and had never completed either. . . . After we won the Mapleton case, we realized that potentially we had the ability to change the world."[47]

In spite of these shifting tactical realities on the part of the environmental groups, the agencies were acting pretty much as they always had, at least from the perception of the outside world. The regional environmental groups had asked the BLM to prepare an EIS on its 1983 agreement with the Oregon Department of Fish and Wildlife about protection of owls on the Coos Bay district. When their request was denied, they appealed the BLM's decision to the Interior Board of Land Appeals who denied the appeal in February 1985 noting that the proper place for such challenges was on timber management plans.[48]

To nongovernmental groups, the FS's Regional Guide also looked like business as usual. While the biologists had been influential in moving the definition of minimum management requirements along between 1981 and 1984, the Regional Guide looked a lot like 1980. According to Andy Stahl: "The 1984 Regional Guide came out including a Spotted Owl appendix which was simply the Interagency plan. Period. Xeroxed. It contained no look at alternative plans, and little justification for why this particular plan was selected. So, I set myself to finding out what was behind the biology of the plan. Within one week, I confirmed one of my hypotheses: that the plan was nothing more than the consensus of a number of agency bureaucrats: midlevel state and federal wildlife biologists who were now in administrative positions. Deputy directors of departments whose jobs were primarily dealing with State legislatures and the like. They were accustomed to political expediency."[49]

The Regional Guide Appeal

The Regional Guide was approved by the Chief of the FS in mid-June 1984. NWF, ONRC, the Oregon Wildlife Federation, and the Lane County Audubon Society notified the FS in July that they would appeal the Chief's decision, and filed their Statement of Reasons on October 18, 1984. The environmental groups described their motives for filing the appeal:

> Appellants awaited the Regional Guide for four years, during which time substantial amounts of old-growth forests were made available for sale and harvest. When finally released, however, the Regional Guide and its accompanying EIS failed to assess in any meaningful way the critical issues relevant to old-growth harvesting and spotted owl protection. In reliance on the "interim" guidance given first in the original Oregon SOMP and now confirmed by the Regional Guide, however, the Forest Service continues to harvest suitable old-growth habitat. Analysis of the impacts of the already implemented spotted owl management guidelines is now to be contained, if anywhere, in individual forest plans. Yet, by their very nature, those plans can not be expected to weigh and analyze the issues central to spotted owl management, for by agency admission those questions are regional in scope.[50]

The Forest Service responded with what was becoming a time-honored refrain of "No Decisions Here," that is, the environmentalists did not understand that the purpose of the Regional Guide was not to make decisions, but to guide the planning process where decisions would be made.[51]

The appellants had a number of specific problems with the final Regional Guide. They first argued that the Regional Guide did not contain an analysis of cumulative or regional-scale impacts of timber harvesting on old growth or owl populations. The FS responded that the Interagency Owl Task Force (the OESTF and the OWIWC) had adequately done this.[52] The environmentalists responded that even if the Owl Task Force had done this, "and appellants, of course, have no way of knowing since the process was not public,"[53] it did not absolve the FS of its duty for assessing regional and cumulative impacts as required by NEPA.

The appellants also argued that the decisions made in the Guide were underlain by substantial scientific uncertainty, and that the SOMP had been constructed by a task force that did not include species viability experts: "the SOMP is based upon little more than a collective guess of several agency biologists as to what might be done to save the Northern Spotted Owl within the constraints of timber production goals."[54] In such a situation, the agency's duty was to either "cure the uncertainty"[55] or prepare a worst case analysis. The Forest Service responded, "There is sufficient biological data available to permit evaluation of the impacts of management activities on the northern spotted owl. The scientific uncertainty is very limited."[56] The agency went on to note that Dr. Michael Soulé had assisted the regional office staff in developing the spotted owl plan, that the approach established by the Regional Guide was peer-reviewed by population viability experts, and that the approach was supported by a monograph on spotted owls that was being written by Eric Forsman and Chuck Meslow.

The environmentalists responded to the Forest Service's claims of limited scientific uncertainty, stating "This is a remarkable assertion." They noted that Forsman and Meslow's monograph contained home range results for six owls, with the smallest amount of old growth present being 1008 acres, and the average old growth available for the six owls was 2264 acres, yet the Regional Guide identified 1000 acres per owl pair as the planning guideline. They also quoted from a paper presented by Forsman and Meslow at a FS symposium on the ecology and management of the owl held in northern California in June 1984, that the management approach adopted by the agency "involves a high degree of risk."[57] The environmentalists pointed out that the scientific opinions on which the Regional Guide approach was founded indicated that the scientists' advice contained considerable uncertainty. For example, an analysis by George Barrowclough and Sadie Coats of the American Museum of Natural History that was used by the FS to justify the Regional Guide approach cautioned that their analysis was a "best case" analysis, and that the agency should "proceed with extreme caution in timber harvesting" until data uncertainties were eliminated.[58]

The FS noted that the minimum number of owl pairs identified in the Regional Guide was not necessarily the actual number of owl pairs that would be managed for on the forests and that: "direction is not so specific and limiting to justify concluding that Forest Plan population levels for spotted owls are, in effect, established in the Regional Guide."[59] Hence, you cannot analyze the impact of the Guide's numbers on owl viability as required by the NFMA regulations, since that will be determined by the forest planning process. The appellants responded by saying this argument was nonsense: "The whole basis for the spotted owl plan and the Regional Guide is that no individual forest can insure viable owl populations; such insurance can only be had, if at all, by adoption of a regional approach. . . . Given that understanding, the Forest Service has itself applied its regulatory requirements to the Regional Guide. Having made that choice, the agency cannot now claim that the regulations do not apply." [60]

Finally, the appellants argued that the FS had relied on and selectively used inadequate and outdated information, and had used the absolute minimums for guiding protection activities from any data set they employed, and called the FS approach a "least-possible-protection" approach. [61] The FS responded that it "considered all data available at the time the Regional Guide was being prepared, and is continuing to gather and evaluate new data. Evaluation of this new information has not revealed any new conclusions that would undermine the validity of the Regional Guide." [62]

The relief that the appellants asked for was the preparation of a regional environmental impact statement on owl management that would involve both the FS and the BLM. They also asked that harvesting of old growth be stopped until the regional EIS was completed. The Forest Service, however, believed that a joint EIS on owl management was "impractical," and that owl habitat management had been "responsibly handled" through the OWIWC. [63] The bottom line for the FS was that:

In the enactment of the National Forest Management Act, the Congress of the United States clearly spelled out the intent that a single, integrated Resource Management Plan was to be developed for each National Forest. A Regional integrated resource management plan was not identified. . . . It is in Forest Plans and project plans, not the Regional Guide, that site specific spotted owl management and timber harvesting decisions are most appropriately made. . . . There are opportunities for interested parties to participate in this analysis process and to raise issues concerning the effects of proposed actions on northern spotted owl habitat where these issues are appropriate. [64]

The environmentalists concluded with their overall assessment:

The Forest Service has in the past authorized, and continues to propose, timber sales in suitable old-growth habitat. Yet nowhere, and at no time, has the agency ever publicly addressed, in compliance with NEPA, the critical environmental questions raised by such harvesting. It has, instead, proceeded either in ignorance of the regional and cumulative impacts of its activities on old-growth habitat and spotted owls (pre-SOMP actions) or it justifies its sales not with analysis, but on blind reliance on the SOMP and the "interim guidance." Appellants, after their 1980 appeal, desisted from challenging such clearly illegal actions in the good faith expectation that the Regional Guide and EIS would assess the environmental consequences of the SOMP. It did not. They can no longer stand by and let the agency daily reduce options for old-growth management before it has complied with NEPA.[65]

The appellants also had a solution in mind. They proposed grounds for a settlement to the Regional Forester in a letter on January 3, 1985. Their settlement proposal was based on two major changes: to increase the acreage of old growth in a spotted owl management area to the average of what researchers had found that owls used, and to account for the possibility that all SOMAs would not be occupied by breeding owls. Their proposal was to increase the amount of old growth protected per pair to 2200 acres, and to increase the number of SOMAs protected to one thousand. A thousand SOMAs would insure a breeding population of 500 owl pairs at 50% occupancy.

The aggregate effect of these proposals would be to increase the amount of old-growth reserved for spotted owls from approximately 400,000 to 2.2 million acres, not an insignificant change. In fact, the impacts would be greater because all of the increase in set-aside acreage would lie in currently unreserved lands, while a good portion of the 400,000 acres was probably going to be located in existing reservations. In addition, the acreage implications were more significant, because the appellants asked for special treatment for two national forests, the Siuslaw because it had so little old growth remaining, and the Olympic Peninsula because its owls were isolated from the rest of the region's population. According to the environmentalists, "We continue to believe that an environmental impact statement of the SOMP is necessary, but, should you accept our proposal, the forest planning process may provide sufficient analysis in light of the increased interim protection the owl would enjoy."[66]

Assistant USDA Secretary John Crowell granted intervenor status to two timber industry groups, the Industrial Forestry Association and the Northwest Pine Association, and both groups filed comments that heavily criticized the Regional Guide, but argued that the environmentalists' appeal should be denied.[67] The industry groups were particularly aggrieved by the establishment of minimum standards through the MMR process. Interestingly, both timber and environmental groups opposed the concept of minimum viable populations, but for entirely opposite reasons. Environmentalists viewed minimums as a ceiling, above which management would most likely never occur, and in their view, the minimums were too little. Timber viewed minimums as a floor, below which management would never fall, even though they felt less owl habitat was appropriate. According to the Industrial Forestry Association, "Establishing such minimums by fiat, across a region, not only taints the planning process, but clearly it is a violation of NEPA."[68]

The timber group also had a number of specific comments on the FS analysis. Their statement on cumulative effects analysis is remarkably similar, and probably even stronger than that contained in the environmental groups' statement, in impugning the work of the interagency owl committee:

[The Chief's statement] seems to indicate that the legal requirement for assessing cumulative effects has been met through the actions of that ad hoc advisory committee. This is a false reliance. The committee had no legal power, lacked multi-disciplinary representation and excluded interested members of the public. We have also seen no documentation that it addressed cumulative effects and we have serious concerns over the legality of Forest Service adoption and reliance on its Draft recommendations. Little or no analysis was conducted on possible resource tradeoffs necessitated by the Spotted Owl Management Plan, a direct conflict with NEPA requirements. Public comments were not solicited or accepted.[69]

The Association also commented on what they viewed were the real motives of the environmentalists in bringing the appeal:

The appeal of the Regional Guide by the environmentalists was predictable. While we were heavily disappointed by the Regional Guide, . . . we viewed the Guide as a somewhat nonconsequential document with limited impact. On the other hand, we were sure the environmentalists would carefully scrutinize the Guide to determine if it contained a legal expedient to achieve their true goal: to reduce or eliminate timber harvest on the national forests. This appeal is a thinly disguised attempt at just that—sharply limit or stop timber harvest—using the Spotted Owl.

The two legal issues raised in the appeal—cumulative effects and worst case analysis—have been carefully developed and cultivated over the years through the courts to become the environmentalist's best tool for stopping any program. (Scratch an environmentalist and he will chime "cumulative effects/worst case" before he even says "hello.")[70]

The timber group concluded that while the Regional Guide was inadequately (and unlawfully) prepared, the appeal should be denied: "The Regional Guide was disappointing, but it is not worth the expense and effort to remand and revise it."[71]

The Decision: A Supplemental EIS on Owl Protection

The responsibility for deciding the appeal fell to Deputy Assistant Secretary Douglas MacCleery, a former Washington lobbyist for the Forest Products Trade Association, and in March 1985, his decision was produced. In an opinion that was largely supportive of what the FS did in the Regional Guide, it nevertheless concluded that "new information has become available since the Regional Guide was prepared which may be relevant to regional direction on spotted owl management." Hence the Chief's decision was reversed and the Regional Guide was remanded to the Regional Forester for preparation of a Supplemental EIS (SEIS) on owl management. The rest of the Regional Guide was considered final and approved. At the same time, MacCleery denied the environmentalists' request for a stay on timber sales in spotted owl habitat.[72]

The remand decision provided extensive direction to the FS on what should be included in the SEIS, and specified that the analysis should consider a full set of alternatives, ranging from preserving all existing suitable owl habitat on national forest lands to no formal measures to protect the owl.[73] The decision also clearly indicated that the Washington, D. C. political leaders were interested in the economic effects of owl protection, including "changes in revenues to counties and the Treasury, effects on dependent communities and jobs, other community impacts, effects on regional timber supplies, opportunity costs (changes in the present net value of the forest) and other appropriate economic and social effects that are associated with alternative levels of spotted owl habitat protection."[74]

MacCleery's decision also set up the way that uncertainty was to be dealt with in the SEIS, along the lines of an adaptive management approach that Hal Salwasser was pushing: make short-term choices based on the best information available, collect new information, and reevaluate your choices regularly in accordance with what you learn. The decision indicated that uncertainty could be dealt with by establishing a monitoring and evaluation process that would generate and analyze information so that plans could be altered in accordance with new information. Hence the task of the SEIS was to set down direction for the next ten or so years, while outlining an approach to provide information needed to evaluate that direction at the end of the planning cycle.

Why did MacCleery remand the owl issue back to the Regional Forester?[75] Some observers felt that he truly believed that what the Region had done was in violation of NEPA and the FS could get in trouble in the courts. Both timber groups and environmentalists were claiming NEPA violations and not having much difficulty making a compelling case. In addition, the environmentalists were threatening a lawsuit, and it was better to keep the decisionmaking process under agency control and out of the courts.

Other observers felt that MacCleery was responding as much to the concerns of timber as to those of the owl preservationists. Timber felt that an honest analysis of the owl situation would support their argument that owls lived in second growth, and were not in the danger the environmentalists said they were. Environmentalists felt that an honest analysis of the owl situation would support their argument that owls were dependent on a lot of old growth, and their habitat was being rapidly diminished. Both sides were betting that they would benefit from a NEPA-worthy analysis of owl management issues.

The most cynical observers argued that timber was willing to go along with anything that would delay taking final action on forest plans. Since the old timber management plans governed forest management until the new NFMA-mandated plans were finalized, and the old plans generally allowed a higher level of timber outputs, industry was benefited by slowing down the NFMA process. To some observers of the events of 1985, MacCleery's actions must have received the political concurrence of timber. Regardless of motive, the remand put the owl issue squarely back in the FS's lap, requiring them to undertake a comprehensive SEIS on owl management that they knew would be scrutinized intensely by all parties.

4

The Forest Service's Last Stand: 1985–1989

The supplemental environmental impact statement (SEIS) analysis was the first time that FS leadership took the owl issue seriously, and considerable resources were put into it. The analysis and evaluation began in the summer of 1985, and a draft impact statement was put out for public review in August 1986. A final record of decision was not filed until more than two years later. The amount of time it took to complete the SEIS reflected how controversial an issue the owl case was becoming. While much of the analysis reflected an honest technical assessment of owl viability under different scenarios, assumptions that were made along the way, and wordsmithing that occurred at the end of the process, were problematic. Some of these problems were inevitable given the nature of the task, and some reflected the longstanding approach of the agency toward the owl issue. The SEIS was the best hope for a technically credible way out of the developing controversy at a time when polarization between the affected groups was not extreme; and the FS failed to provide the needed solution.

The SEIS analysis was innovative in that it laid out and followed a risk analysis approach to scientific decisionmaking, rather than a standard agency exercise in professional judgment, where a group of experts look at the data and come up with a set of conclusions. The decision to follow the risk analysis approach in part reflected Hal Salwasser's interest in having the owl be a test case for using embryonic concepts of risk and population viability analysis. Salwasser had joined the Washington office of the Wildlife and Fisheries staff, and his interest was supported by Jeff Sirmon, Regional Forester in Region 6 at the start of the SEIS effort. The approach that was to be taken was considerably different than what had traditionally taken place. According to Dick Holthausen, a FS biologist who was involved in the SEIS effort:

> What we had been doing up to that point in time had evolved very quickly from just using some rules of thumb, like you need 500 pairs of a species to get past genetics problems, and that sort of thing, into a broader framework that eventually incorporated recognition of genetic risks, and risks from demographic occurrences, and risks from catastrophic occurrences. And it was really just in a year to eighteen months in '84 and early '85 that we began putting all of that together into an overall framework for doing the viability analysis.
>
> [Prior to that time,] you would have gotten a very empirically based literature and data analysis with some astute professional judgment added to it to discuss viability. Frankly, people had [only] considered the question of viability for species that were very, very clearly in imminent danger, and there wasn't a need to make as fine a discrimination concerning viability and risks to viability, because normally the topic just wasn't addressed until a species was clearly in imminent danger. . . . I don't think anyone had ever really tried to forecast risks to viability as we tried to do in this analysis. It was a different beast because we were charged with managing for viability, whereas the Fish and Wildlife Service had always been charged with managing for recovery.[1]

Bruce Marcot, another FS biologist who was central to the SEIS analysis, felt that the risk analysis approach had the potential to change national forest decisionmaking significantly:

> The agency, from district to national forest level to regional level, has typically operated throughout its history by looking to a specialist, such as a biologist, for input as to what would be adequate criteria for providing for a species or a habitat. This view of things as a risk framework would really turn that sideways, in the sense that the specialists would provide their best professional understanding and estimation as to what the likely outcome on the species and on the habitat would be in the short term and the long term. And then they would turn that information over to line officers, who would weigh that outcome against similar sorts of assessments on other interests and issues—the timber, the jobs, and so on—and come up with a final choice.[2]

While this approach would provide decisionmakers with more information than they might have in the expert judgment model, it was not necessarily embraced by the rest of the agency. According to Holthausen:

I think it took a long time for management to even understand what it was that we were attempting to do with the viability analysis. . . . There were people in other staff positions who were leery of . . . the risk analysis aspect—that we would be coming out with an analysis that didn't give an answer but that rather portrayed various degrees of risk and portrayed some of the uncertainties in the information that had gone into the analysis. And there was a lot of concern expressed by several people about the precedent that the risk analysis would set. . . . People said, "Wow, what if we have to do an analysis that shows the uncertainty surrounding timber yield projections? That's going to be a completely different kind of analytical world than the Forest Service is used to dealing with."[3]

The time frame for completing the SEIS was extremely short. A May 1985 schedule for carrying out the work outlined a process that would result in alternatives identified by September 1, 1985, a preferred alternative selected by October 1, and a draft SEIS approved by November 1. A public comment period would be provided by April 1, 1986 and the comments were to be analyzed, and a final SEIS and record of decision approved by May 1.[4] Under any circumstances, this set of deadlines would have been optimistic, but for a contentious and technically complex issue like the owl, it was doubly so.

While the tight deadlines in part reflected a lack of appreciation for the complexity of the issue, they mostly indicated the attitude of the agency toward the need to get the job done and out of the way of more important business. Next to "getting this year's cut out," the most important thing in the life of much of the FS leadership was getting on with forest planning. Forest plans for Pacific Northwest forests had been started in 1979 or earlier, and six years later, they were still incomplete. And the owl issue was seen as a major impediment to getting the plans done. According to Tom Ortman, leader of the FS interdisciplinary (ID) team that put together the SEIS, "We had our thirteen forests with owl habitat, and the release of their draft [forest plans] was contingent on getting a draft SEIS out. And for the release of the finals, we had to have a final SEIS out. So with them nipping at our heels, of course we were hard pressed to get on with the draft preparation."[5]

Everyone involved in the SEIS analysis felt the pressure coming from the forest planning effort. According to Dick Holthausen:

> People were desperate to get those plans out, to finally get something on the street. . . . The motivation on the part of some people was to get the plans out in order to adjust the harvest level [downward]. But with a lot of people, it was simply: number one, professional pride, in that we had said we were going to get a job done. Number two, there was a lot of concern about burn-out on the part of the planning teams, and about how long you could keep those planning teams operating without allowing them to get some sort of final product done. Also we were beginning to operate in a fairly difficult situation where you knew some of the decisions that were coming down through forest planning and wanted to protect your ability to implement those decisions but didn't really have any legal justification for implementing them until you got the plan out on the street. So it was beginning to . . . cause problems for the managers out there. . . . People in the region felt that they were about ready to release final forest plans back in '85, and then [the SEIS decision] was remanded to the region.[6]

While the FS SEIS team tended to work together well, the fast track was difficult for the analysts and in part determined the nature of the analysis process. Bruce Marcot indicated that his interest in establishing a multiparty technical working group to work through some of the data and modeling assumptions went by the boards in part because of the fear that the effort would take up valuable time: "At the start I was interested in gathering the best minds on it to really help me think through the best ways and alternative ways to interpret both theory and data. But that did not happen because of the crisis timeline and framework, so we ended up having to really power through as a rather isolated technical group and ID team."[7]

The tight time frame also meant that the analysis team had to limit the scope of its work in order to get the job done, and that meant using whatever data was already in existence and avoiding pressure to expand the scope of the SEIS analysis. The memo from Washington that provided guidance as to the meaning of the remand decision also indicated that the "supplement will be based on existing and available data."[8] To collect much new information would require time, and time was of the essence.

The short time frame was also one of several reasons that the SEIS focused exclusively on owls, and avoided expanding into the old growth issue. Kathy Johnson, a FS wildlife biologist from the Willamette National Forest, remembered a key decision made in a strategy meeting of the ID team in early July 1985:

> The discussion was real focused on owls, although MacCleery's letter did have a lot of reference to old-growth forests. There was a discussion about were we going to deal with the old growth forest issue, or is this just going to be an SEIS on spotted owl viability? . . . [Regional Forester] Tom Coston basically threw something out on the table, "Well, why shouldn't we deal with old growth forests if that's really the issue?" That was indeed the issue at that time, and I was just quietly watching this. [Deputy Regional Forester] John Butruille's face got all red, and Al Lampi was very uncomfortable. Al Lampi was the director of land management planning.
>
> I guess they kind of let the question be asked and then reacted to it. I believe that Al spoke first. He mentioned that we had forest plans that would be pending the resolution of this issue, that it was important that we move ahead with forest planning, that opening this issue to old growth forests when the appeal was specific to spotted owls only would put the rest of the Regional Guide at risk. There was a perception that we wanted to focus in very narrowly on spotted owls in order to minimize our jeopardy, if you will, on the rest of the Regional Guide so that forest plans could move ahead.[9]

The complexity and amorphous nature of the old growth issue clearly frightened the agency's leadership for a number of reasons, and everyone involved in the SEIS work understood that while old growth was the issue, the SEIS would not address it. For example, Project Leader Larry Fellows (formerly the supervisor of the Siuslaw National Forest) noted:

> We identified early on that the issue was not just the owl, it was the old growth issue. . . . We purposely, consciously made the decision that we would deal with the SEIS as a document to maintain the viable population of spotted owls, recognizing that it had some spin-offs into the old growth thing. . . . There are so many facets to the old growth thing, everything from the intrinsic value of old growth to the ecological value of old growth, that it would have gone far beyond what we intended to do with the SEIS. So we said all along

that the SEIS does nothing more than provide for a viable population of birds.[10]

The SEIS Analysis Begins

Work on the SEIS began in June 1985. A structure was created to coordinate and carry out the work required to complete the analysis. A steering committee was appointed by the Regional Forester to act as an advisory group to the director of planning, who had overall oversight responsibility for the SEIS. The steering committee consisted of the regional directors of fish and wildlife, timber management, and planning. The interdisciplinary team charged with carrying out the analysis and led by Tom Ortman of the regional timber planning staff consisted of Betsy Bailey, a regional office economist, Kathy Johnson, a wildlife biologist from the Willamette National Forest, and Len Volland, a Regional Office plant ecologist. A technical team was created to advise the ID team, with Bruce Marcot and Dick Holthausen as the key members. Others were involved throughout, including agency attorneys, technicians, and writers, and representatives of the Washington office, most notably Hal Salwasser. As the analysis moved toward a decision, the top level line officers in the Regional and D. C. offices (the Chief, his deputies, the Regional Forester, and his deputies) were heavily involved, particularly as a decision was made in the Draft SEIS, and throughout the two years of follow-up activities.

Planning in the Forest Service is an issues-driven process, starting with an assessment by the planning team of the "issues, concerns, and opportunities" facing the unit. Often this list of ICOs comes from a NEPA-generated scoping process that solicits information from concerned members of the public. In the case of the SEIS, the ID team was advised by FS lawyers that NEPA did not require a full-blown scoping process for a supplement. Instead, the team spoke with regional interest groups and developed a short list of issues that were to drive the analysis process. A draft set of issues and concerns was prepared in July 1985 that by and large became the set incorporated in the Draft SEIS. The issues and concerns were that the Regional Guide had not: (1) considered the cumulative impacts of timber harvest and incorporated a "worst case analysis" of what would happen if the agency acted without complete information; (2) incorporated new information into its analysis, nor had it acknowledged disagreements about owl habitat requirements; nor (3) considered adequately the impact of habitat protection on other forest resources and their economic and social effects.[11]

The issues and concerns were partly a product of work undertaken to craft a public involvement process for the SEIS analysis. In the early stages of the work, a number of people inside the FS thought that it might be possible to get all the interest groups together to draft a consensus decision for owl management. The agency hired a consultant, Jerry Oncken from Northwest Executive Consultants, Inc., to help craft and implement a public involvement process for the SEIS. The draft contract between the agency and Oncken indicated that "as a result of this work we see: all parties having an even greater respect for the possibilities of people and groups with diverse interests and viewpoints working together collaboratively; [and] . . . being willing to accept the final supplements as representing their interests to at least an acceptable degree."[12]

After Oncken met with the representatives of the groups interested in the issue, he concluded that the lofty goals set in the initial statement of work were unattainable. In his view, both environmental and timber groups saw the real issue as old growth management, and were far apart in their perspectives and motives:

> The industry people are fearful, defensive, and frustrated. They are defending people's basic needs for making a living and maintaining a standard of living, which they see as being threatened by the environmentalists. They also see the environmentalists as having the upper hand, because the legal system allows them to easily and repeatedly bring things to a halt. . . .
>
> The conservation people, on the other hand, are determined and well armed with information and experts. The needs they are defending are aesthetic. They see virtually no points of agreement with the industry people. . . .
>
> Because of the different need levels that are being defended by the two groups (aesthetic for the conservationists and security for the industry people), I see little possibility of developing any form of consensus with which both sides are happy. Once the management alternative is selected, I recommend that the Forest Service do what it can (within its authority and responsibilities) to help the affected parties offset negative consequences. This can be addressed along the way in the public involvement process.[13]

While some members of the ID team wondered whether they should try a consensus-building process anyway, to others, Oncken's conclusions felt right. It was also clear that a consensus-building process would take a lot of time, with possibly little positive effect. According to Kathy

Johnson, the agency leadership set up the SEIS study as "quick and dirty." Her image of their view was, "we don't have time for all this public participation and negotiations and conflict resolution. We've got to get it done so we can get our forest plans out."[14] Besides, the notion of a negotiated, consensus-building process was nothing that the FS line officers were comfortable with. Speaking in 1989, Kathy Johnson believed that: "I think that was a failure on our part not to question [Oncken's conclusions] more than we did. . . . I think that the issue has gone on so long . . . that people get entrenched in their beliefs. It could have been resolved with the Regional Guide. It should have been. There was enough willingness on the Spotted Owl Subcommittee's part to play that kind of role. We didn't take advantage of the enthusiasm and the energy of that group and we weren't willing to let go of our complete control over the decision. And so we were unwilling to take that risk as an agency, and now look at what we've got. Now we have actually no control over it at all."[15]

Variations on a Theme: Alternative Courses of Action

With an initial cut on the issues and concerns, and ongoing work on acquiring habitat information and developing a public involvement strategy, the ID team moved on to identifying alternative courses of action, and by mid-September 1985, had developed a preliminary set of eleven alternatives. The alternatives ranged from no formal owl protection measures to no further reduction in owl habitat plus the designation of capable acreage as future habitat. In the middle were a set of alternatives that protected different numbers of habitat areas at different amounts of acreage per area, ranging from 375 pairs protected at 300 acres of old growth per pair up to 1000 owl management areas (SOMAs) at 2200 acres per area.[16]

While a fair bit of fine tuning of the different alternatives took place over the next several months, the character of the alternatives stayed pretty much the same throughout the rest of the SEIS effort. On one level, the ID team was working with a full range of alternatives, since the alternatives ran the gamut of no habitat protection to full habitat protection. On another level, though, the biologists associated with the analysis felt that the alternatives were narrow and not creative. Responsible for much of the technical analysis, Bruce Marcot felt that:

An awful lot of at least the Washington staff and certainly our ID team staff effort was spent on devising and assessing variations on a single theme. From the start I had posed that providing for spotted owls in habitat areas that are seen as islands is just one theme, and changing how you space them and how you size them are simply variations on that theme. There are very different ways one can think of that provide for that kind of habitat across a landscape that steps out of that theme.

One is to think of not setting up habitat islands per se but looking at a very, very broad-based sub-basin scale or broader scheduling of projects. . . . Rather, I would want to schedule projects at a much, much longer time frame across a much broader base and thereby know how much of the landscape can or should be kept in various age and cover classes of forests. . . . Another perhaps is to really try to meld landowner goals across many different land ownerships and uses. Because if you let the overall range of the spotted owl tell you where and on what kind of lands it could be found, or has historically been found, it covers private, corporate, state, county, and federal lands. And I would like to think of being able to provide for a species or a set of species as old growth ecosystems, across different land ownerships, and then guide the ownership goals such that that becomes one key focal goal.[17]

In Tom Ortman's view, it was too late to develop protection strategies based on large blocks of habitat, because the land base just was not there:

We couldn't figure out how to cut that kind of acreage out of the landscape. We don't have those kinds of acreages left on the national forest system lands. We have wilderness areas that are unfragmented, we have some roadless areas which have been allocated one way or another, and those probably would represent the best opportunities to do that kind of thing. But we had made an interpretation with Hal's help that "well-distributed throughout a planning area" meant that you went out to the edge of the forest boundaries. And we didn't see a lot of opportunity to do that kind of planning, unless you were going to get into a "close the roads and let everything recover" kind of a strategy.[18]

Bruce Marcot felt that the direction set in the range of alternatives responded to well-established agency norms and historical direction. It was the "inertia of prior planning" that established the range of alternatives to be considered: "The concept of what was sometimes referred to as a grid of spotted owl habitat areas, had been pretty well planted inside the agency. I think perhaps both staff officers as well as line thought if we can simply vary the standards and guidelines on this theme, we can be equally able to provide for species and get on with planning. That there was no real need to go to something more radical."[19] In Marcot's view, setting aside well-marked islands of habitat were attractive from the agency's standpoint for a number of reasons: "It's easy to count up the acres on such sites and to pull that out of the timber base, and it's easy to analyze what sort of an effect that has on overall board-feet, and to relate that to what Congress tells us is the cut each year. How we're set up to start planning, the tools we use, the maps we use and so on fits in. And to think of having to schedule projects across a broad span of space and time, and then to track and try to predict patterns of types, distributions, and quantities of habitat, well that's a heck of a task. Especially when you don't have a good base set of survey data on lots of different kinds of habitats and species."[20]

Selecting a Preferred Alternative

In the next few months, the ID team and associated analysts spent a great deal of time analyzing data, fine-tuning alternatives, developing and running population ecology and economic effects models, and keeping interest groups informed of their progress. The approach to viability analysis was also finalized, including a three-step process of assessing empirical information on the biological and ecological attributes of spotted owls, assessing factors that could cause extinction, and summarizing the results of the assessments in terms of the probability of continued existence of owl populations for each alternative.[21]

The last step was the most important in terms of converting the biologists' and ecologists' analyses into a decisionmaking tool. Viability was defined as the likelihood that a well-distributed population would persist through a specific time period. Rather than attaching numeric probabilities of owl survival to various alternatives, the analysts assigned a five-level rank order scale that represented their assessments of the probabilities of a well-distributed population of owls persisting until various points in time. The time periods that they examined were 0 (current conditions), 15, 50, 100, 150, and 500 years. The five levels of likely persistence were Very High, High, Moderate, Low and Very Low. For example, a "Very Low" meant that "there is a very low likelihood of continued existence of a well-distributed population on the planning area at the future date. Catastrophic, demographic, or genetic factors are highly likely to cause extirpation of the species from parts of or all of its geographic range." A

Moderate level meant that the population would likely survive if nothing unusual occurred.[22]

Operationally, Marcot, Holthausen, and staff had a set of models and analysis procedures that gave them numeric results for a given alternative at a specified point in time, and they had to combine them in some way and translate their meaning into one of the above probability categories. They did this by establishing a set of rules that translated the numeric outcomes of five different indices into the probability categories. The indices described the effects relating to demographic and genetic stability, habitat size, and habitat distribution. For example, an alternative (at one point in time) that evidenced a 95% chance of being stable demographically, with a 0.05 inbreeding coefficient, a habitat suitability index of 0.95, a 95% occupancy rate, and an average spacing of suitable habitat areas of less than 4 miles was assigned a Very High probability of persistence. The table of rules determined the overall score that each alternative was assigned.

By January 1986, the SEIS team had a set of alternatives, and estimates of their ecological and economic effects, and were ready to examine their information and propose a preferred course of action. In the fall, they had defined a set of evaluation criteria to use in selecting a preferred alternative. The criteria specified that each alternative would be evaluated according to the degree to which it:

• Provided for the highest level of viability for the owl;
• Maintained options to incorporate new information on owl biology and habitat preference;
• Maintained or improved the socioeconomic conditions of affected communities;
• Minimized adverse impacts on other resource outputs such as timber; and
• Minimized adverse environmental consequences.[23]

Beyond these criteria, the SEIS analysts had received a lot of input from the Washington office and the agency's attorneys that the FS's obligation was to provide an outcome that balanced the demands of competing needs and interests. Hal Salwasser consistently emphasized the need to find a balanced solution. For example, as described in his meeting notes, Salwasser argued his case in a September 1985 meeting: "My personal concern expressed to the ID team, Lampi, Fellows: The Forest Service 'niche' is to manage forests to provide resources and diversity. Our role is not to set up extremes of protection or intensive tree farms. We must search for a solution that allows dependent industries to adjust and allows for research and managers to learn how to provide for owls and other [old growth wildlife] in managed forests. If we propose in the SEIS to push interests on either extreme too far we will get either a legal or legislative fix or both. Goal of the SEIS should be to propose what we believe meets legal requirements and also provides for resources to flow to local and regional economies. . . . Our interest is in long-term workable solutions, not short-term 'victories' for a special interest that will be reversed by the Courts or Congress."[24]

Salwasser's comments reflected a developing split within the FS staff as to the role of the agency and the need to be sensitive to political variables. For the research biologists, and some of the applied biologists like Bruce Marcot, the primary concern was protecting habitat to insure a viable population of owls. For some of the managers and the policy-level biologists like Salwasser, long-term political sustainability of an agency decision was most important, subject of course to estimates of what the political response would be to any given decision. They argued that if the agency went too far to the preservation side, the political fallout might harm long-term owl protection efforts, and even worse, have a backlash effect on the Endangered Species Act. To critics of this view, the backlash argument was just a shield for continuing business as usual, and maintaining a high level of timber harvest. Regardless of what you believe about motive, the indication from D. C. was that a balanced solution was legally, politically, and technically appropriate, and that perspective influenced the decisions of the SEIS ID team.

The ID team met together with the biological assessment group in January 1986 to select a recommended alternative to forward to the Regional Forester. According to Bruce Marcot:

Ortman brought the ID team into a room, and through the standard ID team process came up with a recommendation for which option would be best given a balance of owls and timber and other resources. . . . What we did was set up on a large board, a table showing for each alternative the outcome for each of a set of evaluation criteria. Such as effects on timber ASQ, effects on owls, likely effects on water quality, effects on jobs, and so forth. We each talked about where we could feel comfortable, in terms of what range of outcomes. That's the economist, the ecologist, biologists, and the timber representative, who was in fact Ortman. And then we looked for a center point or some overlap: which options overlapped most of the people's areas of comfort.[25]

In Tom Ortman's view, the group ended up with a compromise decision:

We had Bruce Marcot, who favored something beyond alternative L, which in my view was the best thing we could do for the owl. . . . You had the operations research analyst and myself—both of us had been in the forest planning effort since '79 trying to blend where our field folks were and what we had established on the ground, and people were finally starting to feel good about, which was the 1000-acre alternative—with the balance on the economic scale. . . .

My personal belief, having lived through mill shut downs in some towns that I lived in in Idaho on two different occasions, and watched what happened when the forests that I'd worked on cut their [allowable sale quantity] ASQ by fifty percent, where you had five mills you ended up with two. . . . On the social side you see people really put in a desperate situation. They have a house payment, they have kids to feed, dad gets drunk, he beats mom up, I mean just terrible stuff. That really biased the way I approached this thing. I could agree with Bruce on what's best for the owl, but to come up with a preferred alternative on that extreme end of the spectrum would not be a good decision.

So at any rate, we argued for a day and a half, and then finally we said "let's look at what's the worst for the owl that anybody could accept, and what's the best." And that's where we came up with an Alternative F, which was 2200 acres in each [spotted owl habitat area] SOHA, and I believe we had something like 550 habitat areas identified, 392 of which were in land suitable for timber and the rest were in wilderness areas. We went ahead and established those, because we wanted to make sure these could all be networked together. So that's what we went back to the Chief with. . . . I think Bruce was reluctant on that. We just put a lot of pressure on him to say yes so we could get the hell out of the door.[26]

Marcot obviously felt that pressure and concurred with the team's recommendation as long as it was tied to the acquisition of more information, and the ability to change direction down the line: "I could go to F if we did substantially increase research. We saw it only as a short-term plan, and if we really had the adaptive kinds of management structure set up to go out, and survey, research, study, monitor, and analyze that set of data, and then feed it back through well-specified criteria and rules back to the planning framework. That is, if we see that habitats are more fragmented than we assumed, perform an analysis. If we see that owls are starting to drop out of sites and so on, feed that back in."[27]

While Alternative F was hard for Bruce Marcot to agree to, it was equally difficult for the Chief and his staff, but from the opposite perspective. Alternative F provided at least 550 spotted owl habitat areas with 1000 acres to be removed from the land base considered suitable for timber production. In addition, to protect long-term options, timber sales were not to be scheduled in an additional 1200 acres per SOHA. To the Chief's staff, Alternative F looked like a set-aside of 2200 acres of owl habitat in 550 locations. Tom Ortman described what happened when the ID Team took their recommendation to the agency leadership: "So we went back, and of course the Chief's staff was aghast because here we had doubled the size of the habitat areas. We were going to completely upset everybody's sales schedules for the next three years. And rattling that through forest planning was going to delay forest planning. So they worked on that all Spring, and finally got to the point where they were ready to take it to the Administration, to the Department. And we worked through them. We had a lot of tug-of-wars about how you display economic effects and effects on counties, and we finally came to a consensus. So in June we got a go-ahead from the Department, Doug MacCleery, to print the draft and distribute it."[28]

The two-volume Draft SEIS (DSEIS) was released for public review in August 1986. The Draft examined the effects of twelve alternative approaches to owl management, most of which looked similar to the original set identified by the ID Team. Except for the extremes of all or no habitat protected, all alternatives called for the setting aside of a certain number of chunks of owl habitat spaced in a defined pattern.

The preferred alternative, Alternative F, was expected to result in the withdrawal of some 313,839 to 690,446 acres from the lands considered to be suitable for timber production.[29] Combined with acreage in already-reserved status, total owl habitat was considered to be 1,413,839 to 1,790,446 acres. When comparing the preferred alternative with Alternative A, the alternative that provided no designation of national forest lands beyond existing reservations and other lands not suitable for timber harvesting, the effect on the volume of timber to be cut annually was about 5%.[30] That is, Alternative F would mean a five percent reduction in timber supplied from the 13 owl forests in the Pacific Northwest, a reduction that the DSEIS translated into a cost of 760 to 1330 jobs in the forest products industry, $18 to 21 million to the U. S. Treasury, and $10 to 11 million to the states.

The result of the preferred alternative on owl viability was mixed, particularly as you moved further out in time. Alternative F was assigned a High probability of persistence at the 15-year point, as were most of the other alternatives with the exception of the two alternatives that provided the least amount of designated habitat. At the 50-year point, Alternative F's probability of providing a well-distributed population was deemed High to Medium. At 100 years, Medium to Low. At 150 years, Low, and at 500 years, Low to Very Low.

The risk approach taken in the SEIS was clearly identified at the front of the Draft SEIS. The document also began to define an iterative approach to decisionmaking, whereby tentative decisions could be made, new information solicited, and the decisions could be altered as needed in the future. This approach meant that a Low probability of persistence at 150 years was not considered to be a problem, because the decision that the SEIS provided was direction for the next ten to fifteen years, and at fifteen years, the probability of persistence was considered High.[31] The Draft SEIS also continued to highlight the longstanding agency perspective that while the analysis might look like a major decision, it really was not: "The standards and guidelines associated with these alternatives do not specify land uses, nor do they specify actions to be taken on any specific land area in the Region. Rather they guide the decisions that are made in individual National Forest Plans regarding what actions are to be taken concerning a specific area. The guidelines will be used by individual National Forests to assure viability of northern spotted owl on a regional basis. The more site-specific consequences of these guidelines will be analyzed in development of the National Forest Plans."[32]

The Interest Groups and the Public Respond

The Draft SEIS was put out for public review in August 1986. While the approach it took had a number of weaknesses, much hard work and thought had gone into the analysis, and agency officials hoped that they had found a point of balance that would achieve some measure of support from some of the affected interests. But the support was not to be found.

The interest groups had been busy pursuing a variety of strategies to attempt to influence agency direction on owl and old-growth protection. Timber interests had been working the political channels fairly hard. Both groups tried to be as active in the SEIS process as they were allowed to be. But both groups, and particularly the environmental groups, began to understand that the playing field on which the owl case would be resolved was one of science and perceived expertise, and they began to muster experts and information that supported their side of the argument.

Andy Stahl at NWF in Portland got a theoretical biologist from the University of Chicago, Russ Lande, involved in examining the spotted owl management plan contained in the Regional Guide, and used Lande's report as the basis for a request for a stay of old growth timber harvest in the Pacific Northwest. The report concluded that "Based on the available evidence, harvest of any of the remaining old forest in the Pacific Northwest poses a potentially serious threat to the long-term viability of the northern spotted owl."[33] Attached to the report were copies of letters from Mark Shaffer at the FWS and George Barrowclough at the American Museum of Natural History, both fairly credible scientists in the spectrum of individuals involved in the spotted owl case. Both supported Lande's approach.

Armed with this "independent" analysis of the spotted owl situation, Stahl moved ahead on two fronts. First, he held a press conference that focused on the Lande study. Second, he used the report as the basis for asking Doug MacCleery to reconsider his earlier decision to deny a request for stopping timber sales in spotted owl habitat.[34] MacCleery's response was to suggest that the FS meet with interested parties and see if they could work something out, while the SEIS work was ongoing.[35] According to Stahl, at that meeting the environmentalists proposed a deal, "We said to [Regional Forester] Jeff Sirmon that we wouldn't go to Court if the Forest Service would agree not to sell six timber sales that were particular thorns in the side of various environmental groups until the SEIS was done. Sirmon agreed and cut the deal. . . . After that meeting, we then had to twiddle our thumbs for a couple of years until the spotted owl SEIS was written."[36]

While the window of formal appeals was closed to environmentalists until the SEIS was completed, other means of acquiring expertise, and the legitimacy that comes with it, were available. Amos Eno, then at the National Audubon Society in Washington, D. C., was concerned about the biased approach taken by the FS on the owl issue, and the "pure advocacy, no holds barred" approach taken by some of the environmental groups involved in the case. He had the idea of assembling a Blue Ribbon Panel of scientists to examine the status of the owl, and provide management recommendations: "to use as a platform for a series of recommendations to Congress and the private sector."[37]

Eno asked the presidents of the American Ornithologists' Union and the Cooper Ornithological Society to recommend members for the Blue Ribbon Panel, with the only constraint being that one panel member had to be a FS biologist so that the panel could not be accused of being divorced from institutional realities. A six-member group was named with William Dawson of the University of Michigan as chair, and their report was published in May 1986. The report endorsed the concept of using the spotted owl as an indicator of the condition of the old-growth Douglas fir forest of the Pacific Northwest, and suggested the following management program:

- Management for a minimum of 1500 pairs of birds (across all land ownerships) in Oregon, Washington, and California;

- Maintenance of an effective distribution of owls through a habitat network like that proposed by the FS; and

- Establishment of per pair habitat areas that included 4500 acres of old growth in Washington, 2500 acres in Oregon and northwest California, and 1400 acres in the Sierra Nevada.[38]

The report also denigrated the notion that a breeding population of 500 was a good rule of thumb for avoiding genetic inbreeding problems. It further recommended that a monitoring system and an extensive research program be established, and federal agency personnel should be made more accountable for their efforts to protect owl habitat. If these measures were taken, listing of the owl as a threatened or endangered species would not be necessary.

The Audubon "Blue Ribbon Panel" report was produced with a good bit of fanfare in the early summer of 1986, at the same time that the FS's Draft SEIS was put out for public comment. Since the Panel's recommendations went beyond those in the preferred alternative in the Draft, they tended to provide a scientifically legitimized alternative that critics could point to in comments on the draft study. And almost no one was happy with the Draft SEIS.

The comment period on the Draft SEIS ran through the fall of 1986, and more than 41,000 responses containing 140,000 specific comments were received by the FS. More than sixty percent of these responses were form letters or clipped coupons from interest group literature. Indeed, the FS staff counted 106 different form letters that were either hand-copied or photocopied and individually signed. Agency staff diligently coded all the comments and organized them for summary in the final SEIS. Of the 41,000 responses, only 344 indicated some support for Alternative F, the preferred alternative.

Environmental group comments generally pointed to the long-term probability of persistence estimates provided in the Draft, and argued that a Moderate to Low probability at 100 years did not satisfy NFMA requirements for protecting biological diversity. Most also noted that the Draft missed the larger issue of old growth and the significance of the owl as an old growth indicator species. They also claimed that the economic effects of habitat designation were overstated, because the allowable harvest under each alternative was compared against Alternative A, which contained an allowable sale level considerably above that which occurred in recent years.

Industry groups argued that the preferred alternative was not a balanced approach and would unnecessarily cause economic chaos in the Pacific Northwest. Most industry comments suggested that the biological basis for owl protection was still too uncertain to justify imposition of serious economic impacts, and that the right course of action was to continue to do more research to find out what action was needed. The comments provided by the Northwest Forest Resources Council in Portland are illustrative:

> We continue to believe that a balanced resolution of this important issue is possible . . . and will assist you in developing a proposal which will not send a devastating shock wave through the recovering economy of the Pacific Northwest. . . . [We believe you should] defer a decision until sufficient information is gathered. . . . To put it bluntly, the Draft Supplement does not provide the Forest Service with a single piece of meaningful information which it did not possess before it began preparing the document. It is a worthless document which provides decision-makers and the public with absolutely no valid information about the spotted owl. . . . In fact it is less than worthless, because through the use of mathematical formulas and scientific jargon, it creates a false impression [that we know more about the owl than is the case].[39]

The FS content analysis team completed their work on the public comments by late January 1987, and set down some summary comments about what they had read:

> During the 90-day review period, interest groups from both sides of the spotted owl issue worked to get their constituents to respond to

the SEIS. Early on, representatives from the timber industry printed and distributed thousands of response forms with multiple choice statements. The respondents checked the statements that reflected their opinions. The forms included several lines so that the respondent could write a personal message. Area timber companies provided envelopes and postage for many of those forms.

Although the ID Team continued to receive copies of that early response form throughout the entire response period, it was replaced by a set of brochures. The set included a four-page brochure entitled, "The U. S. Government's $6,000,000,000 Spotted Owl Plan, Is It Necessary?," and a two-page brochure with a tear-off, mail-in coupon. The coupon contained only one box to check and provided several lines for a personal comment. While the information contained in each brochure was generally correct, it was often incomplete. To illustrate, the brochure's cover states that "The government's plan will . . . designate 1.8 million total acres for the spotted owls." It did not explain that 1.1 million of the 1.8 million acres were in established wilderness and reserved areas.

It appears that the respondents who had not read the SEIS often based their comments on the information presented to them in the brochures. One respondent, who was an employee of the timber industry, attended a company-sponsored meeting concerning the spotted owl issue. She said after later reading the SEIS, she felt that the company had given inaccurate information, and was concerned about the standpoint of other employees who may not have read the document. . . .

The concern of the wood products industry was emphasized by the various forms of encouragement they purportedly offered their employees to write a letter, sign a form letter, or mail a coupon in response to the SEIS. Respondents mentioned compensations such as $18 worth of Green Stamps or a night on the town.

The environmental and wildlife interest groups have also encouraged their members to respond to the Draft SEIS. In many instances, however, these groups have not been so much in support of the spotted owl as in support of preserving the remaining old-growth timber. Once again, the information provided to the public was generally accurate but incomplete. . . .[40]

Overall, the FS analysts were not encouraged by what they had read, either in its implications for the agency or the image it provided of the public. In their view, "Ninety percent of the letters indicated the people did not read the SEIS, and were not well informed on the issues." In addition, the "Forest Service lost in the credibility game in the PR area. People did not think Forest Service had done [its] homework enough." As to the SEIS itself, it was "not a readable document for the average reader. [They were partly] misinformed because the documents are so difficult."[41]

Moving from Draft to Final SEIS

As the FS moved from the Draft to a final SEIS, issued for public review in April 1988, a number of changes were made in the analysis and how it was presented in the document. Some of these changes were substantive, some were cosmetic and responded to the widespread confusion on the part of the public in their reviews of the draft, and some were strategic. The character of the decisionmaking process also changed, as it increasingly involved the active participation of top-level agency officials.

While the Draft SEIS had discussed twelve alternatives, the analysis team was encouraged by agency lawyers and the Washington office to simplify and streamline the analysis, and the number of alternatives were reduced by half. In addition, the recommendations framed in the Audubon Panel report were crafted into an additional alternative, labeled "Alternative M." Hence, the Final SEIS looked at seven alternatives, ranging from no additional owl protection to all owl habitat protected, though in terms of timberland that would be affected, most of the alternatives were packed into the lower end of the range of possibilities.[42]

In addition, the team began modifying the Draft preferred alternative, Alternative F, by looking at the effects of varying the amount of habitat set-aside per owl pair depending on how far north the SOHA would be, a change that coincidentally would make Region 6 planning consistent with Region 5 (California) planning. The group also began working on different ways to show the effects of the analyses. A decision was made not to highlight the number of designated SOHAs, as had been done in the Draft analysis, but rather to focus on the number of owl pairs that could be supported by the different alternatives. Hence, nowhere in the final draft does the number of SOHAs to be set aside get established.

Other changes were also made that tended to minimize the perceived effects of owl habitat requirements on timber production. The base for evaluating effects was changed, moving from a fairly old and optimistic projection of potential timber harvest to the estimates provided in the draft forest plans' preferred alternatives. Since the draft plans for the thirteen owl forests in the region would be setting aside commercial timberland for other purposes (such as recreation), it was not necessary to consider those reservations a result of the owl issue. Substantively, this seemed a reasonable change, since the forest plans would generate the most realistic base for analysis, but it also resulted in the perception of less economic impact due to owl protection, and that was a good strategic move.

A great deal of effort went into defining and clarifying the concept of adaptive management. While long-term adaptation to new information was implicit as a decisionmaking approach in the Draft SEIS, the Final used graphics and descriptive language to indicate that the decision that was being made was only a 10–15-year decision. New information would be provided by ongoing research, and it could be incorporated in the next round of forest plans (should the current round ever end). This emphasis on short-term decisionmaking was reinforced also by dropping the 500-year point in the analysis of viability. While those who promoted the concept of adaptive management in the agency viewed it as an effective management approach given the existence of uncertainty, others viewed it as a "brilliant" strategy for putting off a tough choice.

As the analysts worked on the Final SEIS, the decisionmaking process also changed significantly. One decision criterion, the effects on the Allowable Sale Quantity, became extremely clear: no more than a five percent reduction in the ASQ. Everyone on the analysis team understood this to be a constraint in making a decision on the Final SEIS. According to Dick Holthausen: "One of the hard decision criteria was that the decision should not trade off more than five percent of the timber harvest. That was a pretty clear criterion. And to my knowledge, pretty openly stated. . . . That was very much a Chief's and a Regional Forester's decision. . . . It was pretty clear in the selection of an alternative for the final that there was going to be very strong consideration given for maintaining, at least over the short term, something close to existing harvest levels. . . . I'm convinced that the Washington office was absolutely convinced that it was necessary."[43]

While the criteria were clear to the team, they were never specified in the Final SEIS. According to Bruce Marcot:

Interestingly, those criteria did not seem to show up clearly at all, either in the final plan nor in the ROD, the final Chief's record of decision. They were fairly well specified in the Regional Forester's slide show talk that the ID team crafted for him, but they seemed to drop out of written documentation, or were written in a way that at least I had a hard time telling the results of the biological technical assessment from the decision criteria.

Part of the problem with the final was that with those lines being fuzzed over, it was very hard for folks outside of the inner circle here—folks in special interest groups, publics, and even higher up forest research—to tell the difference between decision criteria and the science end of it. And it looked like the science end was stating that the Chief's decision was scientifically sound, and scientifically the best and so on for owls. Whereas in truth, that was only one piece of the decision criteria.[44]

As the team worked on the Final SEIS during 1987, the process also became more closed to the outside world. According to Dick Holthausen:

All of the work on the analysis leading up to the final EIS was done in a much more closed-door atmosphere than the work that had gone on getting ready for the draft. Because the agency had gotten so much comment on the draft, the majority of it negative from one camp or the other, . . . it caused the agency to simply kind of draw in upon itself concerning the issue. In my opinion, one of the major false steps that we made in this whole process was making it too much of an in-house process. It was really closing the doors on it, going from the draft to the final. And I think that the kinds of reactions we got to the final were fairly predictable, given the amount of interaction that had not been going on with any of the interested agencies or groups between the draft and the final.[45]

While individuals from outside the agency were less involved, top level leaders of the agency were intimately involved in the crafting of the Final SEIS. Grant Gunderson, a FS wildlife biologist who was involved in the analysis between Draft and Final SEISs, describes some of the interaction between the analysts and the Chief: "We kept going back to the Chief. We'd have 'here's where we are' sessions, and the Chief would say 'we need a little more of something.' Our perception was that something would be some way to do more for owls and more for timber. Find this magical answer, find this thing that's going to solve everybody's problems and make everybody happy. . . . Everybody here understood [that there was no magical answer], but I don't know if the Chief understood that there wasn't a way out."[46] Gunderson indicated that the level of involvement of agency leaders reached fairly deeply into the details of the analysis: "It was unbelievable the scrutiny that some of this got at those levels. . . . When we took the Mt. Hood map back and showed [the Chief], he was looking at individual SOHAs saying, 'Why is that one there?' and 'I don't like that one there because you're right next to a wilderness area.' And we got the word loud and clear that he did not want to see those there. He was involved in the actual SOHA locations. In fact, [Congressman] Les AuCoin was involved in . . . the actual SOHA locations . . . on the Siuslaw, and questioned why we needed three up on the Hebo District." [47]

Bruce Marcot also described the intense involvement of the Chief, the Regional Forester, and their staffs, noting that they: "worked very, very hard to try to really find that very, very narrow optimal point in there that would satisfy interests from all sides. . . . What I saw in common was trying to find the answer, trying to find some way to craft standards and guidelines for owls and still provide for timber access principally. Of course with an eye on other things such as effects on water quality and so on. But there seemed to be many, many hours spent on how can we re-craft the standards and guidelines in order to define some very narrow wedge inside of this broad spectrum. It seemed over time that if we could only find the correct formula for standards and guidelines—how much habitat, where, and how to set it up, and so on—that that's the answer. I saw that as a theme in common from start to finish." [48]

All of this took a lot of time. The adjustment to using the forest plans' preferred alternatives as the base required a lot of effort. New maps and better documentation took time to produce. In addition, the draft copies of the Final SEIS went through extensive editing from all levels of the agency. Dick Holthausen suggested that part of the time involved in getting the Final SEIS out was due to the inherent difficulty of the political choice that the Chief had to make: "I suspect that there was a lot of time consumed just in top line becoming comfortable with a decision that could be released. . . . I think by that time everyone was convinced that we were facing an unwinnable situation, based on the comment we'd gotten on the draft and on the continuing controversy over the management of the spotted owl. It was clear that the public just consisted of factions that were diametrically opposed, that there truly was, at least as far as we could see, no middle ground that was going to make any elements on either side very happy. So I think the agency was faced with what looked like a lose–lose situation." [49]

Tom Ortman remembered a key meeting at which the Chief endorsed the concept that went into the final choice of alternative: "We went back with this F Adjusted. I did not have hopes that we'd get a decision. I thought they'd have to mull it over, because there were still subsequent economic impacts, and we had a lot of controversy surrounding the thing. And by that time, we'd had a suit against the Fish and Wildlife Service over their decision to not list [the owl as a threatened or endangered species.] . . . I can remember yet the Chief saying—several of his staff group were giving advice, different directions—'Tell me why I shouldn't just go ahead and make this decision and let the forces that shape this country react? It's going to go into the court system; it's going to go into the legislative system; that's the way our country is supposed to work.' He went ahead and endorsed the Alternative F Adjusted. So we came back knowing that that was going to be it."[50]

The final two-volume SEIS (FSEIS) was released for public comment in April 1988. Once again, the document noted that the decisions described within it were not really final decisions, since "standards and guidelines do not specify land uses, nor do they specify actions to be taken on any specific land area in the Region. Rather, they guide planning in individual National Forest Plans."[51] The FSEIS also noted that the direction was only relevant to the current cycle of forest planning, which meant that it would only guide management for ten to fifteen years. The need to address old-growth forest management was added as a new issue in the FSEIS, but the agency's response to it was a traditional one: "general issues of old-growth management are left to Forest Planning as part of meeting overall multiple-use objectives."[52]

The FSEIS's preferred alternative specified that SOHA size would vary by physiographic province, ranging from 2700 acres in the Olympic Peninsula to 1000 acres in northern California, but did not specify the number of SOHAs that would be set aside. 347,700 acres of suitable timber lands would be designated as owl habitat, which would yield a High probability of persistence for the entire subspecies (including those in northern California) at 15 and 50 years, and Moderate probability at 100 and 150 years. Subpopulations would not fare as well. For example, the Olympic Peninsula population was given a Moderate probability of persistence at 15 years. The economic analysis in the FSEIS estimated that the plan would cost approximately 163 million board feet of timber annually, representing—not surprisingly—5% of the ASQ. Job impacts were estimated at 455 to 910, loss of timber receipts to the U. S. government were projected to be $18 to 21 million, and loss of timber receipt payments to counties was estimated at $8 million.

When compared against the Draft SEIS, the FS had seemed to win the battle of finding something better for both timber and owls. Harvest impacts were reduced from 189.9 to 163 million board feet, though both translated to a 5% reduction in ASQ. Job impacts fell from an average of 1000 in the draft to somewhere around 700 in the final. Lost revenue to counties declined by a quarter. Even though timber and economies seemed to fare better in the Final, so did the probability of success of the owl, rising to Moderate from Moderate to Low at the 100-year point. While the FSEIS seemed to be a better alternative for both timber and owl interests, in the broader scheme of things and as seen by the outside world, the changes did not look very dramatic at all. Two years of considerable agency effort had been spent, and while the Final SEIS was very nice to look at, the substantive outcomes did not appear to be much different than the earlier draft.

Other Attempts to Maintain Control

Agency leaders looked forward somewhat uneasily to the reaction to the Final SEIS, though other ongoing events had as much to do with the nature of the reaction as the SEIS effort itself. While the SEIS work had been ongoing, the FS, other federal and state agencies, and the affected interest groups had been hard at work on other owl-related activities. The activities of the FS and the Fish and Wildlife Service had been aimed at avoiding the listing of the owl as a federally recognized endangered or threatened species. For example, the FS had created a Spotted Owl Research, Development, and Application Program (RD&A) as an interunit program involving Regions 5 and 6, and the Pacific Northwest and Pacific Southwest Research Stations in January 1987. While its formal mission was to develop and apply improved information, its informal mission was to demonstrate adequate effort and concern about the owl on the part of the FS to avoid a decision to list the owl. Since the owl was threatened primarily by FS activities, the more action taken related to owl management, the better.

This approach was supported by the FWS leadership, which was pursuing the Reagan administration's policies of minimizing the constraining effects of environmental regulation on resource development. Having avoided the need to take action for a number of years, the FWS was dragged into having to take action one way or another on the owl issue by petitions from several environmental interest groups. Up until this time, the environmental groups had debated, but avoided, pushing a decision on the status of the owl. Some felt that they would lose in court because the FS's judgment would appear to be the most legitimate basis for a judge's decision, and the implied threat of going to court had more impact on the course of owl and old-growth management than having to carry out the threat and lose. Others felt that no matter what actions the courts took, they would be overruled by the political process.

A petition by Green World, a tiny environmental group operating out of a phone booth in Massachusetts, forced the hands of the environmental groups and the FWS. Andy Stahl, who moved from NWF to the Sierra Club Legal Defense Fund (SCLDF) in January 1987 and dragged them into the owl issue, explains:

> We knew in 1985 that we could stop every timber sale in old growth and that we could get the owl listed. We decided that we shouldn't do it, because public opinion was not developed well enough. The Green World petition forced our hand. They were not the first out-of-the-blue petition, by the way. The Beckwitt family in northern California had written to the FWS. The Sierra Club freaked that there should be a petition on this, by no means a novel thing, and I found the Beckwitt's and talked them out of it. They withdrew their petition.
>
> As to Green World, you can't find them and talk them into anything. But besides, we were ready by that time. The foundation had been laid. To lay more foundation, we had to push the issue to get newsworthy events. At the time of the Green World petition, we were already drafting a petition. It just took us so long to get the national groups on board.[53]

Green World sent a letter to the FWS in October 1986 that cited the Audubon Panel report as justification for a request that the FWS list the spotted owl as an endangered species under the ESA. Not wanting to deal with the issue, the FWS determined that the letter was not a "petition" because it was not explicitly identified as a petition. Green World added the phrase "Petition to List a Species as Endangered" on the top of their October letter and resubmitted it to the FWS in January 1987. This time, the FWS took the letter more seriously, and started a ninety-day review period in which it had to determine if the petition "presented substantial evidence." Exactly ninety days later, the FWS Regional Director in Portland determined that the Green World petition met the test.

Three months went by before the agency announced in the Federal Register that it would initiate a "status review" of the northern spotted owl and requested public comments.[54] Faced with an upcoming decision by the FWS, the other environmental groups took action, and SCLDF

filed a petition on behalf of 29 environmental groups requesting the FWS
to list the northern spotted owl as an endangered species. According to
Andy Stahl, "When Green World submitted its petition, there was the
feeling that if there is going to be a petition, it had better be a good one.
So we wrote our own."[55]

A FWS review team worked through the fall of 1987 on its status review, and put out a draft for peer review in October. In response, FWS
biologist Mark Shaffer commented that, "the most reasonable interpretation of current data and knowledge indicates continued old growth harvesting is likely to lead to the extinction of the subspecies in the foreseeable future," and argued strongly for listing.[56] Shaffer had solicited the
perspectives of conservation biologists Michael Soulé and Bruce Wilcox.
In Soulé's view, "I can't see how the Service can come to any other conclusions than that this species is endangered." From Wilcox's perspective,
"there is little doubt in my mind that the owl is at least at imminent risk
of becoming endangered."

In spite of the set of internal and external biological judgments that the
owl warranted listing, the agency denied the petition on December 17,
1987, indicating that listing of the northern spotted owl was not warranted at that time. The FWS explained that it intended "to continue to
utilize all of its existing authorities to help to ensure that the species does
not decline to the point where listing would be warranted. The majority
of the owl's habitat occurs on Federally managed lands and the Service
has been working with these resource agencies for some time to monitor
and protect the northern spotted owl. The Service believes that through
proper management a viable population of northern spotted owls will be
maintained throughout its present range."[57]

One of the ways that the FWS and the FS leadership chose to demonstrate their joint commitment to owl protection, and avoid the horror of
having to list the owl under the ESA and suffer the consequences, was by
signing an interagency agreement regarding owl management and research activities. According to FWS Director Frank Dunkle, the Interagency Agreement, signed on December 1, 1987, "will assure the protection and perpetuation of the spotted owl. I see this as another milestone
in cooperative approaches that assist the United States Government in
carrying out its obligation to protect the endangered species of the
World."[58] The interagency agreement was expanded to include the BLM
and the National Park Service in August 1988.

But the approach taken by the federal agencies began to fall apart in
1988. In January, acting under encouragement of the state department
of wildlife (WDOW), the Washington Wildlife Commission declared the

northern spotted owl a state-listed endangered species, and directed the WDOW to prepare a recovery plan for the subspecies. Given this status, and the active lobbying of several internal experts who had become advocates for the owl, the state agency became much more active in opposing federal management direction. In April, they filed an administrative appeal against the Bogy II timber sale on the Soleduck Ranger District in the Olympic National Forest, and in May, joined SCLDF in petitioning the federal district court for a temporary restraining order on the sale. In July, they filed an administrative appeal of the Loner Elk timber sale on the Olympic National Forest, and in August, requested that the Mt. Baker-Snoqualmie National Forest defer sales in owl habitat until additional surveying could take place, a request the FS subsequently denied.

Throughout the period, the WDOW experts became more vocal opponents of federal and state forest management action. WDOW owl expert Harriet Allen made the case for listing the owl as a state endangered species at the Washington Wildlife Commission's January 15, 1988 meeting. Addressing FS owl management plans, she argued that: (1) an inadequate number of SOHAs were being set aside; (2) those that were being set aside were not in the right place; (3) the SOHAs designated in the state of Washington were too small to be effective as owl habitat; (4) there were no provisions for replacement sites if individual sites were lost due to catastrophic events; and (5) there was inconsistency in the way that national forests were implementing owl guidelines. In summary, she stated:

> The [FS] management strategy strongly favors timber harvest over owl protection. SOHAs are often drawn around planned timber sales. In most cases, everything outside the SOHA boundary that is not otherwise protected is planned for future harvest or has already been logged. The SOHAs will become small islands of habitat. There is no assurance that this is going to sustain owls over time.[59]

Other activities in the region also undermined the approach to the issue set up by the FS and the FWS. Partly reflecting the increasingly activist Washington state wildlife experts, but suggestive of the concerns of many of the state and federal owl experts, the Spotted Owl Subcommittee of the Oregon–Washington Interagency Wildlife Committee updated their 1981 plan, and produced new draft guidelines early in 1988. The guidelines called for a more extensive program of owl habitat protection than was being thought about by the FS. They included a "landscape approach which combines large block reserve areas containing a high density of adjacent pairs with the network approach of small blocks of habitat, usually for single pairs."[60] All known sites populated by owls were to be maintained in areas where population levels were considered to be low, and no further timber harvest would occur in selected areas, including the Olympic Peninsula. SOHAs were to contain at least 3800 acres of owl habitat in Washington, 2200 in Oregon, and 1600 in northern California.

The FS leaders were not pleased by the activities of the Subcommittee. On the cover of a draft set of the guidelines forwarded to the agency in October 1987, Hal Salwasser noted: "Is this what you expect from the subcommittee? What about options and their biological consequences? . . . This is only one option with no disclosure of relative biological merits of meeting law and policy." Deputy Chief Lamar Beasley wrote to FS Wildlife staff director Bob Nelson: "Bob, I don't understand this at all. Why is this group drafting management guidelines at this time?" Nelson replied: "Basically the biological community does not support what we are doing. Since this group includes our own researchers, I assume they don't either." Hal Salwasser's note on the routing slip put the agency's reaction more succinctly: "Out of control out there."[61]

Other events also signaled the unraveling issue. In mid-January 1988, the National Audubon Society indicated that it would seek endangered species status for the marbled murrelet, an old growth-nesting seabird. In May, the Ninth Circuit Court of Appeals issued an order prohibiting the BLM from selling any old growth forest trees over two hundred years old.[62] The suit had been brought by a number of environmental groups who were protesting the mid-1987 decision of the BLM Oregon State Director not to prepare a supplemental EIS on the impacts of the BLM's timber program on spotted owls.

The course of action taken by the FS and the FWS was perhaps affected most by events that grew out of the FWS's decision not to list the owl as a threatened or endangered species. Acting on behalf of 25 environmental organizations, SCLDF filed suit in federal district court in Seattle on May 6, 1988, contesting the FWS action. The environmentalists argued that the reason for the agency's decision not to list the owl was concern for the economic consequences of listing, a consideration that is not allowed under the provisions of the ESA. On November 17, 1988, U. S. District Judge Thomas Zilly remanded the case back to the FWS, finding that the agency was "arbitrary and capricious" in its earlier decision not to list the owl as threatened or endangered. To everyone involved in and observing the case, it was clear that the FWS leadership had acted in disregard of the advice of their internal experts, and that politics had played a significant role in their statutorily required biological decision. The agency was given ninety days to provide further analysis of its decision on the owl. Not surprisingly, the agency indicated in December that new information had come to light, and that it would reopen the status review.

Response to the Final SEIS

The Chief of the FS signed the Record of Decision for the Final SEIS on December 8, 1988. By that time, it was pretty clear that there was not much else to do, though agency leaders still tried to find that narrow point of balance between the different interests. In the eight-month period between issuing the Final SEIS and the Record of Decision, the FS received almost a thousand letters in response to the Chief's request for public comment. The primary change in the Record of Decision was to increase the amount of habitat to be protected in a SOHA on the Olympic Peninsula to 3000 acres within 2.1 miles of a nest.

The final decision on the SEIS was not viewed favorably by many of the internal agency experts. While they understood that the decision was not intended to be based purely on biology, they still were uncomfortable with the final choice. Dick Holthausen describes the reaction of the biologists:

> I think the biologists as a group were fairly uncomfortable with it, because it appeared to us that it was going to defer a hard decision on managing for spotted owls, rather than taking it on in the current round of planning. And we didn't perceive that that decision, or the social or political climate for that decision, was likely to get much easier. . . .
>
> I think that any of the biologists, if they were to tell you their pure druthers from a biological standpoint, would say we really ought to be maintaining all the habitat we have out there right now. But people recognize that this is not purely a biological decision. That was really the thinking behind doing it as a risk analysis in the first place. We recognized that there was risk and uncertainty involved on all sides, and that the process of making a decision was going to be a very complex one that involved not just biology, but economics and politics. . . .
>
> My biggest fear [about the decision] was that it would allow the maintenance of a given harvest level for the next five to ten years, and then if people still had the same consensus about what was needed for spotted owl management, it would have caused that har-

vest level to decline very quickly. And I've been convinced all along that someone needed to be dealing with a strategy for a gradual adjustment in that harvest level, rather than a precipitous one, so that the economic and social system had some opportunity to adjust to that change. I was concerned, I know some others were concerned, that it was just going to leave us facing the same nightmare five or eight or ten years down the road . . .[63]

The fact that few of the agency biologists supported the Chief's choice was a major concern within the agency, particularly having seen what happened to the FWS. Agency lawyers must have laid awake at night worrying about what would happen to their case if they got into court. They tried hard to push for a decision that the biologists could support, and harder to get the biologists to agree to support the Chief's decision once it was made. According to Bruce Marcot:

[The lawyers] were trying to push the steering group, the project leader, and the ID Team into having the biologists stand by the Chief, and we just very simply stated what our role was spelled out to be from the very start. That is, to simply analyze effects on certain aspects of the issue and to state that as cleanly as we can. And if we're asked for our personal views, we will give them honestly. And that's in large part why I have not been called up on a stand so far, and will luckily not be, at least by the agency. Because they see that my answer to those kinds of questions couldn't help the agency.

I think that the Chief picked something that he honestly thought was a good balance. . . . There were many, many things going on behind the scenes, talks at the highest level between Department of Interior and Agriculture; among Park Service, BLM, and the Forest Service; between Fish and Wildlife Service and the Forest Service early on . . . If I was the Chief, would I make the same decision? Why don't you give me a year or two to think about that. I have a very, very hard time putting myself in his shoes with all the political pressure.[64]

While the agency biologists might have had reason to be organizationally kind, though professionally opposed, in assessing the validity of the Chief's decision, the interest groups had no trouble indicating what they felt about it. On January 24, 1989, a timber industry coalition and a coalition of sixteen environmental groups filed separate appeals of the Chief's decision. On February 3, Assistant Secretary of Agriculture John Dunlop denied the appeals. Inside the federal bureaucracy, there seemed to be no point in delaying the inevitable court suits.

A few days later, the lawsuits were filed. On February 7, the Northwest Forest Resources Council filed suit against the FS in the Portland Federal District Court. The next day, Seattle Audubon and five other environmental groups filed suit in Seattle. At the same time, the Western Washington Commercial Forest Action Committee filed in Portland against the FS decision, and on March 2, the Washington Contract Loggers Association filed in Seattle. In mid-March, Seattle District Court Judge William Dwyer granted a temporary restraining order on 139 planned FS timber sales, and extended the injunction indefinitely in May.

On February 28, 1989, more than four hundred loggers staged a rally on the steps of the state capitol in Olympia, Washington, and marched on a "friendly" Washington State Senate committee to demand passage of anti-set-aside legislation. The Associated Press reported that there were scuffles and shouting matches between the loggers and "greeners" or environmental activists. More than 150 truck rigs clogged the area around the capitol, horns blaring. State Senator Amondson hoisted his young son aloft and "received a hero's welcome," asking "Is it spotted owls or kids? Our kids come first."[65] And in the summer of 1989, tensions continued to rise and the amount of civil disobedience increased, leading to threats even against Smokey Bear and Woodsey the Owl, as described in the first paragraphs of the introduction to this book. From the Forest Service's perspective, an agency that liked to keep its business off the front pages of the newspaper, all hell had broken loose.

5

All Hell Breaks Loose: 1989–1993

In the events that followed the response to the Chief's decision on the SEIS, the FS largely lost control over the direction of the issue. Its expertise and information were challenged, the credibility of its leadership was weakened, and numerous members of Congress and others began talking about fundamental changes in the direction and mission of the agency. Throughout the next four years, technical studies prepared by the agencies and interagency groups generally pointed toward the need to protect a significant amount of habitat, the administration and the Pacific Northwest Congressional delegation tried to craft a more moderate course of direction, and the courts and the environmentalists tried to hold the government to the direction identified in the studies. Both timber interests and preservationists continued a battle for the hearts and minds of the American public on an issue that had achieved national notoriety. And the politics of the 1992 Presidential and Congressional elections, which focused largely on the health of the domestic economy, tended to polarize the positions of elected officials and the public. Through four years of blood, sweat, and tears, the spotted owl issue was tossed back and forth between numerous political and administrative hands, consuming a lot of human energy and generating a considerable amount of anxiety without producing much understanding, much less a stable outcome.

Perhaps nothing had more impact on the overall level of anxiety associated with the controversy as the decision by the FWS in April 1989 to propose that the owl be listed as a threatened species under the Endangered Species Act. It is sometimes easy to forget that all of the controversy up until this point revolved around an animal that was not on the federal list. Listing under the ESA loomed large as a threatening unknown, whose consequences were feared by most of the involved parties. For the Forest Service, listing the species would at a minimum mean that Fish and

Wildlife Service staff would have their hands in Forest Service business, and would also probably mean more single purpose set-asides, fewer timber receipts and timber staffers, and political problems. For the Fish and Wildlife Service, listing meant immersion in a political controversy that everyone understood was a no-win situation.

The listing decision was equally threatening to other groups. For timber, listing meant that their opposition would be given an additional lever into on-the-ground management decisions; even if overall timber supplies ultimately were not affected, the uncertainty of the situation was hard to deal with. For many environmental groups, listing meant having to live with their great fear that they may win the owl battle but lose the endangered species war. For the courts, it increased the likelihood of unending, unproductive litigation. For Congress, it suggested the possibility of having to face the politically unattractive proposition of making a decision that would be perceived as a choice between endangered species (a clear and long-standing public value) and economic well-being (an equally compelling public value). In the spring of 1989, these fears, having circled above the fray for some years, descended almost visibly on most of the individuals and groups involved in the owl case. Even if the threat posed by the listing decision was not fully realized, it raised the level of anxiety and reduced everyone's decision space considerably.

A proposed listing for the owl was published in the *Federal Register* on June 23, 1989.[1] It proposed listing the owl as a·threatened species throughout all of its range in the Pacific Northwest and British Columbia. According to the proposed rule, the owl was "threatened throughout its range by the loss and adverse modification of old-growth and mature forest habitat primarily from commercial timber harvesting."[2] Clearly the leadership of the FWS did not want to propose listing the owl, but they made the only credible choice they could make. A February 1989 report by the U. S. General Accounting Office had made it clear that politics, and not science or law, had been responsible for the earlier FWS decision not to list the owl.[3]

With the likely listing of the spotted owl and injunctions on FS and BLM timber sales, the pendulum of power and influence over forest management in the Northwest had clearly swung toward the environmental groups. But by the late 1980s, strategists in the environmental movement were seasoned enough to understand that a judicial strategy is only as good as the politics underlying it. To succeed at shifting the balance of management of Northwest forests toward protection of old growth and associated animals and plants, the environmental groups had to neutralize the power of the timber lobby acting through the Pacific Northwest Congressional delegation, and offset the inherent power of the agencies. To accomplish the first task, they continued developing a strategy of nationalizing the old growth issue, pulling it out of the arena of regional politics where they were doomed to failure.

To win on the national stage, the environmentalists had to find Congressional sponsors willing to challenge powerful members. In the House of Representatives, Bruce Vento (D-MN), Chairman of the Interior Committee's Subcommittee on National Parks and Public Lands, and Jim Jontz (D-IN), a member of the Forestry Subcommittee of the House Agriculture Committee, helped move the issue along. In the Senate, Patrick Leahy (D-VT), Chairman of the Senate Agriculture, Nutrition and Forestry Committee, became a patron. All three got involved partly because of staff members who were interested in the issue and had solicited their boss's support. All three had significant pro-environmental constituencies and little to lose in their home districts by taking a stand on old growth protection in the Pacific Northwest. Vento and Leahy had institutional turf concerns as well. For many years, the Appropriations Committees had been setting substantive national forest policy by establishing an Allowable Sale Quantity that promoted cut levels that increasingly conflicted with the policies established by the authorizing committees. But the fact that all three individuals were willing to take on the power of the Pacific Northwest Congressional delegation suggests how far the environmentalists had come in legitimizing the issue as a national concern, and generating enough constituent support to provide protective cover to members of Congress willing to take action.

To counterbalance the power of the agencies, the environmental groups worked hard to develop parallel sources of expertise, and challenge those of the FS and BLM. Their efforts involved soliciting the support of outside experts and hiring new staff scientists and economists, and the information they brought into the debate increasingly was seen as more credible than that of the land management agencies. For example, mapping and inventory work by The Wilderness Society and National Audubon was showcased in a Congressional hearing held in June 1989, setting up a battle over numbers between the environmentalists and the FS. The Chief continued to hold to the FS's estimates that some 6 million acres of old growth remained, while the environmental groups claimed that the real figures were closer to 2.5 million.[4]

Pushing the Pendulum Back: The 1989 Timber Summit and Its Aftermath

The Oregon Congressional delegation pushed the pendulum of power back somewhat by orchestrating a "timber summit" held in Salem, Oregon on June 24, 1989. Organized by Senator Mark Hatfield (R-OR), Representative Les AuCoin (D-OR), and Oregon Governor Neil Goldschmidt, the one-day summit was billed as a day for information exchange, in the hope of establishing a dialogue that would lead to a compromise that would let the disputing interests get past the short-term crisis atmosphere established by the timber sale injunctions. In fact, the summit became a masterful means of shifting the force of initiative from the environmentalists back to the pro-timber Congressional delegation.

The summit involved three representatives of environmental groups, three representatives of industry groups, the regional or state directors of the FS, BLM, and FWS, and most of the Oregon Congressional delegation. Members of Congress from Washington were not included, a strange exclusion that affected the ultimate effect of the summit in Congress. None of the nongovernmental groups were certain about the objectives of the summit. Most of the environmental groups thought it would be a media show, with little substance, and some environmental staffers argued that they should not attend. But they really had little choice: To stay in the debate and appear reasonable, and not the extremists that their opponents claimed they were, the environmental groups participated in the summit, and as it turned out, bore the brunt of the Congressional questions and comments.

Worse from the environmental groups' standpoint was that the summit forced them into a reactive mode, by providing a platform at which Les AuCoin and Mark Hatfield presented what was billed as a compromise proposal to deal with the short-term timber supply problem. Following a closed-door luncheon of the Oregon delegation, Hatfield and AuCoin presented a four-part proposal: Timber supply levels (the Allowable Sale Quantity) would be prescribed at 8 billion board-feet off FS lands over a two fiscal year period (1989–1990); BLM sales would be 2 bbf for the same period; and the agencies would be required to prepare and sell at the 10-bbf level. The proposal also sought to provide certainty for industry by requiring the environmental groups to ask the courts to drop the injunctions on sales, and precluding them from filing judicial and administrative appeals of the sale program. Cutting would be directed away from significant old growth areas, through the inclusion of language requiring nonfragmentation for "significant old growth Douglas Fir stands." And the proposal was to be a short-term action that would not prejudge decisions on future sale levels or land management plans.[5]

The proposal was presented as a three-day offer. The groups had until Tuesday to respond, or the proposal would be withdrawn, and the "normal" appropriations process would generate the allowable sale level. Both the agencies and the timber industry responded with qualified support for the proposal, but the environmental groups were thrown into a tizzy. The alliance of environmental groups was an uneasy coalition of different interests, perspectives, and capabilities, and the response to the timber summit exacerbated their differences rather than bringing them together in the face of a common enemy. Having viewed themselves as finally holding a strong hand in influencing the direction of forest management in the Pacific Northwest, the environmental groups found little of substantive merit in the Hatfield and AuCoin proposal. While the ASQ levels were nine percent less than those in 1987 and 1988, they were considerably more than those provided by the draft forest plans, and roughly equal to actual harvests averaged over the past five years.[6] The nonfragmentation language would recognize the value of old growth for the first time in federal law, and that would be an important symbolic victory, but the thought of giving up their right to appeal a decision was anathema to groups grounded in the view that their fundamental mode of action was to challenge governmental decisions.

The environmentalists responded on June 27 with a letter that was intended to buy time until a counterproposal could be developed. They noted that the timber supply levels were moving in the right direction, but did not go far enough. Similarly, the proposal's call for nonfragmentation of old growth was positive, but specific language was needed to implement it. Finally, the groups were unable to support limits on judicial review, in part because of "our belief in citizens' fundamental right to seek justice in the courts."[7]

On the afternoon of Friday, July 14, the environmental groups presented their counterproposal.[8] It supported a timber supply level of 9.6 bbf for the two years, but noted that the level could not be sustained over time. It contained language intended to give the nonfragmentation requirement some teeth, by outlining a set of screens that would limit what could get cut. For example, no timber sales were to be allowed in habitat areas that the FWS determined might contribute to the owl's viability. On judicial review, the environmental groups made what they viewed as a significant concession: they would agree to sign legally enforceable stipulations in federal court that would require them not to challenge timber sales conducted in accordance with the other guidelines of their proposal. Such a process would bind the groups in this specific case, without creating the dangerous precedent of having Congress limit review of forest management decisions. Finally, the counterproposal noted that the agreement must be a first step toward a longer term process to deal with the issue in a more comprehensive way.

The response to the Ancient Forest Alliance's counterproposal was rapid and almost universally negative. In a meeting later that evening, the Congressional delegation told the environmental groups that they "can't live with this."[9] On Monday, both the timber industry and the Forest Service responded to the proposal. While industry's view was expected, the speed and nature of the FS response surprised the environmentalists, particularly because it came without any discussions with the groups. The agency deemed the environmentalists' proposal as "unworkable."[10] According to Audubon staff member Brock Evans, FS Chief Dale Robertson told them, "we can't live with this, you're pushing us off a cliff."[11]

The FS was particularly aggrieved by the screens the environmental groups would establish to minimize fragmentation. In the FS's view, the timber supply levels agreed to by the environmentalists were a sham, since they could never be achieved if the nonfragmentation screens were applied. The constraints established by the screens would be "as stringent as the current court injunction imposes on timber sales."[12] In addition, the screens would rule out many of the areas already prepped for sale by the FS, and the lead time involved in prepping other areas would mean that little timber would come on line during the next two years. The agency also commented that the provisions the environmental groups outlined to insulate the sale program from court delays would not work. The bottom line for the FS was that, while the proposal seemed to allow a 3.9 bbf timber sale level for the agency in fiscal year 1990, it would actually result in a sale program similar to that allowed under the court injunction: about 1.5 to 2.0 bbf in 1990. The agency continued to support the Oregon Congressional delegation's proposal.

In the aftermath of the timber summit, and the resulting negotiations over appropriations bills in the Senate and the House, the divisions in the environmental community festered through hundreds of hours of conference calls, strategy sessions, and negotiations on Capitol Hill. Some groups, particularly the regional and local groups located in the Northwest, wanted to go for broke and work toward a major floor fight in the House and Senate. Others, such as the Sierra Club, were leery of angering powerful members such as Hatfield and AuCoin who were sometimes supportive on other issues, and sought a compromise solution that built on the Hatfield/AuCoin proposal. Most of the Washington, D.C.-based environmental staffers fell somewhere in between, estimating the likelihood of winning the short-term battles as extremely low. In their view, a compromise on the short-term crisis would give everyone some breathing room, and would allow the preservationists to continue their longer term strategy of building political support at the national level that would yield victory down the line.

At times throughout the negotiations, these differences weakened the hand of the environmental coalition and so absorbed the time and energy of staff members that very little other work was done. Hard feelings developed between groups and individuals that took some time to work through after the battles were over. One environmental staffer commented that if Mark Hatfield had known what the impact of the summit and proposal would have been on the environmental community, he would have done it several years earlier.

The final appropriations bill (called the Hatfield/Adams bill) was passed in October 1989, and built broadly on the parameters of the original Hatfield/AuCoin proposal. It prescribed a timber sale program for fiscal years 1989 and 1990 of 9.6 billion board feet (7.7 bbf FS; 1.9 bbf BLM). To provide more certainty to timber interests, the bill did several things: First, it provided incentives to the environmental groups involved in lawsuits against the FS to free up 1.1 billion board feet held in litigation bondage, and do so quickly. Within two days after enactment, the FS was to provide the environmentalists with a list of timber sales ready to be offered for sale during 1989. If the groups identified sales totaling 1.1 bbf within fourteen days after receipt of the list, the other sales on the list would not be offered through the end of fiscal year 1990. Second, it streamlined the administrative appeals processes, and proclaimed the FS's Final SEIS and accompanying Record of Decision as adequate to govern management of the affected areas, and not subject to further judicial review. In addition, it prohibited the courts from issuing a temporary restraining order or preliminary injunction on fiscal year 1990 timber sales, though it did not touch the court's power to issue a permanent injunction if a decision on a pending timber sale was arbitrary, capricious, or unlawful. Finally, it prescribed deadlines for judicial review and authorized the use of Special Masters to expedite the processing of lawsuits.

For the preservationists, the appropriations bill precluded timber operations in known owl areas and gave them some involvement in the selection of areas to be cut. Areas identified as SOHAs under the FS FSEIS, and some 110 BLM habitat areas that had been set aside under an agreement with the Oregon Department of Fish and Wildlife, were to be off limits to cutting. Forest-level advisory boards, including an equal number of environmental and business representatives, were to provide input to the FS in determining which remaining areas would be cut. The FS was directed to minimize the fragmentation of "the most ecologically significant old growth forest stands," and to revise its 1988 record of decision (attached to the Final SEIS) based on new information, some of which was to come from the activities of an Interagency Scientific Committee that had been set up by memorandum of understanding in the previous year.

In the conference report accompanying the final bill, Committee members made clear their dissatisfaction with the effectiveness of prior agency actions:

> The managers have agreed to this provision because a large portion of the Forest Service's and Bureau of Land Management's (BLM) fiscal year 1989 timber sale programs on the thirteen national forests in Oregon and Washington and five BLM administrative districts in western Oregon known to contain northern spotted owls have been interrupted due to legal challenges, and because these challenges have raised serious concerns about the adequacy of planned actions of the agencies with regard to managing habitat for the northern spotted owl and minimizing fragmentation of old growth stands. The managers are extremely concerned that the agencies did not pursue avenues to resolve the conflicts or more adequately address the issues raised in these legal challenges, thereby requiring the inclusion of this section. The extraordinary measures included in this section, particularly with regard to judicial processes, have been reluctantly agreed to because of the failure of the agencies to take steps on their own to resolve these matters in a manner which could have prevented the current situation.[13]

No one was entirely satisfied with the outcome of the appropriations battle. Timber got the promise of a short-term federal timber supply at fairly high cut levels, but received it at a price of unprecedented environmental group input into the timber sale process and the explicit recognition of old growth in federal law, a change that was sure to come back to haunt them in the debate over the long-term management issues. The environmental groups largely defined their success as having staved off a much worse outcome. Cut levels were lower than initially proposed, they received a role in choosing what stands would be cut, and made symbolic progress on the old growth issue, but they lost control of the issue, at least temporarily, and got few concessions of significant meaning in the short term. The limitations on judicial review were particularly odious, leading to court suits that the environmental groups eventually won and then lost. The Ninth Circuit Court of Appeals ruled in September 1990 that the constraints on review outlined in the appropriations bill violated the separation of powers doctrine contained in the Constitution, but by the time the ruling was handed out, the issue was moot since it only applied to sales through September 30.[14] The Supreme Court reversed the appeals court decision in 1992, leaving open the possibility that future legislation might incorporate restrictions on judicial review.[15]

Other groups received a mixed outcome from the appropriations battle. Mark Hatfield and the rest of the Congressional delegation got some breathing room, and regained control of the issue, but the fact that a compromise with the environmentalists was necessary highlighted their waning influence over the long-term management decisions. The agencies perhaps fared the worst. The legislation provided for unprecedented outside involvement in their day-to-day operations, and further undermined their credibility as land management experts. And it highlighted their continuing lack of control of the issue.

Attempting to Regain Control: The Interagency Scientific Committee

The FS's attempt to regain the high ground of technical credibility was to convene an interagency group of scientists who would create a scientifically legitimate plan for owl management. Under the authority of an interagency memorandum of understanding signed a year earlier, an Interagency Scientific Committee was chartered in October 1989. Its existence was further acknowledged and blessed by language in the Hatfield/Adams appropriations bill. Headed by one of the FS's most credible scientists, Dr. Jack Ward Thomas, Chief Research Wildlife Biologist at the Forestry and Range Sciences Lab at LaGrande, Oregon, the ISC consisted of six individuals including four FS, one FWS, and one BLM employee. To link their efforts to agency management and nongovernmental groups, the Committee developed a set of designated agency manager contacts, and a set of "knowledgeable observers" who would represent the interests of the three states, the National Park Service, the forest products industry, and the environmental community. In many ways, the ISC looked a lot like the much earlier Oregon Endangered Species Task Force and some of the same people were involved in both. But besides a significant increase in the availability of technical information and approaches, the major change lay in the political environment. Unlike the earlier interagency efforts, the product of the ISC was sure to have an impact on the course of forest management.

The mission of the committee was the "development of a conservation strategy for northern spotted owl management and cooperation."[16] While the scientists recognized the broader issues of old growth and biodiversity protection, they kept their focus on the owl. After an extensive set of discussions with governmental and nongovernmental experts, field inspections and literature review, and peer review by a set of scientific associations, the scientists concluded, "that the owl is imperiled over significant portions of its range because of continuing losses of habitat from logging and natural disturbances. Current management strategies are inadequate to ensure its viability. Moreover, in some portions of the owl's range, few options for managing habitat are left, and options are inexorably declining across its range. Delay in implementing a conservation strategy cannot be justified on the basis of inadequate knowledge."[17]

Published in April 1990, the report of the ISC called for the protection of 8.4 million acres of forestland in the Pacific Northwest, including acreage in parks and wilderness areas and state and private lands. The scientists abandoned what they called a "flawed system" of small (1 to 3 owl-pair) spotted owl habitat area (SOHA) set-asides, in favor of larger blocks of habitat, which they called Habitat Conservation Areas (HCAs).[18] The HCAs were designed to house a minimum of 20 breeding pairs of owls. After the FY1990 timber sale program was completed, no timber was to be sold out of the HCAs, and the committee suggested that a fire management policy be developed to protect the set-aside areas. The committee recommended that at least half the forest outside the HCAs be maintained in timber larger than 11 inches in diameter (dbh) and with at least 40% canopy closure—trees that are roughly 30 to 50 years old. Their strategy was expected to result in a breeding population of 2200 pairs of owls before the end of the twenty-first century (a 30% increase over 1990 levels) though a 30 to 40% decline was likely in the near term.

The outcome of the ISC could not have been particularly surprising since a number of the committee members including Eric Forsman had been recommending protection for the owl for years, but that reality did not make their report any less controversial. It became the jumping off

point for another round of political handwringing, administrative and judicial activities, and Congressional initiatives, although the dynamics of
the discussion had somewhat changed. Increasingly, the debate was about
what kind of balance between economic impacts and owl preservation
was appropriate, and less about the scientific justification for owl protection. While the timber industry continued to argue that spotted owls reproduced successfully in second growth forest, particularly in northern
California, and letters to the editor of various newspapers contained reports of owls allegedly nesting in shopping malls and high rise apartment
buildings, the Interagency committee report tended to put such arguments to rest. According to the scientists, the experience in the redwood
forests in northern California could not be extrapolated to the predomi-
nantly Douglas fir forests in Oregon and Washington. In their view, protective action was warranted, and without it, the owl would likely become extinct.

The ISC report was unveiled at a hearing conducted jointly by three
Congressional subcommittees on April 4, 1990. At the hearing, FS Chief
Dale Robertson said that implementing the Committee's proposal would
reduce harvest levels on national forests in Oregon, Washington, and
northern California by some 1.0 to 1.3 billion board feet, representing an
overall reduction of 30 to 40% in harvest levels on Oregon and Washington national forests. An equal percentage reduction was foreseen for harvests from BLM lands.

Neither the FS nor the BLM had undertaken detailed analyses of the
economic impact of the ISC recommendations, but the timber industry
and its supporters were ready with predictions. According to Mark Rey,
executive director of the American Forest Resource Alliance, a newly organized coalition of pro-timber groups housed at the National Forest
Products Association in Washington, D. C., the plan would result in "the
immediate loss of 9,000 to 12,000 timber industry jobs with additional
job losses in northern California and in supporting industries." Based on
timber prices in Oregon and Washington, the industry estimated it would
cost $95 million to protect each owl pair. According to Rey, "The spotted
owl scientific plan . . . offers a stark choice between people and owls."[19]

Members of the Pacific Northwest Congressional delegation focused on
issues of balance and fairness. For example, Oregon Representative Bob
Smith (R), whose district includes ten national forests, noted that "The
question we should consider is whether or not Pacific Northwest families
should be entitled to the same consideration as spotted owls. Families
have as much a right to the forest as owls. . . . We've already set aside 53

percent of all the forests. They will never be touched, never be harvested."[20] Sid Morrison (R-WA) asked the central question, "Half of the
federal lands we have to work with is not enough to protect the species.
How much is enough?"[21]

The Interagency Scientific Committee recognized that the ultimate decision was one involving tradeoffs: "Conservation problems cannot be
solved through biological information alone, nor from applying 'scientific
truth.' Rather, solution comes from a combination of considerations that
satisfy society's interests. A strategy that has any chance of adoption in
the short term and any chance of success in the long term must include
consideration of human needs and desires. To ignore the human condition in conservation strategies is to fail."[22] While the Committee noted
the importance of an explicit assessment of the economic and social impacts of implementing their strategy, it highlighted the complexity of the
ultimate decision: "Adoption of the conservation strategy, however, has
significant ramifications for other natural resources, including water quality, fisheries, soils, stream flows, wildlife, biodiversity, and outdoor recreation. All of these aspects must be considered when evaluating the conservation strategy. The issue is more complex than spotted owls and timber
supply—it always has been."[23]

While publication of the ISC report did not end the spotted owl controversy, it did give weight to the proponents of preservation and helped
legitimize new Congressional and administrative action. The preservationists lost no time taking advantage of the report. On the same day as
the Congressional hearing, Congressman Jim Jontz (D-IN) introduced a
bill that would establish a national ancient forest reserve system carved
out of existing FS and BLM lands.[24] By this time, most people understood
that Congressional action was probably necessary to craft a final solution
to the owl controversy, and Jontz, joined by 24 cosponsors, none of whom
were from the Pacific Northwest, started the formal process of Congressional posturing. The proposed Ancient Forest Protection Act was envisioned as part of a larger package that included economic assistance and
compensation for affected workers and communities. Indeed, members
of the Pacific Northwest delegation were busily working on legislation
that might include low interest loans to promote diversification in timber-
dependent communities, job retraining for displaced workers, financial
assistance that would enable mills to retool for second growth logs, and
compensation for all affected parties.

Further restricting raw log exports was also a likely part of the ultimate
solution. It could provide supplies to mills to ease the transition to

smaller harvests off federal lands, and was supported overwhelmingly by the public, but was opposed by the Bush administration because of its impact on trade relations with Japan. The fact that the larger timber companies such as Weyerhaeuser were the primary beneficiaries of log exports reinforced the administration's opposition, and influenced key members of Congress, not the least of whom was House Speaker Tom Foley (D-WA). Foley's staff had made it clear in the negotiations over the Hatfield/Adams appropriations bill that log exports were not on the negotiating table.

The Interagency Scientific Committee report also gave new legitimacy to the FWS decision whether to list the northern spotted owl as a threatened species throughout its range. Having proposed listing on June 23, 1989, the FWS by law had one year to make a final decision, and no one was surprised by the agency's decision to do the scientifically "right thing" and list the owl as a threatened species. The final listing was published in the *Federal Register* on June 22, 1990. In announcing the listing decision, FWS Director John Turner indicated that "The biological evidence says the northern spotted owl is in trouble. We will not and, by law, cannot, ignore that evidence. . . . But I strongly believe there is room in the world to protect both owls and loggers. Our intent now . . . is to find ways to protect the owl with the least possible disruption to the timber economy of the Northwest."[25]

The Administration Seeks "Balance"

Between the listing decision and the ISC conservation plan, a scientific base was established that could have been used as a baseline from which to craft a package of compensation and mitigation, but the direction set in the scientific studies was not politically palatable, at least in the view of the administration. What followed was a period wherein technical studies indicated one direction, the administration tried to craft another, and the courts and the environmental groups tried to hold the administrative agencies to the direction indicated by the technical studies. Members of Congress introduced a variety of pieces of legislation that together would no doubt eventually become a legislated solution to the conflict. Everyone sought an appropriate "balance," but since there is no absolutely correct answer to the question of how much is enough, decisions are made only when the politics stabilize and one side amasses enough power to force a decision out of the system and hold everyone to it. For the next two years, political instability continued to be the norm as environmental and timber interest groups lobbied for their interests, and national electoral politics began to play a role.

The Bush administration was clearly opposed to a significant amount of land protection for the owl, but rather than the obvious and sometimes illegal perversion of scientific judgment that had taken place during the Reagan years, the Bush approach was to delay taking action, attempt to craft strategies with fewer economic impacts, try to change the rules of the game by seeking changes in the ESA, and swing with the politics of the moment. Secretary of the Interior Manuel Lujan, Jr. and President Bush established the tone of their response to the FWS listing decision in the month preceding the decision. In an interview with the *Denver Post* that was widely reported around the country, Lujan, charged with enforcing the ESA, stated his personal opinion that the ESA is "just too tough an act, I think. We've got to change it." Citing the case of the Mount Graham red squirrel, a listed subspecies that affected construction of an observatory in Arizona, Lujan said, "The red squirrel is the best example. Nobody's told me the difference between a red squirrel, a black one, or a brown one. Do we have to save every subspecies? Do we have to save [an endangered species] in every locality where it exists?"[26] President Bush emphasized a consistent theme of his: balance. On a West Coast trip raising campaign funds, Bush stated that "I'm interested in the owl, but I'm also interested in jobs for the American family. . . . I want to be known as the environmental President, but I also want to be concerned about a person's ability to hold a job."[27]

There was considerable debate at the highest levels of the administration as to what to do in response to the listing decision. The FS wanted to adopt the recommendations laid out in the ISC report, while White House officials, led by domestic policy adviser Roger Porter and under pressure from Republican lawmakers such as Mark Hatfield, felt that its economic impacts were too high. According to one FS official, "We're getting a lot of heat from the White House . . . as to why we haven't come up with something that provides more timber. Believe me, we've tipped the world upside down looking for something that would provide more timber and protect the owl."[28] Indeed, while the administration had planned to release its plan for further action simultaneously with the listing decision, the ongoing internal administrative battle forced a delay.

On Tuesday, June 26, the administration announced its "Five-Point Plan to Preserve Owl and Protect Jobs." It called for the following actions in response to the listing of the owl as a threatened species:

- The BLM would implement a strategy different from that identified in the ISC report, and allegedly would protect more owls at a cost of 1000 jobs, rather than the 7600 jobs estimated to be a cost of the ISC plan.

- For fiscal year 1990, both the FS and BLM timber sales would follow the provisions of the Hatfield/Adams provision of the 1990 appropriations act.

- For FS lands past FY1990, the administration would convene a high-level interagency task force, chaired by the Secretary of Agriculture, to begin work on devising a forest management plan for fiscal year 1991, and submit a report to the President by September 1, 1990.

- The administration would seek to convene the Endangered Species Committee, the so-called "God Squad," should a federal agency receive a jeopardy opinion from the FWS on a proposed timber sale. It would also press the Congress to broaden the mandate of the Endangered Species Committee to "allow it to develop a long-term forest management plan for Federal lands."

- The administration announced its support for bills that would ban the export of raw logs from state lands, an activity that reportedly would reduce job loss in the Pacific Northwest by roughly 6,000 jobs by the year 2000.[29]

Clearly the forces promoting a balance point different from that contained in the ISC report had won. Equally important to the direction set by the administration was the timing of the announcement and its relationship to electoral politics. The five-point plan was announced in a news conference held directly after another news conference at which administration officials announced the President's decision to curtail offshore oil drilling off California, Oregon, Washington, Florida, and New England's Georges Bank. No one believed that the two decisions were unrelated. Rather, they appeared to give one to the environmentalists and one to the developers. Equally important was the relationship of the decisions to the future of important Republican politicians.[30] The President's stand on the drilling issue was thought important to gubernatorial candidates Pete Wilson in California and Robert Martinez in Florida. The owl issue was of obvious importance to Senator Mark Hatfield and other Republican lawmakers up for reelection in the fall of 1990.

Having declined to use the ISC report as the foundation of a solution to the owl conflict, the administration took on the burden of finding something better, and failed miserably at the task. The interagency task force, comprised mostly of top level political appointees, spent three months conducting hearings and negotiating among the major political players in the controversy, and produced a three page news release on September 20, 1990. The *Seattle Times* called the administration plan, "so sketchy that lobbyists and government spokesmen alike were unable to describe some of its key features."[31] The primary recommendation in the plan was that the 1991 allowable timber sale quantity be set at 3.2 billion board feet, roughly 0.6 bbf more than that envisioned in the ISC report. When asked where the additional timber was to come from, task force spokespersons were unable to respond. The task force also called for timber sales to be insulated from environmental laws, and the ESA to be amended so that economic and social factors could play a greater role in species decisions.

The task force's recommendations contained little that was new and impressed no one on either side of the owl issue. According to Kevin Kirchner of the Sierra Club Legal Defense Fund, "There was a sense of dismay that after three months, the most that the task force could come up with was a 2½-page press release."[32] Timber lobbyist Mark Rey commented that "at least no one will ruin a weekend studying it."[33] A House Merchant Marine and Fisheries Committee staff member summarized the views of most involved parties: "So Congress and the Pacific Northwest carefully awaited the recommendations of this Delphic-like task force, only to find out later, two weeks before time runs out, that these folks really have nothing to say."[34]

While the interagency task force produced little of substance, it did have an impact on the course of events. It demonstrated what little slack was left with which to craft a solution. After spending three months, top level political appointees were unable to find a solution. Indeed, at one point in their deliberations, task force members reportedly were thinking of endorsing the ISC report recommendations, but the administration told them to try again.[35] The task force also consumed time valuable to crafting a long-term solution. At the end of the appropriations negotiations a year earlier, everyone knew that more difficult negotiations would be necessary to find a solution to the conflict, or a crisis would blossom forth each year over the coming year's timber sale program. Indeed, at the end of 1989, conditions were fairly ripe for a negotiated compromise that would attract most of the groups' support. The interagency task force refocused the hopes of pro-timber groups, and defused the energy that might have been put into negotiations. If nothing else, it squandered the time necessary to develop a compromise over the next year's timber sale appropriations.

For its part, the Forest Service tried to walk a line between the politics of the administration and Congress, and the scientific credibility of the ISC report. Having been directed by the Hatfield/Adams bill to incorporate the ISC recommendations and new information into agency plans for owl management by September 30, 1990, the Department of Agriculture gave notice on September 28 that the FS was vacating the December 1988 Record of Decision, and that it would manage timber sales in a manner "not inconsistent with" the ISC report.[36] At the same time, the FS proposed a fiscal year 1991 timber harvest of 3.2 bbf. In order to cut at that level, the agency proposed to accelerate sales in roadless areas, tracts previously set aside to protect single pairs of spotted owls, areas already severely fragmented, and areas that were rejected for cutting during the previous two years by the Hatfield/Adams-created advisory boards.[37]

Environmental groups were horrified by the FS proposal. Even though the roadless areas proposed for cut had not been designated as wilderness in earlier wilderness bills, environmentalists never stopped viewing them as wilderness and hoped to protect them in future legislation. Even more upsetting to the environmentalists was the proposal to cut some 181 million board feet from areas that they had guarded from sale during 1989 and 1990. According to Charlie Raines, timber specialist for the Sierra Club and a member of the Mount Baker-Snoqualmie National Forest timber advisory board, "I'm angry. We're going to fight them on this. We tried to work with the system. What they're saying is we got snookered. They're going to go back and take what we thought was protected."[38]

The Environmental Groups Press Their Advantage in the Courts

The reaction of the environmentalists in the Pacific Northwest was to continue to press their case through the courts, and the actions of the administration gave them numerous opportunities. In the fall of 1990 and winter of 1991, they went after the FS for scheduling timber sales without having a lawful spotted owl plan in place. The September 28, 1990 notice that they would operate in a manner not inconsistent with the ISC report was viewed as arbitrary and capricious, having been made without undergoing the normal administrative procedures including a public hearing and an environmental impact statement. Indeed, FS officials testified that they had started working on a new plan early in 1990 (following passage of the Hatfield/Adams appropriations bill), but had been stopped while the administration's interagency task force did its thing. According to Associate Chief George Leonard, it was expected that the task force would develop an option that would be the basis for a new FS plan, but "they never did."[39] The court was not sympathetic: The FS was enjoined temporarily from selling any more timber in habitat suitable for the owl until such a plan and an environmental impact statement was in place. The agency had until June 15, 1991 to provide the court with a timetable for completion of the plan, and had an ultimate deadline of March 5, 1992 to complete the plan.

The environmentalists also continued to lean on the Endangered Species Act lever. In the FWS's June 1990 listing of the spotted owl as a threatened species, the agency had declined to list habitat critical to the owl's survival because it was not "determinable" at the time.[40] Yet the Endangered Species Act makes it clear that listing and critical habitat designations are to occur simultaneously except when habitat designation is not determinable because of limited information, or not prudent because it might enable collectors to locate and collect endangered or threatened organisms. In the owl listing, the FWS claimed that they lacked the resources to adequately assess critical habitat by the June deadline, and that the ISC report's publication in April did not give the FWS enough time to respond.[41] The court disagreed with the agency's reasoning, stated that the agency had abused its discretion and was acting in a manner that was arbitrary and capricious. District Court Judge Thomas Zilly gave the FWS until April 30, 1991 to propose critical habitat.

The agency published its proposed critical habitat designation on May 6, 1991, and the extent of land area described as critical habitat surprised everyone: 11.6 million acres were identified as critical habitat, including 6.5 million acres of FS land, 1.4 million acres of BLM land, 0.7 million acres of state and tribal lands, and 3 million acres of private land.[42] Not included in the 11.6 million acres were lands already reserved from timber harvest, such as wilderness areas or national parks, estimated at 1.9 million acres, because such areas were already protected and did not need any special protection. Hence the total acreage of lands affected by the designation was much larger than previous administrative pronouncements that usually relied on existing reserved lands as core habitat areas.

The FWS relied substantially on the ISC analysis and strategy for owl conservation. Critical Habitat Areas (CHAs) went well beyond the acreage identified as Habitat Conservation Areas (HCAs) in the ISC report for several reasons: The CHAs had to be identified as legal entities that could be mapped and recorded in the *Code of Federal Regulations*, and that meant that CHAs were identified by section lines, while the ISC boundaries were less defined and generally corresponded to natural system boundaries. The CHAs contained acreage of currently unsuitable habitat to facilitate the development of large contiguous blocks of habitat. While the ISC report was a plan for the management of existing federal lands, the FWS proposal also identified nonfederal lands that were considered an important part of an overall habitat matrix, including sections of land that were intermixed with old growth parcels (the BLM checkerboarded O&C lands were primarily of this type) and corridors to facilitate dispersal between habitat areas. Indeed, the proposal stated that "there may be significant impacts on private and other non-Federal lands."[43]

In publishing the proposed habitat designation, the FWS attempted to forestall the panic that agency officials knew their proposal would set off. First, the proposal noted that the habitat designations by themselves did not prescribe any particular management regime for a given area; rather, they meant that the FWS would be reviewing proposed actions under the Section 7 interagency consultation process. For private lands included in the critical habitat proposal, the notice indicated that proposed actions were only affected by the designation if they required a federal permit. The proposed rulemaking also made it clear that the areas proposed were simply an opening move in what was expected to be an iterative process of proposals, analysis, public review, and modification.[44] The proposed areas had not been screened adequately for their economic impact as required by the ESA. Since the Secretary of the Interior can exclude any area from critical habitat if he determines that the benefits of exclusion outweigh the benefits of inclusion (unless the exclusion will result in extinction), and Secretary Lujan had made his biases known, the proposed rule was to be seen as giving notice to affected parties: Make an argument as to why your favorite area should not be included, and maybe you will win the great owl sweepstakes.

The proposed critical habitat designation got everyone's attention, including that of private landowners such as Weyerhaeuser and other large timber companies who did not rely on federal lands for a timber supply. Just as visible was U. S. District Court Judge William Dwyer's decision two weeks later to make the injunction on FS timber sales in owl habitat permanent until the FS had completed a new spotted owl management plan and environmental analysis. The judge's decision came after a week of testimony that was widely reported by the media, and seen as one of the broadest forums on the owl/old growth issue that had yet been held.

The hearing included testimony from biologists, government officials, timber industry officials, and representatives of local communities. A major point of debate was over the economic effects of a permanent ban on logging in owl habitat. Estimates went as high as the loss of 11,688 direct jobs and 25,500 jobs overall.[45] The FS estimated the impact at 3,000 direct jobs with sales dropping from 3.7 billion board feet to 0.8 bbf. Agency officials argued that they should be allowed to harvest timber consistent with the ISC report recommendations. Ironically, the timber industry found itself supporting the ISC report recommendations as providing something better than an outright ban on logging in owl habitat.

Judge Dwyer was not impressed by their arguments, and put the blame squarely on the FS and the administration:

> The problem here has not been any shortcoming in the laws, but simply a refusal of administrative agencies to comply with them. This invokes a public interest of the highest order: the interest in having government officials act in accordance with law. . . . This is not the usual situation in which the court reviews an administrative decision and, in doing so, gives deference to agency expertise. The Forest Service here has not taken the necessary steps to make a decision in the first place—yet it seeks to take action with major environmental impact. . . . Had the Forest Service done what Congress directed it to do—adopt a lawful plan by last fall—this case would have ended some time ago.
>
> More is involved here than a simple failure by an agency to comply with its governing statute. The most recent violation of NFMA exemplifies a deliberate and systematic refusal by the Forest Service and the FWS to comply with the laws protecting wildlife. This is not the doing of the scientists, foresters, rangers, and others at the working levels of these agencies. It reflects decisions made by higher authorities in the executive branch of government.[46]

While Judge Dwyer recognized his obligation to weigh the various public interests affected by the injunction, he was unconvinced by the predictions of dramatic economic effects. His decision restated many of the environmental groups' counterarguments about the magnitude of economic impact and concluded: "To bypass the environmental laws, either briefly or permanently, would not fend off the changes transforming the timber industry. The argument that the mightiest economy on earth cannot afford to preserve old growth forests for a short time, while it reaches an overdue decision on how to manage them, is not convincing today. It would be even less so a year or a century from now."[47]

The Dwyer decision and the FWS-proposed critical habitat designation got everyone's attention, and everyone joined in the feeding frenzy on the FS and the administration. For example, stating that the Bush administration had "botched" the handling of the owl issue, Washington Governor Booth Gardner called for Congress to grant the Pacific Northwest states or a new regional authority the power to make decisions over the future of the area's old growth forests.[48] His special assistant for timber issues, Rich Nafziger, claimed that "There is no leadership from the top. The result is that the administration does not appear to have the credibility with either the courts or the environmentalists to broker the kind of balance that is so sorely needed."[49] Congressman Bruce Vento (D-MN) argued that "We are losing jobs because the past two administrations have refused to allow the Forest Service and Bureau of Land Management to follow the environmental laws."[50] Jim McDermott (D-WA) claimed that "The administration has been willing to systematically violate the law to lose one court decision after another, rather than help find solutions."[51] Norm Dicks (D-WA) called for the administration to find a defensible way to move forward on the issue: "I want the Secretary of Interior and the Secretary of Agriculture to get together and come up with an approach that's consistent on federal lands, and is scientifically credible. Because without that, we're going to get killed in the courts."[52]

For its part, the administration was faced with a deteriorating state of affairs. Its technical credibility was gone, and its politicization of science challenged. Under court orders, both the FWS and FS had to come up with new owl protection plans, and it seemed clear that the recommendations contained in the ISC report were probably the minimum possible.[53] Nine months after the owl was listed as a threatened species, a recovery team was named to develop a protection strategy. While the administration reportedly had delayed naming a recovery team so that any recommendations would be postponed until after the 1992 elections, the February court decision on critical habitat forced their hand. But the composition of the team was controversial and reflected the administration's efforts to influence the outcome of the recovery plan. Normally recovery teams consist solely of scientists with expertise in the biology of the listed species. In contrast, the owl recovery team included economists and political appointees as well as biologists. Marv Plenert, FWS regional director, was named team leader, an unusual event indicating the political sensitivity of the issue.

The environmental groups and scientists complained about the unusual composition of the committee. Chuck Meslow, director of the FWS's Cooperative Wildlife Research Unit at Oregon State, said "I think it's pretty general knowledge that the group of biologists on there are pretty concerned about not being misused."[54] Syd Butler, The Wilderness Society's vice-president for conservation, had a more dramatic metaphor for the committee: "The wolf politicians and the lamb biologists may lie down together, but the lamb isn't going to get much sleep."[55]

The Train Wreck and God Squad Strategies

With the courts pushing the agencies to act, the administration and the pro-timber forces in Congress and industry began to sense that their best shot at getting relief was through an exemption from the Endangered Species Act. Members of the Northwest delegation had introduced their own version of ancient forest legislation to counter the protectionist Jontz bill, and Congressional gamesmanship was underway, with Don Young (R-AK) introducing legislation that would declare Jim Jontz's northern Indiana district a national forest.[56] But it was clear that the only outcome possible from the Congress was a compromise containing a set of new protected areas at the cost of a significant economic assistance package.

A number of opponents of new actions to protect endangered species, including some in the administration, saw the situation as a toehold into more fundamental changes in federal endangered species policy. The Endangered Species Act was up for reauthorization in 1992, which could provide a window of opportunity for amending the basic policy to require more balancing of economic impacts against preservation benefits. By this time, the spotted owl was seen as the tip of the endangered species–development conflicts iceberg. Controversies over the status of the delta smelt in northern California, the California gnatcatcher in southern California, the marbled murrelet and the Pacific salmon in the Pacific Northwest, and the red-cockaded woodpecker in the southeast, among others, rose to media consciousness, and appeared to support the arguments of opponents of the ESA. "They're everywhere. They're everywhere. It's not just loggers; it's all of us." A deteriorating economy made a weary public more willing to listen to anti-preservation arguments, and a reorganized timber lobby and a growing "wise use" movement, a coalition of commodity interests including both corporate and community-level groups, were more effective at bringing their perspectives forward.

Many critics of the administration believed that administration strategists were following a "train wreck" strategy for dealing with the owl issue.[57] They believed that the administration, under the leadership of White House Chief of Staff John Sununu, a longstanding opponent of environmental regulation, was doing as little as it could to find a solution to the owl controversy, so that the controversy would run out of control and result in pressures to fundamentally change federal policy on endangered species protection. The endangered species train would crash and burn under the force of its own controversy.

The train wreck strategy depended, however, on Congressional action to weaken the ESA, and such action was not at all likely given the mood of the Congress. Senator Bob Packwood (R-OR) had tried earlier in the year to change the act to jump start the "God Squad," but Senate Majority Leader George Mitchell (D-ME), one of the original sponsors of the act, gave an impassioned speech late at night in defense of the act. Packwood's amendment was soundly defeated. It was unlikely that other powerful members of Congress would be willing to take on the Act. Senator Mark Hatfield had been rocked by allegations of a financial scandal that came to light in May.[58] The alleged scandal damaged Hatfield's ability to take controversial stands at least in the short term, and just as important, it drew his time and attention away from policy matters.

While fundamental changes in the ESA were unlikely, it was speculated that it might be possible to get some set of timber harvest activities exempted from the provisions of the Act. When faced with an irreconcilable conflict, the Endangered Species Committee could exempt a project from the provisions of the ESA. The so-called God Squad committee had been added in 1978 as a political pressure valve in response to the snail darter–Tellico dam situation in east Tennessee, and it seemed a likely source of relief. From the perspective of top administration officials, the committee was a particularly logical course of action because it included seven individuals, six of whom were top-level administration officials, and the seventh was an individual appointed by the President to represent the affected state(s). The main barrier to be overcome was that there were substantive tests that a case for exemption had to meet, and no exemption had yet been given. Action by the committee to grant an exemption would be seen as a major political and moral statement with little precedent.

The administration decided to try the God Squad strategy not with FS lands, but with proposed timber harvest on BLM lands. FS leaders had chosen to implement the provisions of the ISC report on the ground, and were receiving a fair bit of cooperation from the FWS. Under the guidance of Director Cy Jamison, the BLM had gone a different route. Instead of following the provisions of the ISC report, the BLM had come up with a different plan that purportedly would protect owls at a lower economic cost. The Jamison Plan avoided harvest in Habitat Conservation Areas, and protected owl territories that had been identified under the Hatfield/ Adams 1989 appropriations bill, but did not comply with the provisions that regulated harvest levels outside HCAs. The agency estimated 1000 jobs lost as a result of the Jamison strategy, rather than the more than 7000 forecast under the ISC approach. When asked what would happen if the FWS determined that the BLM approach would cause jeopardy to the owl, Director Jamison indicated that the agency would ask for an exemption under the God Squad provisions of the ESA.

When the BLM submitted its fiscal year 1991 timber sale program to the FWS for consultation review, only a third of the proposed sales were cleared. Fifty-one were held up with proposed modifications to protect owls, and another 52 sales were rejected as likely to jeopardize the continued existence of the spotted owl. From these 52 sales, the BLM chose 44 and submitted an application for exemption from the ESA.[59] Secretary of Interior Lujan determined that the BLM had met the three preliminary qualifying tests identified in the ESA, and accepted the application on September 30, 1991. The God Squad strategy was underway.

The ESA contains fairly explicit guidelines on how and when a proposed exemption has to be considered. The Secretary of the Interior has to certify that: there are no alternatives to the proposed action, action is in the national or regional public interest, appropriate mitigation measures will be taken should an exemption be granted, and the agency proposing action has acted in good faith, that is, it has refrained from taking irreversible actions while consultation and the exemption process is underway. The seven-person Endangered Species Committee can only grant an exemption if it is satisfied that these same tests have been met, and that the benefits from acting outweigh the costs to a species. The Committee itself consists of the Secretaries of Agriculture, Interior, and Army; the Administrators of the Environmental Protection Agency and the National Oceanic and Atmospheric Administration; the Chair of the Council of Economic Advisors; and a representative of the affected state(s) appointed by the President. A decision to grant an exemption must be made within 170 days following acceptance of an application, and must receive five of seven Committee members' votes.

While the 44 BLM timber sales were fairly small potatoes, most observers of the process felt that they had much larger significance. According to Mark Rey, executive director of the American Forest Resource Alliance, "It may have a greater bearing on the fate of the Endangered Species Act than on the owl and the timber industry."[60] Sierra Club lobbyist Jim Blomquist also viewed the proposed exemption as a step toward bigger things: "They're attempting to see what the public mood is. If they can get away with an exemption, they will come forward with much larger actions."[61]

Most environmentalists and many administration critics felt that Lujan's convening the God Squad was a continuation of the train wreck strategy. In their view, Lujan wanted the committee to grant an exemption and have it overturned in court in order to change the political climate for amending the ESA.[62] If the exemption process failed to provide relief, the Act could be portrayed as inflexible and needing change. In March, Congressman Peter DeFazio (D-OR) noted, "I don't think the train wreck theory can anymore be denied. It's either an intentional train wreck, or a train wreck where there's just no one in the engineer's cab. One way or another, we had the wreck and the indifference, incompetence, and intransigence of the administration caused a lot of casualties."[63]

The exemption process went forward amid controversy alleging administration manipulation of the process. Not only was testimony taken from expert witnesses, but an open public hearing was held in Portland, surrounded by hundreds of loggers and logging trucks. According to Kevin Kirchner, an attorney with the Sierra Club Legal Defense Fund, "They are manipulating the process, presumably for political purposes."[64] Kirchner cited conflict of interests among Interior Department staff, the withdrawal of the EPA as an intervenor supporting the FWS's opposition to the timber sales, and the withdrawal of a FWS brief contending that the BLM had not consulted in good faith, one of the preliminary tests that had to be met prior to convening the God Squad. Whether or not any of the environmentalists' charges were true was beside the point. What they guaranteed was that any outcome of the exemption process would be appealed. And when five of seven committee members voted to grant an exemption on May 14, 1992 (the EPA Administrator and the representative from Oregon voted against the exemption), no one was surprised when the environmental groups went to court to contest the decision.

In the summer and fall of 1992, the dispute was still in full bloom. Court suits against BLM, FS, and FWS continued; timber harvest injunctions were in effect; ancient forest, economic assistance, and ESA reauthorization bills were being debated by the Congress; and the FWS-appointed spotted owl recovery team was working on one recovery plan while an administration task force sought to devise another parallel, less-costly recovery plan.[65] For its part, the FS continued to work on an owl management plan that would bear up to court scrutiny, having produced a plan in January 1992 that was deemed inadequate by Judge William Dwyer in May.[66] All of these activities were affected by a festering economic recession and the politics of a national election year, wherein key lawmakers from the Pacific Northwest were seeking reelection or election to different posts, and a weakly supported president and administration cast about for a message that would keep them in office. While a huge amount of individual and organizational time and energy had been spent on the issue, an endpoint was not clear.

The Clinton Administration Begins to Take Action

As a new administration began its term of office in January 1993, the political dilemmas represented by the spotted owl conflict presented themselves to a president who ran on a platform centered on improving the domestic economy, and who had a weak environmental record in his home state.[67] The choice of Al Gore as vice-president, who had authored a book on the global environmental crisis, was seen as a strong statement in favor of environmental protection, and high-level administration appointments appeared solidly in the environmental camp: former Arizona Governor (and League of Conservation Voters president) Bruce Babbitt named Interior Secretary; former Wilderness Society head George Frampton named Assistant Secretary over the FWS and the NPS; and former House Agriculture Committee staffer (and SAF resource policy expert) Jim Lyons named Assistant Agriculture Secretary over the FS. And the early rhetoric from the administration appeared to set a direction different from that of the previous twelve years, with Babbitt calling for a multiagency, ecosystem approach to resource management grounded in a scientific assessment of biological resources to be carried out by a new National Biological Survey.[68]

At the same time, other early moves appeared to respond to the longstanding political realities of natural resource management. Visits to the Pacific Northwest by Babbitt and Agriculture Secretary Mike Espy appeared to be designed to avoid environmentalists and old growth forests, focusing instead on loggers and millworkers. Babbitt shocked the environmentalists by stating that the administration was ready to "lift the edges of the injunctions" in order to get timber flowing into local mills.[69]

In March, the administration again surprised environmentalists by back-pedaling on several key promises made during the campaign. Efforts to deal with below-cost timber sales, low grazing fees, and minuscule mining royalties had been proposed as part of the administration's budget package but were dropped in the face of a great deal of political pressure. In response, NWF President Jay Hair said "what started out like a love affair" between the environmental community and the administration "is turning out now to be more like date rape."[70] In the early days of the administration, it was clear that President Clinton needed all the political capital he could muster to promote a variety of initiatives, including deficit reduction, economic stimulus, and health care reform, and early political fiascoes presented the image of a weakened and disorganized administration that was very willing to compromise its objectives in the face of political opposition.

Yet the magnitude of the political dilemma facing the administration was matched by an opportunity to forge the political will to settle the owl dispute. While President George Bush had derided the preservationists as those "spotted owl people," candidate Bill Clinton pledged to convene a multiparty working group to resolve the controversy within the first 100 days of his administration,[71] and he was true to his word. Scheduled for April 2, 1993, the forest conference represented the first step by the Clinton administration to legitimize a compromise solution to the controversy.[72]

All parties viewed the conference warily. Armed with court injunctions, a more environmentally inclined Congress and White House, and remembering the outcome of the 1989 timber summit, environmentalists in particular wondered how they could benefit from such a meeting. According to ONRC conservation director Andy Kerr, "It reeks of smoke-filled rooms. These are national forests, and we need to have a national review of the issue. They want a timber-supply summit. We'd like to have a national-forest summit. If it's a timber summit, I ain't coming."[73] Rick Nafziger, head of outgoing Washington Governor Booth Gardner's timber team noted that the elements of a settlement were not difficult to define, but the polarization of the stakeholders would be difficult to overcome: "The components are simple. The problem is that the interest groups are so far apart."[74] One congressional aide indicated the difficulty of the situation: "This is going to make setting up the Middle East peace talks look like a Sunday picnic."[75]

Held in Portland, Oregon, the forest conference was remarkable in several ways. It showcased the President and Vice-President of the United

States, along with three cabinet secretaries, sitting around a conference table for a full day, talking domestic policy with those who ostensibly would be most directly affected by any course of action, while the rest of the nation had the opportunity to watch the proceedings on national television. Reflecting the lauded style of an earlier televised conference on the economy held in Little Rock, Arkansas, the forest conference suggested that the top levels of government took the issue seriously, and viewed it as a complex problem needing a serious solution to get beyond the gridlock that had prevailed for the preceding few years.

Just as remarkable was the set of individuals who were not visibly participating in the dialogue at the conference: the agencies and the Northwest Congressional delegation, some of the major historic players in the dispute. The symbolism was unmistakable: Here was a conference focused largely on the future of national forest management in the Pacific Northwest, and the chief of the FS and the elected representatives of the region's population were not at the table. While the agencies and members of Congress were present as invited members of the audience, the team planning the event orchestrated it to focus on a small set of experts and a larger set of Pacific Northwest residents who were personally affected by any outcome. This approach reduced the amount of grandstanding at the conference, offset the need for many participants to defend past actions, and focused many of the presentations on the kind of personal stories cherished by the President and loved by the media. From the strategic perspective of the administration, it also had the potential of focusing follow-up activities on solving a recast problem rather than on justifying past positions. The handpicked panelists included twenty-one representatives of timber (split approximately evenly between labor and management), four fisheries groups, nine environmentalists, six scientists, a handful of local and state government representatives, two economists, two sociologists, one vocational counselor, and the Archbishop of Seattle.

The conference also was notable for explicitly reframing the controversy as something bigger than the spotted owl issue. The problem was defined as: how to protect a broad range of environmental values within the old growth ecosystem while dealing humanely within a regional economy that was undergoing a normal process of transformation. While the needs of some 480 old growth dependent species were aired (and the threat of 480 more spotted owl controversies described), more important was a discussion of the impacts of timber activities on salmon and other fisheries. The Congressional delegation had tried in preconference negotiating sessions to keep the salmon issue off the table, but presentations at the conference made the very logical connection that one segment of the economy may well benefit from changes in other segments of the economy: Salmon stocks could improve as logging declined and management practices changed.

The discussion of economics focused not just on the impacts of the logging restrictions due to the owl lawsuits, but on a broader understanding of regional, national, and international economic forces. A Congressional Research Service study issued in the weeks preceding the conference had indicated that rising lumber prices were more the result of economic recovery than owl-caused supply constraints, and that the real cost of lumber was no higher than it had been in the 1970s.[76] Other studies marketed by the environmental groups in the weeks before the conference highlighted automation and the investment behavior of multinational corporations as key components in the decline in timber employment in the Northwest.[77] In closing one of the day's sessions, the President reiterated the theme that had been central to his campaign platform: the unavoidable need to change in the face of a changing world. While he was sympathetic to the human costs of change, he noted that the average 18 year old in all parts of the nation will change the kind of work he or she does seven or eight times in a lifetime, and highlighted the changes in defense-dependent industries. In Clinton's words, "I cannot repeal the laws of change."[78]

In concluding the conference, President Clinton called on his cabinet secretaries to craft a "balanced and comprehensive long-term" plan to end the stalemate and to complete it within sixty days. He reiterated the theme of compromise: "I don't want this situation to go back to posturing and positioning. To the politics of division that has characterized this issue in the past. I hope we can stay in the conference room and out of the courtroom."[79] Clinton established five principles to guide the construction of a plan:

- The needs of loggers and timber communities had to be addressed.
- Forest health had to be protected.
- The plan had to be scientifically sound.
- A sustainable and predictable level of timber had to be provided.
- The government had to speak with a single voice.

The FEMAT Report and Option 9

To respond to the President's directive, the administration created three interagency working groups, which began working around the clock to complete their tasks by June. Jack Ward Thomas led the forest ecosystem management assessment team (FEMAT), which focused on strategies for managing the federal public lands in the Northwest. Other groups addressed labor and community assistance and interagency coordination. In addition, agency staff were asked to investigate some of the side issues that had been kept separate from the owl controversy in previous years: regional economic concerns, salmon and fisheries management, log exports, regional timber supply, and the health of the forests on the east side of the Cascade Mountains in Oregon and Washington.[80]

Any plan produced by the working groups had to conform to the normal procedures established in federal administrative and environmental law, and an interdisciplinary team was established jointly by the FS and BLM to construct a draft environmental impact statement that would examine the environmental effects of the FEMAT options and identify a preferred course of action. The EIS would supplement the 1992 plans that Judge Dwyer had found inadequate and would define management direction for FS and BLM lands that hopefully would meet NEPA and NFMA requirements. Judge Dwyer had given the agencies a deadline of July 16, 1993 to provide a new plan, and the SEIS effort was framed to adapt the substance of the FEMAT report into NEPA-defensible terms.

The FEMAT was the centerpiece of the post-conference effort, and it was directed to "take an ecosystem approach to forest management and . . . address maintenance and restoration of biological diversity, particularly that of the late-successional and old-growth forest ecosystems, maintenance of long-term site productivity of forest ecosystems; maintenance of sustainable levels of renewable natural resources, including timber, other forest products, and other facets of forest values; and maintenance of rural economies and communities."[81] Its mission was to identify and analyze a range of alternative courses of action for the management of FS, BLM, and NPS lands within the range of the northern spotted owl. Biological diversity considerations were defined more broadly than a single species like the owl; the team was directed to consider the needs of the marbled murrelet, which by that time had been proposed for listing as a threatened species, other old growth associated species, and anadromous fish, and to consider the connection of various ecosystem segments. Options had to be provided that would yield medium to very high probability levels of species viability.

The charge to the FEMAT provided the basis for an analysis that could lead to what had been needed for years: a scientifically credible, multiagency, ecosystem-oriented management plan that had the blessing of the top levels of the administration. While it did not consider the problems and opportunities involved in the management of nonfederal lands, and did not involve stakeholders in a way that might build support, adding those objectives at that time probably would have doomed the effort.

The FEMAT report, produced in draft form in early June, yielded the groundwork for the administration to chart a course through the political negotiation and maneuvering that was sure to come. The report discussed ten options, most of which were based in part on strategies framed in previous studies, particularly those identified by the so-called Gang of Four report.[82] Each option allocated the more than 24 million acres of federal land within the range of the owl to one of seven categories, four of which were primarily land set-asides for protection purposes:

- *Congressionally reserved:* Roughly seven million acres of land that had been previously designated by the Congress as nonmanaged wilderness, national parks, wild and scenic rivers, and the like.

- *Late-successional reserves:* Areas that were protected from most management activities so as to maintain a functioning old-growth forest ecosystem, though some level of silviculture might be permitted to enhance the development of old-growth characteristics.

- *Administratively withdrawn areas:* Areas already removed from timber harvest because they were designated for recreation, visual protection, or certain other administrative objectives, or they were not technically suited for timber production.

- *Riparian reserves:* Buffer zones along streams, wetlands, and lakes.

Areas in which timber harvest and active management might go on were allocated among three categories:

- *Managed late-successional areas:* Areas in which silviculture and fire hazard reduction would go on to avoid catastrophic loss due to fire, disease, or infestation, and to maintain appropriate levels of late-successional and old-growth stands on a landscape scale.

- *Adaptive management areas:* Areas ranging from 84,000 to 400,000 acres in which experimentation would occur. "Their purpose is to provide areas where managers can use innovative approaches, perhaps at a landscape scale, to achieve management objectives. These areas will also provide a laboratory for innovative social mechanisms for managing federal lands and areas of mixed ownerships in a more cooperative and interactive fashion."[83] Only Option 9 included areas allocated for adaptive management.

- *Matrix:* All federal lands outside the above six categories. All scheduled timber harvest would take place within matrix lands, and guidelines were prescribed for management of the matrix. Most of the alternatives applied the ISC 1990 strategy called the 50–11–40 rule for management of lands outside reserves so as to provide dispersal habitat for owls.[84]

The amount of land allocated to these last three management categories ranged from 2.8 million acres in Option 1 to 8.6 million acres in Option 7, which was fairly close to the "no action" alternative: management prescribed by current FS, BLM, and FWS plans. In the team's assessment of biological consequences, however, only Options 1 through 5 and 9 appeared to satisfy the objectives of protecting the set of old growth–associated species, including the owl and the murrelet. The economic effects of these options were dramatic: The "probable sale quantity" of timber ranged from 0.2 bbf in Option 1 up to 1.0 bbf in Option 5, with forecasts of timber industry employment in the next decade ranging from 112,900 jobs in Option 1 to 118,600 in Option 5. These levels represented significant reductions when compared with historic data on timber sold and regional employment: the 1980–1989 averages were 4.5 bbf with 144,900 jobs; the 1990–1992 averages were 2.4 bbf with 125,000 jobs.[85]

Option 9 appeared to do a little better. While covering the ecological bases, this option was forecast to yield 1.2 bbf per year with 119,800 jobs in the region, and it was to accomplish this through the magic of adaptive management. Option 9 was added in a second round of option development under the encouragement of administration officials who sought to find something better than 1.0 bbf of timber, without compromising environmental objectives. Dubbed the "lean and mean" approach by forest ecologist Jerry Franklin,[86] the option relied on ongoing experimentation in ten areas dispersed across the two-state region. Representing roughly a quarter of the managed lands under Option 9, adaptive management areas would employ "technical, administrative, and cultural/social" innovations, including methods of "new forestry," public–private partnerships, interagency cooperative management, and the like. While a major objective of the areas was described as "learning to manage," they were also expected to produce timber, and the specific areas were located adjacent to "adversely economically impacted communities [so as] to provide opportunity for social and economic benefits to these areas."[87] Option 9 also jettisoned the 50–11–40 rule for management of matrix lands, which presumably would provide managers more discretion in how they laid out timber sales.

The Clinton Plan

While the interagency groups were developing their reports, life did not stop for the other players involved in the ongoing controversy, and all continued their efforts to influence the ultimate outcome. Both timber representatives and environmentalists placed ads in national newspapers, sought favorable editorials, and attempted to mobilize grassroots support. A horde of lobbyists descended on Congress, and a horde of Congressional representatives attempted to influence the administration.

All sides sought to define the framework through which a presidential decision would be made and understood. For timber, the early indications were that the options considered by the FEMAT would result in dramatic reductions in timber harvest levels, and they began a campaign to define a higher bottom line: 2.5 bbf, approximately the average harvest levels during 1990–1992.[88] Environmentalists responded that they had already lost 90% of old-growth forests, and hence any harvest level represented a major concession. And the administration did its best to lower expectations about harvest levels: By letting it be known that harvest levels associated with the ecologically viable options considered early on by the FEMAT were no more than 1.0 bbf, administration officials sought to frame any decision that would result in greater harvest levels as a plus for timber.

Clearly the pressure coming from timber's Congressional supporters was considerable, and the realities of making a decision that would cost jobs for a president who ran on a platform of improving the domestic economy were tough, and by many accounts, Option 9 was constructed at the end of the FEMAT process as a way to respond to these pressures.[89] In addition, its adaptive management approach appealed to agency officials because it offered greater discretion to on-the-ground managers and allowed researchers to try out new forestry approaches. Assumptions made about how discretion would be exercised resulted in forecasts of higher timber harvest levels while providing adequate biological diversity protection.

Given the "something better for all" character of Option 9, it was not surprising when it became the preferred choice of the President's advisors. In a draft memo to Clinton from an Executive Office committee headed by Katie McGinty, director of a newly created Office of Environmental Policy, the President was urged to adopt Option 9.[90] The memo stated that the overriding goal was to cut as much timber as possible while still complying with environmental laws. Option 9 would do this by "front-loading" a ten-year harvest program with higher yields in early years. While it would average harvest levels of 1.2 bbf over the ten-year period, it would provide 2 bbf in the first year and 1.7 bbf in the second. The memo suggested that the higher harvests in early years would help gain the support of Pacific Northwest lawmakers, though some scientists might oppose the plan.

If the President's advisors thought they had found middle ground by promoting Option 9, they were mistaken, and as the details of the plan began leaking out, timber, the Northwest delegation, and environmental groups came unglued. In a heated meeting between the Congressional delegation and McGinty, an angry House Speaker Tom Foley blasted the administration for mishandling the decisionmaking process and allowing the environmentalists to define the debate over the plan.[91] According to Foley, it was highly misleading to claim "that the administration had chosen an option that expanded or maximized the timber cut. In fact, the plan reduces timber operations by 80 percent from historic levels. It means huge job losses in communities; it comes close to meeting the goals of people who want to end timber operations."[92]

The initial reaction of environmentalists to the likely decision was equally extreme. According to Brock Evans, vice-president of the National Audubon Society, "This is war. It's political science, not biological science."[93] Adaptive management had been viewed as the equivalent of smoke-and-mirrors ever since the FS's SEIS effort in the mid-1980s, and while some of the agency players had changed, the environmental groups felt no more trusting of the agencies. In action alerts and phone calls to members, the groups sought to mobilize opposition to the approach taken in Option 9. One computer-generated action alert sent to Sierra Club members asked them to write or call the President. According to the alert,

The timber industry recently panicked when it was scientifically determined that timber cutting levels must be dramatically reduced to protect species and ecosystems. In order to sidestep the scientific data and continue timber cutting, the new option is based on so-called "New Forestry" techniques—techniques which are completely untested and would unravel the entire old-growth ecosystem. This "experimental" alternative has the timber industry licking its lips and sharpening its chain saws. But we cannot afford to let the timber industry and Forest Service experiment with the precious remains of our Ancient Forests. . . . We must tell the President that the timber industry–inspired solution is not acceptable. And we must tell him today.[94]

In spite of intense lobbying from all sides, when the President announced his preferred course of action in early July, no one was surprised that it was based on Option 9, including a cut level of 1.2 bbf per year, the creation of buffer zones around sensitive riparian areas, and the creation of adaptive management areas.[95] His plan also targeted $1.2 billion in new funding over five years to offset the economic impacts of the change in management. (He claimed that the plan would directly affect 6000 jobs in 1994, but would create more than 8000 jobs and fund 5400 "retraining opportunities.") Domestic milling would be encouraged by eliminating a tax subsidy for companies that exported raw logs. Timber salvage operations in the forests on the east side of the Cascades would be accelerated. The President's plan also sought to break the judicial gridlock and provide timber for harvest: the administration would seek a lifting of the injunctions on timber sales, and if necessary, seek Congressional action to limit judicial review. Officially, the President's plan was produced as a draft supplemental EIS, with the FEMAT report attached as an appendix.[96]

Following release of the Clinton plan, the political battle began in earnest, with both sides unhappy with the direction set by the plan, though the environmental groups appeared to soften their rhetoric. For the environmentalists, the adaptive management zones and the possibility of salvage logging meant that some of the areas that they had sought as inviolate reserves would be open to management, and they did not trust the agencies to do the right thing. According to Sierra Club executive director Carl Pope, "With this proposal, the fate of the forests will stay in the hands of the very agencies that pushed them to the point of collapse. It is a sincere attempt to protect the remaining ancient forests, but the way it is crafted won't accomplish that goal."[97] In addition, the groups were also highly concerned about any limits on their rights of judicial review.

For timber interests, the Clinton plan resulted in dramatic reductions in timber supplied from federal lands when compared with levels in the recent past. Warning that 85,000 not 6000 jobs would be lost, industry and labor leaders viewed the plan as a "betrayal" of the President's promise to create a balanced solution. According to Mike Draper, executive secretary of the Western Council of Industrial Workers, the plan would "devastate the wood products industry."[98] Regardless of the ultimate outcome, it seemed clear that the plan might not end the stalemate that had pervaded federal forest management for five years. According to Mark Rey, vice-president of the American Forest and Paper Association, "the Clinton administration has spent 90 days developing a solution to the crisis that will leave us exactly where we were when it began: court-ordered gridlock, with no timber being sold for the foreseeable future."[99]

As a ninety-day public comment period opened on the Draft SEIS, and close to 100,000 comments poured in, the political posturing, interest group strategizing, Executive Office deliberations, and battle for public approval continued. Coffin-carrying logging trucks circled downtown Portland, while a flood of action alerts sought to mobilize citizen involvement, if only to send a form letter to the President. Bill Clinton had promised a final decision by the end of 1993, but what would come next was anyone's guess. Was Congressional action necessary or likely? Could the administration act unilaterally, and would its action hold? House Speaker Foley continued his vocal criticism of the President's plan, yet it was unclear whether his message was that the plan had to change, or that the President should keep the decision within the administration and take the heat for it.[100] A Congressional battle over old-growth forests would be divisive and leave the lawmakers from the Northwest exposed to treacherous political crosswinds.

In the fall of 1993, more than twenty years after the spotted owl became a consideration for forest management in the Pacific Northwest, the controversy raged on. While the Clinton administration had mustered the political will to buy into a solution that was generally viewed as scientifically credible, it remained to be seen whether others would allow a settlement to take hold. Timber interests already had gone to court to challenge the legitimacy of the FEMAT report, arguing that the FEMAT process violated provisions of the Federal Advisory Committee Act. Environmentalists battled with the administration over how much timber could be released from court injunctions as a short-term supply measure while the SEIS process was completed, with administration officials threatening to seek restrictions on judicial review if not enough timber was released.[101]

As a broader set of resource policy issues began to be pushed by the administration, including the reauthorization of the Endangered Species Act, the prospect of the owl and old-growth dispute becoming enmeshed in these other issues was discouraging. At the same time, changes in the top leadership of the Forest Service in December, with Jack Ward Thomas appointed as Chief, signified a window of opportunity for public forest management that was ripe with possibilities. While there was no question that the spotted owl controversy had helped to fundamentally redefine the objectives and politics of federal resource management, more so than any other modern day environmental issue, its ultimate effect was yet to come.

TWO

Learning from History

6

Tough Choices: A Difficult Issue under Any Circumstances

While the long history of the spotted owl controversy has many meanings, the case provides significant insights into the character of natural resource and environmental disputes, and the ability of our decisionmaking processes and agencies to resolve them effectively. The prime benefit of history is as a learning process for social change, where ideas and methods are tested, evaluated, and either embraced or thrown away. The owl dispute persisted in part because:

- the nature of the issue itself promoted the controversy;
- the underlying characteristics of the American policymaking process did little to resolve it;
- the incentives influencing the behavior of numerous actors in the policy process exacerbated the underlying problems in policy processes;
- the policies that had been crafted to manage natural resources and regulate environmental quality tended to perpetuate the dispute; and
- the qualities and norms of present-day management agencies reinforced it.

The next five chapters explain the evolution of the owl case as a function of these five sets of variables.

For reasons that are explored in Chapter 7, even simple, uncontroversial policy issues are often difficult to resolve, and the spotted owl issue was neither simple nor uncontroversial. The character of the issue itself; the intensity of the stakes held by numerous individuals and groups; uncertainties about science, economics, and policy; and the fact that seemingly little room existed to craft a solution by the time the parties were serious about finding one, all made the controversy extremely difficult to resolve. Even if we had a decisionmaking system that routinely provided opportunities for effective, informed problem-solving, and a set of participants who truly wanted to settle the issue, the spotted owl controversy would have tended to persist. This chapter explores those characteristics of the issue that contributed to its persistence.

Separate but Interconnected Subissues

Natural resource and environmental disputes generally involve a multiplicity of subissues, so that seemingly simple choices become battles over a variety of other substantive, organizational, and political objectives. Indeed, natural resource decisionmaking is one of the few places where explicit choices are made about the economic and environmental future of a region of the country, and hence, decisionmaking often takes on a significance beyond the immediate issue. Since American institutions are not very good at long-term planning (that is, imagining a desirable future state and designing a strategy to bring it about), decisions about the future are made by reacting to events or proposed activities in an ad hoc way. The future emerges, shaped mostly by the hundreds of choices made by individuals without regard to their meaning for the broader society, but influenced at least somewhat by the many fewer decisions made as collective choices by our policy processes. The essence of many natural resource controversies is that they are collective choice decisions about the future state of the landscape that are not made any other way. As a result, resource policy issues are often more serious and difficult to resolve than they otherwise might be.

The spotted owl controversy was in fact the confluence of a number of different issues, each of which represented distinct policy streams. Sometimes when tributary issues combine into a raging controversy, they can be disaggregated and solved separately. In this case, the multiplicity of issues tended to muddy the waters, clouding the possibilities for resolution.

While some have suggested that the owl was purely a surrogate for other issues and concerns, the long-term survival of the northern spotted owl clearly was an issue that motivated the involvement of a number of participants in the controversy. The owl's preference for habitat features generally associated with older, unmanaged forest types—large trees, broken tops, standing dead trees, fallen decayed trees, and a multilayered canopy—did not bode well given the historic and predicted decline in such areas mostly due to timber harvest.[1] Fragmentation of what remained, further threatened the owl population, as younger forest areas promoted predation by great horned owls and competition by barred owls. In addition, entire portions of the owl's range were isolated due to natural barriers such as the Columbia River Gorge or development corridors such as the Willamette River Valley, raising concerns about the effects of genetic isolation on long-term viability.

Even if there were no spotted owls in Oregon or Washington, the future of the remnants of the old growth forest would have been hotly debated. Part of the concern was over other species that were thought to be associated with the old growth system, including some four hundred or more species reported as being dependent on or associated with old growth forests.[2] In addition, a rebounding interest in regional scale planning and ecosystem-based management in the end of the 1980s led scientists, interest groups, and some policymakers to be concerned about larger units of geography, and the old growth system of the Pacific Northwest was a logical focus of this concern. The old growth issue underlay much of the spotted owl controversy, yet few decisionmakers in federal land agencies or on Capitol Hill knew how to deal with it, and few tried or wanted to try. In many ways, inattention to the developing old growth issue doomed efforts to solve the owl controversy in a stable and enduring way.

Since the turn of the century, another theme had been tied up in the politics and management of public lands, the preservation of wildlands, and this issue too was a portion of the owl controversy. Prior to passage of the 1964 Wilderness Act, preservation proponents pushed both for single-site set-asides as national parks or protected areas, and administrative management of national forest and other federal lands as primitive areas, protected from commodity or other development. With enactment of federal wilderness legislation, interest group strategies shifted toward Congressional designation of land set-asides. In the Pacific Northwest, some 4.7 million acres of federal lands were established as designated wilderness, with major additions made through the 1984 Oregon and Washington wilderness bills. Following their passage, wildland preservation advocates looked to other policy strategies, and protection of endangered species habitat was a logical successor strategy.

One of the tricky things tied up in these three issues is that, depending on which you attempt to solve, they potentially lead you to different land allocation solutions. That fact complicated attempts to resolve the spotted owl controversy. The best strategy for enhancing the survival prospects of the owl may not be the best approach to protecting the old growth ecosystem. For example, it may be possible to "grow owls" through habitat modification or population management, but these approaches probably will not be the best way to protect old growth as inviolate reserves. Achieving consensus on land set-asides for primitive recreation and wilderness values may include protection of favorite backyard wildlands, that is, isolated fragments of old growth located close to population centers, but this approach is unlikely to protect owls or old growth ecosystem characteristics. The intermingling of these three issues in the spotted owl controversy made resolution considerably more difficult than it would be if the only policy and management concern had been ensuring the long-term survival of the owl.

Other issues were also tied up in the debate. Concerns over the direction and implementation of land management policy by the Forest Service, the BLM, and state and private landowners in the Northwest had been longstanding, and the owl was a logical focal point for these concerns as well. For years before the owl issue achieved national prominence, a debate had been ongoing about the purposes of national forest lands, and the methods used to achieve them. Battles over timber cut levels, below-cost timber sales, roading, clear cuts, and wilderness review had been ongoing for many years. Similar controversies existed over the direction of BLM lands in western Oregon, state- and county-owned lands in Oregon and Washington, and forest practices on private timberlands.

Issues related to public land management priorities and practices were grounded in more fundamental questions about the efficacy and values of the federal resource management agencies themselves. Survivors of resource management disputes increasingly questioned the motives, competence, and trustworthiness of the Forest Service, the BLM, and the Fish and Wildlife Service. The critics also believed that, while the science base was expanding significantly, preparing the way for new patterns of creative resource management, the agencies were resistant to change and out of date. Some questioned whether Forest Service leadership was at all committed to nontimber resource values and whether the agency's long-standing image as a professional, technically based bureaucracy was more appearance than reality. The BLM and the FWS were easier targets for criticism, because while they emulated the independence and image of professionalism maintained by the Forest Service, they lacked the depth of tradition and barriers to explicit politicization held by USFS.[3]

If nothing else, these ongoing concerns about the public land agencies had resulted in well-established visions of agency misdirection and ineptness. By encouraging the resource management agencies to retrench toward their core technologies—commodity production for the USFS and BLM, and game animal management for the FWS—the Reagan administration helped set up the agencies as clearer targets, as decisionmaking became more politicized and development-oriented, in spite of laws and public values to the contrary. By the mid- to late-1980s, agency critics had dug their trenches deeply, and when the owl dispute came onto the battlefield, it was viewed as one more example of agency mismanagement, confirming preconceptions that had been framed for a number of years. As a result, whatever actions the agencies took in the spotted owl dispute were likely to draw immediate fire, and they did.

Other broader environmental policy issues were involved as well. Court action in the 1970s established the seriousness of federal endangered species policy, and pro-development forces argued that the ESA's absolute mandate to protect was inappropriate, for it appeared to preclude balancing various social objectives against endangered species objectives. The 1980s brought a federal administration that was sympathetic to the development interests' arguments and who tried to find ways around the ESA's mandate.[4] These actions, and the dismantling of the FWS endangered species bureaucracy through budget cuts and staff decentralization, mobilized advocates of continued or enhanced endangered species protection. The fact that endangered species issues increasingly seemed to involve larger areas of real estate and more than one species had the effect of mobilizing both sides, as the developers perceived an even greater threat, and the preservationists understood the higher stakes in the outcome.[5]

The owl issue came to roost in a region that was undergoing a significant economic transformation. While habitat set-asides for the owl clearly would have an impact on the regional economy, other more fundamental forces were at work long before the owl became a visible public policy issue. Some individuals had begun to raise questions about the long-term viability of the timber industry in the Pacific Northwest. Market forces were clearly having an impact on the industry: The recession in the early 1980s reduced new housing starts nationwide with significant implications for the wood products industry. More than eighty mills in Oregon and Washington closed in the early 1980s because of the recession,[6] and similar effects were evident in the recession of 1990–91. Industry sources claimed that the downsizing in the early 1980s was in fact a positive force, shaking marginal firms out of business, and encouraging the survivors to implement productivity improvements such as laser-guided saw rigs. In their view, the future of timber in the Northwest was bright, not fading.

Nevertheless, many in the region worried about the overall dependence of the region's economy on timber. Curiously, both economic planners and ecologists share a belief in diversification as a long-term survival mechanism, yet for most of its history, the Pacific Northwest's economic base had been grounded in timber. Even if current activity had resulted in a broadened economic base, many perceived the region as an economic monoculture and feared the regional effects of economic downturns that seemed to be particularly associated with heavy industry. Historically, timber-based regions saw booms and busts as timber was logged out, and some viewed the Northwest as the final act in a slow-moving timber play that starred the Northeast in the 1700s and 1800s, and had featured the Midwest in the 1800s and early 1900s.

Even if the region itself could weather a shift from a timber-based economy, some communities would be hurt by any transformation. The long-term economic and social health of communities dependent on national forest timber was thus very much at issue in the spotted owl dispute. Towns like Forks, Washington; Molalla and Sweet Home, Oregon; and numerous others were dependent on national forest timber as a source of jobs in harvesting trees and milling them into wood products.[7] And these communities were the front-line in swallowing the economic impacts of any policy or industry-led change. They were aggrieved equally by changes in national forest timber supplies or changes in employment due to automation in the industry, though the former was easier to blame. The ability of the communities to absorb layoffs was slim, as alternative jobs in other industries were generally unavailable. Jobs created in the much-vaunted tourism industry generally were low-paying and fairly subservient, and hence unattractive to workers used to good pay with a fair bit of autonomy.

These same communities were in double jeopardy because their local public services were underwritten heavily by federal timber receipts, and an additional issue that became tied to the spotted owl controversy was the long-term funding mechanism for local public services such as roads and schools. One-quarter of revenues from all timber receipts paid to the Forest Service for national forest timber is returned to local governments, and one-half is paid from BLM O&C lands. In recent decades, this arrangement has been a major source of county revenues, offsetting the need to raise money to provide needed public services in other ways. But a shift in the funding arrangements for public services was likely as old growth resources were logged out on national forest lands. At the national level, forest policy critics argued that the timber receipt formula was inappropriate anyway, because it created a set of local interests dependent on federal public resources, and in so doing, contributed to political pressure for inefficient forest policies.

Concerns about who was benefiting from the extraction of Northwest natural resources extended into the international realm. Since old growth Douglas fir yields clear, high quality lumber, it was heavily in demand overseas in countries like Japan that had long since depleted their indigenous sources of high quality timber. As a result, a thriving market in raw log exports sent huge volumes of Pacific Northwest timber overseas, carried in the hulls of large cargo ships, serviced in Pacific Northwest ports such as Astoria, Oregon, and Tacoma, Washington. Since much of the value extracted from a log comes in its milling into finished products, Northwest economies were receiving only a small portion of the economic benefits associated with a chunk of wood. Since most of the corporations who were major players in the log export market were large, multinational corporations, some critics claimed that regional economic benefits were thin, even where national or global benefits were large. And although it was contrary to federal law to export raw logs harvested on federal lands, in an interconnected log market, a log harvested from private lands and taken from domestic milling simply creates enhanced pressure for additional cutting on federal lands to feed the domestic mills.

The issues of raw log exports, the magnitude of regional and local economic benefits from timber, and the long-term diversification of economic activities in the Northwest were significant issues on the plate of regional decisionmakers long before the owl appeared. They would have been difficult issues to deal with regardless of the spotted owl controversy. But the controversy and the economic issues became interspersed in a fairly counterproductive way. For firms seeking to improve their competitiveness within the industry, wise capital investments were hard to make because the owl issue increased the level of uncertainty about raw material supplies. On the other hand, the losers in the economic transformations of the 1980s and 1990s could use the owl as a scapegoat, deflecting public attention from more fundamental questions that should have been addressed. Neither timber nor sustainable development interests were well-served by the deflection, and it contributed to the tenacity of the owl issue.

Finally, all sides in the owl issue agreed that important international ramifications and messages were inherent in how the spotted owl controversy was resolved, and the global dimensions of the issue contributed to difficulty in resolving it. Since wood products are traded in an international market, U.S. forest policy has implications for patterns of international trade. Administration and industry officials claimed they opposed export controls on raw logs, for example, in part because they feared reprisal from importing countries in related or other markets. Some critics of these trade policies suggested that since the Japanese and others were major underwriters of the U.S. public debt, that administration officials were loathe to do anything to alienate their benefactors. Others claimed that the trade issues were largely a smokescreen for corporations that saw huge profits in international trade in natural resources.

Besides trade issues, everyone agreed that important messages were being sent overseas by the direction taken in resolving questions over long-term use of the forests of the Pacific Northwest. Proponents of preservation argued that if the United States did not protect its old growth forests, and the biological diversity contained within them, how could it effectively argue that Brazil and other countries should stop harvesting tropical rainforests? Industry officials counterargued that if the U. S. did not satisfy demand for wood products from American forests, they would be supplied by increased harvests in the tropics, where biological diversity concerns are much more serious due to the much higher level of diversity per unit area. Proponents of sustained yield management of the timber resources of the national forests argued that if we could not manage timber in Pacific Northwest forests where the soils and climate allow for long-term sustainable production, and where the governmental system was better at monitoring and minimizing environmental impacts than developing countries, where could we?

This long set of substantive issues underlay a dispute that on the surface appeared to be simply about owls and jobs. As discussed below, other issues dealing more with political power and organizational turf were also tied up in the dispute and contributed to its resistance to resolution. In addition, because some of the subissues were more ripe for settlement than others, the policy process had a hard time dealing effectively with the owl issue because of its implications for issues that were less ripe for settlement. If the future of the owl was all that was at stake, it is likely that the dispute would have been much easier to resolve.

A Diverse Set of Interests and Motives

Since environmental and resource controversies are multifaceted, and often involve real choices about the future, multiple issues translate into multiple interests, and a variety of individuals, groups, and agencies usually are involved in the debate. In the owl case, the multiplicity of subissues guaranteed that many parties would be affected by any attempt to resolve the controversy, and by the end of the 1980s, a large set of groups with diverse interests, motives, and styles had mobilized to voice their views in the decisionmaking process. At minimum, that meant a lot of parties around the negotiating table. In reality, it also meant an extraordinarily complex set of negotiations among individuals and groups operating with different information, varying motives, fluctuating power, and different levels of interest in seeing the issue resolved.

The environmental groups included a diverse set of national, regional, and local organizations without unanimity of interests or tactics. Their motives reflected the subissues identified above. Some groups, like the National Audubon Society, were truly concerned about the status of the owl and old growth-related biological diversity. Others, like The Wilderness Society, were primarily focused on the need to protect the wildlands that old growth represented. Still others, like the Sierra Club, were primarily concerned about protecting opportunities for wildlands-based outdoor recreation, while others, like many of the local Audubon chapters, were mostly concerned about protecting the fragments of old growth forest in their "backyards."

Some groups, like the Oregon Natural Resources Council, were vehemently anti-Forest Service, while others, such as Headwaters, were strongly anti-BLM, and many were opposed to the aesthetics and lifestyles associated with large-scale, mechanized timber operations. Some, such as the National Wildlife Federation, could be seen as fairly moderate in the spectrum of political possibilities, while others, such as the direct action-oriented EarthFirst "non-organization," advocated fundamental changes in social organization and political power.

As the controversy wore on, differences in style and objectives appeared between the regional groups and the nationals (and in some case, the Northwest and Washington, D. C. offices of the same groups). For example, leaders in the battles of the Northwest, such as the Oregon Natural Resources Council, were bit players in the halls of Congress. While ONRC worked hard to nationalize the old growth issue, in doing so, they gave up a fair bit of control over the direction of the issue. After the controversy landed in Washington, D. C., the regional groups increasingly felt the national enviros were too accommodating and overly inclined to negotiate. On the other side, the nationals felt that the regionals were not at all cognizant of political realities and likely to lose more than they could gain by holding out for more than was possible. Other differences, such as revenue-base, size, type of staffing, and quality of leadership, were also important and contributed to diversity of style and motive in the environmental group community. Even these observations, however, are overly simplistic since individuals and subunits within organizations often had differing objectives and opinions about tactics.

It is important to understand the texture within the environmental community because it contributed to the multiplicity of interests involved in the owl controversy. The owl dispute was not one of environmentalists against loggers, but rather one that involved a diversity of individuals and groups with some common and some divergent interests. At times, the fragmented nature of the environmental community made it difficult to find common ground and very difficult to bind all parties to any agreement.

The set of timber interests were similarly diverse. The concerns and motives of contract loggers and small mills that were highly dependent on national forest timber were very different from those of mills that were not. Less affected mills included companies that owned their own forest lands, drew on other public or private timber sources, or had converted to handle smaller logs or other kinds of wood products. Large industrial forestry concerns such as Weyerhaeuser and Plum Creek were more concerned about log exports and international trade issues and less affected by reductions in national forest timber supply. Indeed, firms that controlled other sources of raw material could well reap a windfall profit from rising stumpage prices due to reductions in supply. Nevertheless, all of these interests were concerned about the precedent that an adverse decision on the spotted owl case could provide, and the shift in political and economic power that it would represent.

The set of small, timber-dependent communities scattered throughout the Northwest can be considered to be a part of the timber lobby, but their concerns were somewhat different. Most of the mills were not in business to provide jobs and community stability; rather, they provided jobs as a byproduct. Mill workers and loggers had seen corporations close down mills in places like Valsetz, Oregon, and they understood that their concerns about maintaining their communities and lifestyles were quite different from those of most corporate planners. The wood products industry in the Northwest paid fairly good wages, and any shifts in the communities' economic bases were likely to result in a lower standard of living. In the wake of mill closures often came a significant unraveling of the social fabric of the communities, generating higher rates of crime, alcoholism, and domestic violence.

These different sets of timber-based interests were reflected in numerous associations and lobbying groups organized in the Northwest. While the National Forest Products Association was the lead lobbying group in Washington, D. C. for most of these industry components, and the Northwest Forest Resources Council in Portland attempted to provide regional representation across the spectrum of interests, various industry associations represented distinct segments of interest, including such organizations as the National Independent Forest Manufacturers Association, representing the small mills, the Washington Citizens for World Trade, representing the large companies, the Western Washington Forest Action Committee, based in Forks, and the Washington Contract Loggers Association. The Yellow Ribbon Coalition and the Oregon Project were organizations grounded in the concerns of dependent communities.

By the end of the 1980s, a wide-ranging set of federal agencies also were involved in the owl controversy. At times this included the active involvement of the White House and several Cabinet-level officers, all trying to minimize the political fallout of a decision on owl management. Faced with federal deficits of enormous size, a philosophy of government that tended to promote resource development, and a national economy that at the end of the 1980s and beginning of the 1990s was anemic at best, high level involvement sought to minimize the amount of timber locked up by owl protection.

The interests and motives of the Forest Service will be described in Chapter 10, but it should be noted here that there is no one Forest Service. Rather, the agency is an assemblage of numerous components, individuals, and professions with different values and attitudes about the owl issue. The spotted owl case involved activity on the part of literally hundreds of individuals in the agency, affecting personnel on sixteen national forests in two regions, and numerous regional and Washington office staff members. In addition, many other individuals in the agency were affected by the issue, both directly through effects on personnel, and indirectly by impacts on the agency's public image.

As in any collection of diverse individuals, differences exist in interests and attitudes about all things, and the owl issue tended to reveal rather than hide these differences. Over the course of the controversy, differences in approach and values were apparent between the line officers, individuals with significant decisionmaking authority such as District Rangers, Forest Supervisors, etc., and staff, such as wildlife or planning personnel. Clearly there were also differences between the professions, with the wildlife biologists, who traditionally have been fairly isolated in the organization, tending to value owl protection more than their colleagues in timber management. The researchers tended to be further down a continuum of levels of protection than were the management-oriented biologists. Washington office staff were more aware of the organizational turf and political dimensions of the issues, and field level staff appeared more cognizant of the difficulties of implementing policy on the ground.

Similarly, the Fish and Wildlife Service and the Bureau of Land Management should be understood as organizations with diverse interests that in part reflect staff heterogeneity. Neither organization has a strong tradition of nongame animal protection, though both employ a set of biologists who are actively concerned with the management of sensitive wildlife species. The endangered species staff of the FWS and the wildlife biologists operating out of the BLM Oregon state and district offices were consistent advocates of owl protection, while their management did as little as they could to promote activities that would result in diminished timber harvest. Through much of the issue chronology, the FWS and BLM leadership were solidly reflective of the Reagan and Bush administrations' policies favoring resource development on federal lands.

As the efforts of the FS, the FWS, and the BLM to contain the developing issue failed, and interest groups from all sides mobilized to take action, numerous members and committees of the U. S. Congress became involved in the controversy. At times, the Congressional battles could be understood by contrasting the values and interests of the Representatives and Senators from Oregon and Washington with those of a set of pro-environmental members from eastern states. At other times, the battles could best be understood by examining the turf battles between the appropriations committees, who set agency budgets including the FS's timber budget, and the substantive, authorizing committees who have oversight authority over statutes like the National Forest Management Act and the Endangered Species Act. And at other times, examining the interests of Congressional leadership in simply maintaining control over the flow of activities in the Congress help explain what happened as the owl issue moved into the Congress.

But even these characterizations are too schematic, for there was significant diversity among these different Congressional groups. For example, some members of the Pacific Northwest delegation, such as Jolene Unsoeld and Peter DeFazio, were more willing and able to seek a middle ground, and others, such as Bob Smith and Denny Smith, were unyielding and hard line. And there were some Congressional representatives from eastern states who are generally supportive of things environmental, who were not out in front of the pack on this issue.

There were numerous other players in the spotted owl drama, and they provided a great deal of the texture that was evident throughout the chronology. State and local agencies in Oregon, Washington, and California, elected officials at the state and local levels, University-based researchers, professional associations, local chambers of commerce, other federal agencies, and a variety of others all played a part in the developing owl controversy. Just as different spatial scales of analysis provide different kinds of insight into natural processes, so do different levels of administrative and political analysis. The dynamics within the Washington state government (for example, between a natural resources department that was statutorily mandated to raise revenues from state lands, and philosophically oriented toward harvesting trees, and a department of wildlife that uneasily housed both game animal interests and a vocal set of non-game wildlife biologists) are fascinating and help to explain the development of and response to the owl issue.

Making an effective and durable set of policy choices on the owl issue was difficult not only because these multiple groups would have to buy into a solution, but also because the implementation of an agreement would continue to require their concurrence and active participation. We know that the difficulty of implementing public policies rises exponentially as the number of involved parties increases.[8] To succeed in implementing any resolution of the owl controversy, hundreds of individuals and organizations would have to act appropriately. Researchers, land managers, corporate executives, small mill owners, loggers, school superintendents, road commissioners, federal agency executives, interest groups and elected officials in Washington, D. C., and throughout the Northwest, and many others, would be involved in implementing any agreement.

Given their diverse motives and interests, it would obviously be difficult to get them to act in concert. More important, it was hard for disputants seeking agreement to sign on because they did not trust the other parties to implement any agreement. Thus, even if the parties had interests that could be accommodated in an agreement, their fears of problems arising in the implementation of any agreement reduced the chances of getting an agreement to start with.

Extraordinarily Large Stakes

Most natural resource issues are characterized by multiple subissues and multiple players and yet do not fester as long and as painfully as the spotted owl dispute. What distinguished the owl controversy was the intensity of the stakes that various participants had in the way it was resolved. By the end of the 1980s, what was at stake for the various groups substantively, bureaucratically, and politically was extraordinarily large. For many of these groups, the owl dispute was not just another policy controversy, but rather a battle for their future.

At the start, we are talking about the future of at least six million acres of land, a lot of land by anyone's calculation. While other policy battles pitted species preservation against economic concerns, the owl dispute was fundamentally different in the magnitude of involved acreage and level of economic impacts. In passing the Endangered Species Act in 1973, the few members of Congress that thought much about what was seen largely as a symbolic act[9] envisioned fairly small scale, localized impacts incurred for small remnant populations numbering in the tens or hundreds of individuals: perhaps altering a project slightly, setting aside small areas as a refuge, including a fish ladder on a dam, or other minor changes to everyday agency activities. Even the celebrated snail darter case, at its most threatening point, would have required that a fairly small reservoir not be completed in east Tennessee. But here we had a species that ranged over millions of acres of land with the potential for economic impacts that were far from localized (and hence affected numerous Congressional districts), and frankly, was not down to its last few individuals drawing their final breaths.

For many, but clearly not all, of the affected timber interests, these forests were the only way to continue logging and milling operations in the style to which they had become accustomed, at least until second growth stands came on line in twenty or thirty years. Loggers and wood products firms had seen their national forest land base whittled down over time from what it had been in the 1950s and early 1960s with almost complete access. Then came wilderness set-asides, first administratively and then by Congress. This was followed by additional lands being placed "off limits" in the forest planning process. Finally, along came the spotted owl issue which, in their view, threatened most of what was left. Individuals in towns like Forks and Sweet Home had seen mills close in similar communities in the early 1980s and watched as the towns dried up and blew away. Land reservations for the owl appeared to directly threaten their everyday existence. Hence, the battle was a fight for the survival of their families and neighborhoods—threats that are highly mobilizing.

For the Forest Service, the national forests in the Pacific Northwest are an unusually valuable segment of their national land portfolio. In 1988, for example, approximately 5.5 billion boardfeet of timber was produced from the national forests in Region 6, representing about 44% of all national forest timber cut that year. (Harvest levels from 1980 to 1989 averaged 4.5 bbf.) What is just as important as the volume of timber cut is that the forests of the Pacific Northwest are generally money-makers for the USFS. In 1988, for example, the U. S. government lost money on the timber sale programs of 74 of 120 national forest units.[10] Indeed, 37 national forests made less than fifty cents on the dollar invested. Timber in Region 6, on the other hand, yielded an average of $1.78 on the dollar in the same year. In a time of huge federal deficits and a philosophy of government that promoted government operations that generated revenue, the USFS's Region 6 was its showpiece of timber operations, and hence made the demands of the owl even more threatening to many interests within and dependent on the organization.

For the Forest Service as an organization, the owl dispute came to embody threats to its organizational values and control, criticisms that were abundant in the 1970s and 1980s, but were captured particularly well in the owl controversy. Whether you view the logging of the old growth forests in the Northwest as the last bastion of cut and run forestry, or the necessary conversion of overmature forests to managed, multiple use, sustained yield forests, it was clear that the cycle of clearing the virgin native forests that began in the Northeast in the 1700s and swept through the Midwest in the 1800s was drawing to a close in the Northwest in the late 1900s. Yet agency leadership in Region 6, supported by Washington office leadership and encouraged by Congress, continued to plan for more of the same.

Even leaving aside changes in public attitudes toward federal forests and other economic forces at work, completing the cycle of liquidating what for all practical purposes is a nonrenewable resource meant that the days of timber as absolute ruler were probably over. For Forest Service leaders who had staked their Region 6 wagon to the timber star, that meant that the owl dispute was even more threatening. At minimum, fewer timber sales meant lower budgets and fewer staff. Indeed, in the late 1980s and early 1990s, much of the agency leadership's concerns were focused on the impact of the dispute on workforce size and morale. At maximum, the spotted owl dispute could mean that direction for the national forests increasingly would be set by forces outside the agency, pursuing objectives that were not within the longstanding traditions of agency culture and values.

The political stakes were equally high. As a region with outstanding environmental amenities and well-established resource commodity industries, the Pacific Northwest housed strong political forces in favor of environmental protection and timber production. Even members of Congress like Senator Mark Hatfield, who some viewed as the patron saint of the timber industry, simultaneously promoted environmental quality measures, at least in election years. The spotted owl dispute landed smack dab in the middle of these two sets of interests and offered them a stage on which basic questions of political dominance could be settled. What happened in the owl case had meaning for who would hold political power in the region, and how it would be exercised, and more fundamentally, who would determine the region's long-term future.

Finally, the policy implications of the spotted owl issue were unusually significant. Both pro-development and pro-preservation interests came to view it as a test case. More extreme preservation interests believed that success over the owl issue would yield a precedent for federal endangered species policy even more powerful than the snail darter victory. Moderate preservationists feared a potential backlash attending victory in the owl issue, suspecting that winning the owl battle would cause them to lose the endangered species policy war. If a decision could be crafted that protected owls and economies, offsetting a political backlash, it would be a great precedent for future conflicts. Development interests believed that the owl issue could result in a redefinition of the absoluteness of the Endangered Species Act, and would be a test case that they would win because, in their view, no one in their right mind would set aside millions of acres of land and put significant numbers of people out of work for an owl.

Uncertainty about Science, Economics, and Policy

The magnitude of what was at stake guaranteed that the owl issue would have been a hard fought battle under any circumstances, but resolution of the owl dispute was made more difficult by the lack of key pieces of information and clear scientific direction at many points throughout the controversy. Technical uncertainty is almost always present in environmental and resource issues. It clouds choices, and invites delay, as we seek to know what we know we should know. Uncertainty also provides an out for decisionmakers who do not want to make hard choices and take the heat for doing so. We can always study the situation one more time. Occasionally such studies provide new insights, bound the inherent uncertainties, or build a shared understanding of a problem among differing interests. More often, they provide political leaders with a great way to give the appearance of action while buying time, hoping that the underlying politics of the matter moves one way or the other.

Uncertainty also gets tied up with the value differences that underlie the differing interests of stakeholders, as these differences get expressed in the debate. The planning and regulatory processes that have evolved to make natural resource management and environmental protection choices are defined as technical processes, yet most also involve value choices that are hidden by the technical framing of decisionmaking. Technical uncertainty allows value choices to be expressed as differing technical viewpoints. Science becomes advocate science as different values are transformed into different technical perspectives, and the debate plays on and on through lengthy and repetitive agency studies and courtroom battles.

In the spotted owl case, while the dimension of technical uncertainty was exaggerated as an underlying problem in reaching settlement, decisionmakers were uncertain about a set of biological and ecological issues, economic realities and effects, and the role and obligations of government and interest groups. For example, while a considerable amount of human effort was put into studying the owl in the 1970s and particularly the 1980s, decisionmakers were faced with a significant amount of uncertainty about the population size and habitat needs of the owl. Besides an inherent bias against gathering information about nongame species, the wide-ranging nature of the owl made information collection difficult. To count owls, researchers go into the woods and call to them. The fact that this is best done at night (in the dark) at regular intervals through often-steep, old growth forest makes this research labor-intense, demanding, and dangerous. Given that the owl could potentially live in millions of acres of such forest, makes counting owls even more difficult. As a result of these inherent difficulties, estimates of the size of the known owl population varied considerably.[11]

While several parameters of the owl's population biology were unclear throughout the debate (including its reproductive success, dispersal distances, and survival rates of young birds) one of the most researched and debated issues was its habitat needs. Was it old growth dependent? Could it survive in second growth stands? Why was it found in northern California and southern Oregon on some second growth stands?[12] If it was old growth dependent, just what was it that the owl needed from old growth? How large was the bird's home range, and how much old growth had to be contained within the home range?[13]

Even if home range size had been established definitively earlier in the case chronology, policymakers would have been hampered by a lack of information about the extent of old growth habitat on the ground. As discussed below, this lack of information was partly because the Forest Service did not routinely collect and map information on forest type in a way that would allow an accurate estimate of suitable habitat, and partly because they were not very good at keeping their inventory data up to date so that it reflected on-the-ground realities. In addition, part of the problem was definitional. An "interim" Forest Service definition of old growth published in 1986 was still considered a draft as late as 1989,[14] and it was still a question as to whether old growth acreage equaled suitable owl habitat acreage. And even though agency experts recognized the need to map old growth acreage at the first meeting of the Oregon Endangered Species Task Force in 1973, estimates of old growth in 1989 still ranged from an estimated 2.4 million acres on all lands in Oregon and Washington, according to The Wilderness Society,[15] to 6.2 million acres on national forest lands alone, according to the Forest Service,[16] a large gap by anyone's standards.

The lack of conclusive information about the owl's population size, demographics, and status was compounded by a lack of understanding of the effects of environmental change on the long-term health of the owl population. The lack of baseline data made long-term forecasting inherently difficult, but uncertainty about the effects of specific changes in the owl's habitat made long-term assessment even tougher. For example, one key question that nagged researchers and made the definition of suitable habitat even more difficult was the effect of habitat fragmentation. Clearly the sum of all old growth acres did not equal acreage of effective habitat because the acreage was not contiguous and independent of activities occurring elsewhere on the landscape. Hence the spatial relationship of different chunks of land became important, and key questions had to be asked. At what size did a patch of old growth forest in a sea of clearcuts become effective habitat? At what distance could patches of habitat be located from each other? What would be the effects of predation and competition as the landscape changed over time? Late in the story, a hybrid cross between the spotted and barred owls appeared and was labeled the "sparred" owl. As barred owls were able to invade the spotted owl's range, was hybridization a threat to the genetic identity of the spotted owl?

Even if we had in hand good static baseline information and an effective understanding of the effects of environmental change on owl demographics, the manner in which all this information would be put together in a systematic way was an evolving science through much of the story. Approaches to analyzing long-term viability of an endangered species population were experimental and debated throughout the development of the owl dispute, leading to considerable scientific debate about their interpretation and effectiveness for making policy choices. For example, effective models for forecasting population viability were not in existence in the 1970s and early 1980s. Population viability analysis and theories and methods of conservation biology appeared in the 1980s, leading to considerable and sometimes virulent debate about specific models and approaches. Were rules of thumb based on fruitfly experiments applicable to owl populations? Was there an absolute minimum number of individuals in any population below which extinction was assured? The fact that most analytic models required empirical data that was not in existence meant analysts had to make assumptions in order to move forward, and those assumptions were open to question.

Finally, an additional set of questions and issues needed to be resolved in order to make good scientific choices in the owl case, and these dealt with the ability of land managers to actively manage to improve the likelihood of owl survival. Could habitat be actively manipulated or silvicultural practices be adopted that would promote owl survival or at least minimize the negative impacts of timber harvest? Was an owl protection strategy limited to setting aside specific areas of old growth habitat? If so, what should the pattern of habitat areas look like? Would the owl survive by reserving circles of old growth located in a 12-mile grid across the Northwest as was proposed in the Forest Service's 1988 SEIS? Should the habitat areas be clustered more tightly, as was the plan in the Jack Ward Thomas report in 1990? How would a protection strategy generate more old growth-like forest in the future in order to deal with the inevitable loss of habitat through blowdown and climate change?

As in most technically complex policy choices, more questions than answers were evident about the owl's current and future status, let alone a related but bigger and less articulated set of issues about the biology of other old growth-dependent species and their interrelationships. Such complexity and attendant uncertainty is not a problem when faced with little need to take action or a fairly low stakes choice. Indeed, researchers thrive on the medium of uncertainty, just like so many molds on the petri dish of federal and state dollars.

But given the intensity of the stakes and the need to act without a firm resolution of many key questions, the biological uncertainties were problematic. Decisionmakers were faced with moving targets rather than absolutes and this made their choices incremental and difficult. In addition, because they rarely understood the reasons that key pieces of information changed in midstream, often policymakers tended to be distrustful of the motives of the messengers of new information. In less controversial situations, experts would meet, decide on their collective professional judgment, and lay out their wisdom for society to follow. In this case, expert judgments were questioned within organizations and in the courts and the political process, and the inherent and apparent uncertainties left policymakers without a firm basis to offset the strong economic and political arguments at play.

Questions about the biology of the owl were not the only uncertainties underlying the controversy; an equal and not quite opposite set of uncertainties prevailed about economic realities and futures. While the overall issue that needed to be addressed was what would be the impacts of a change in management practice to protect the spotted owl, the way in which different groups analyzed it, and the assumptions that they made, heavily affected the answers they provided to this question. The approaches that most of the analyses used were fairly similar in method; what differed was their strategic choice of assumptions, with dramatic effects on outcomes. All in all, the various and conflicting economic projections were difficult for decisionmakers and citizens to sort out. As one 1992 analysis of the economic analyses, conducted by the American Forestry Association's Forest Policy Center, concluded:

> Several major studies have been conducted in recent months to estimate the effects of protecting habitat for the northern spotted owl in the old-growth forests of the Pacific Northwest on employment in the region's forest products industry. For many policymakers and concerned citizens, these reports have been the source of as much confusion as information. The projections of decreased employment that can be expected in the remainder of this decade range widely— from 12,000 job losses in one study to more than 147,000 job losses in another.[17]

Underlying these vastly different projections were several issues: One of the most confusing differences between these studies was the use of different baselines against which impacts were judged. Reports by the FS and The Wilderness Society used the 1990 timber sale levels established by the revised forest plans. Since the new forest plans established harvest levels considerably below actual levels in the 1980s, use of these as a baseline tended to minimize the effects of changes due to owl management. In contrast, estimates by the American Forest Resource Alliance (AFRA), a timber industry group, used an average for the period from 1983 to

1987, when the timber harvests were considerably higher than those called for in the revised forest plans. Using historic harvest levels as a base tended to overestimate the actual impacts of owl protection on the regional economy.[18]

Other assumptions that differed among economic projections included the land base that was considered to be affected by owl protection and the choice of multiplier used to forecast economic effects.[19] How the forecasts separated impacts due to owl management from the impacts of changing forest plans,[20] and how they dealt with the effects of technological change in the forest products industry also differed. While many of these assumptions could be identified through a careful reading of the various studies, few of the decisionmakers and media officials had the opportunity to do so. By reporting a "bottom line" estimate of economic impacts, each group could put whatever spin it wanted on the data, and the media was ineffective at sorting out differences. Unfortunately the uncertainty surrounding the diverse forecasts, fed by advocate economics, added another layer of complexity to the owl situation. Not only was there uncertainty as to the benefits of owl protection measures due to scientific questions, there was considerable murkiness about the costs that would be incurred by whatever strategy that was followed.

Uncertainties about biology and economics were matched by several uncertainties in underlying policy questions, and these added their share of complexity to the owl controversy. For example, since the advent of the modern day welfare state embodied by the New Deal in the 1930s, and expanded in the 1960s, there has been continuing uncertainty about the role and obligations of the federal government in relation to other levels of government and the private sector. Not the least of these uncertainties was a lack of clarity about the federal role in predicting, avoiding, and mitigating the effects of regional economic transformations. It is easy to say that there is no obligation of the public sector to deal with declining economies, but the reality is that society ends up picking up the tab as economic impacts become human impacts. Welfare and other social service spending rises as the social safety net begins to catch losers in economic competition. Heightened violence and domestic problems cause stresses that generate long-term social costs.

In the area of natural resource policy, booms and busts had always characterized economies based on the extraction of natural resources, and were not necessarily an overriding public sector concern when they involved privately owned resources. But what happens when the engine of the economy was publicly owned land? What is the obligation of the public sector to maintain these economies after having contributed to establishing them? In many ways, these questions went to the heart of a longstanding schizophrenia in American natural resource policy: Was the purpose of our policy to promote economic development in support of the current generation, or was it to protect and steward the resources to insure their transmittal fairly intact to future generations? This overall question of direction had permeated public lands policy since the creation of the nation, and is replicated in conflicts of purpose in the resource management agencies.

In many ways, the Forest Service and BLM are most subject to problems caused by these policy uncertainties. Their multiple use mandates legitimize a wide variety of objectives, including community stability concerns. Whether their dominant clients are local or national interests is generally decided by District Rangers and Forest Supervisors who are both members of the local communities and staff of a federal agency. Conflicting direction from Washington adds both to the confusion and the opportunity at the ground level to go whichever way they want. In the owl case, the fact that strong endangered species preservation mandates and large allowable sale quantities of timber were simultaneously flowing from Washington caused a considerable amount of uncertainty on the ground. When no one was watching, ground-level managers would resolve the uncertainty in their own way and get on with their jobs. Increasingly, though, groups were looking over the managers' shoulders, watching what they were doing, unhappy with the balance point the managers provided, and able to stymie decisionmaking.

The roles of these nongovernmental interests were also in question. Were the set of interest groups that appeared and argued over an issue, the totality of public interests in a decision? Should they have unlimited opportunity to challenge agency decisionmaking? Should public involvement in decisionmaking be limited at some point to overcome the public impotence problem that resulted from too many cooks in the kitchen? These underlying governance questions had bothered Alexander Hamilton and Thomas Jefferson at the birth of the nation, and they continued to haunt the decisionmaking process in the 1980s. For example, one of the thorny questions facing members of Congress in the 1989 appropriations battle that would determine the timber sale program for fiscal year 1990, was whether or not limits should be placed on judicial review of agency decisions to insure that the timber sales went forward. The question caused a major response from interest groups, and members of Congress who had not previously been involved in the debate, because of fears of a tactical and philosophical nature.

In many ways, the spotted owl dispute raised fundamental questions about the nature of our federalist system, for it concerned a question of the primacy of different geographic interests: local, state, regional, or national. Should an apparent national interest take precedence over conflicting local or state interests, and if so, should the losers be compensated for their loss? Should losers in a subnational political competition be able to "appeal" the regional outcome to the national level, as was done consciously by ONRC and other regional environmental groups by nationalizing the spotted owl issue?

The scientific, economic, and policy questions underlying the spotted owl debate added uncertainty to agency and Congressional decisionmaking. Choices were delayed in order to deal with these lingering questions, and strategic delays occurred, benefiting both environmental groups, who needed time to build a strong political coalition, and timber interests, who were benefited by cut levels established by the older forest plans. Finally, the presence of uncertainty made decisionmakers wary of selecting courses of action that could leave them out on a limb without adequate support, and cognizant of the fact that there would be grounds to challenge their decisions, no matter what they decided.

The Nature of the Policy Choices

Besides the multiplicity of issues and actors, and inherent questions underlying the debate, other characteristics of the spotted owl dispute made it a particularly difficult issue to resolve. To make a choice, decisionmakers were asked to balance short-term tangible economic and organizational impacts against long-term intangible goals, and these kinds of choices are particularly hard for our public decisionmaking processes to make. The intent of the Endangered Species Act is to bind ourselves to what most would agree is an admirable set of long-term goals. Protecting endangered species rates up there with motherhood and football (formerly baseball) as enduring American symbols that receive, on the surface at least, fairly unanimous support.

The dilemma in carrying out policies aimed at achieving benefits that will be evident over fairly long periods of time is that decisionmakers are still faced with short-term choices that affect our ability to pursue the longer term goals. Rational short-term action may well overtake rational and even necessary long-term objectives. Just like keeping New Year's resolutions is difficult when faced with that one, incredibly small, piece of chocolate cake, foregoing short-term objectives to pursue alternative goals that will pay off sometime well down the road is difficult, particularly to a society used to fairly immediate gratification. Just like balancing the federal budget by cutting services or raising taxes is individually and hence politically difficult, the economic and political costs of owl protection are hard to swallow, in part because the benefits of protection largely will be evident sometime in the future.

The dilemmas posed by short-term choices with long-term benefits were exacerbated in the owl case, because the benefits of protection are largely intangible. Most of the reasons proponents advance to protect rare plant and animal species, including protection of genetic diversity and biotic stability, and preservation of aesthetic and moral values, are hard to compare with the dollars and cents tangibility of lost jobs in affected communities. Decisionmaking is always easier when the metric evaluating benefits and costs is common across decision alternatives. Converting owl benefits to dollars can be done, but only artificially, and lack of agreement on what such a conversion means limits our ability to convert intangibles into tangibles in a binding and enduring way.

The owl dispute was also difficult to resolve because under its surface were clashes of fundamental human values, that is, differences of opinion that are deeply held and unresolvable on objective grounds. It is hard to find middle ground between the underlying premises of someone who argues owls should be protected at any cost, and those of someone who argues that the benefits of owl protection should be balanced against those of other potential human activities in owl habitat. Similarly, it is difficult for advocates of wilderness preservation and proponents of sustained yield forest management to agree on their underlying attitudes and values. Or for those who view clearcuts or tree plantations as ugly, and those who view well-done timber management activities as a thing of beauty, to find much common ground. Whether it is the smell of napalm in the morning (as one gung-ho Army officer valued in the 1979 movie "Apocalypse Now"); the sight of newly cleared, burned, and replanted clearcuts; or the dank silence of the old growth forest; different individuals evaluate aesthetics and desirable courses of action in divergent ways, reflecting their own underlying value sets and orientations. Both sides can argue until they are blue (or green) in the face about morality, aesthetics, and human obligations without being able to agree on a common set of values on which to ground decisionmaking.

In the spotted owl case, not only were fundamental differences in the value of owl protection inherent in the dispute, but related philosophical differences clashed as well. Preservation versus conservation, a philosophical debate in existence since the turn of the century, was also at play in the owl dispute. Should wildlands or other natural features be protected as inviolate sanctuaries for a variety of reasons or is an unmanaged forest inappropriate and wasteful? In addition, other philosophical differences underlay the owl dispute: Should societal choices be based in a homocentric or biocentric set of values or perspectives? Should we be risk-averse or risk-taking about actions taken today, and how they will affect future generations? Will technology protect us from losses of genetic diversity or losses of economic infrastructure so as to guarantee our future lifestyle against the choices we make today?

In a related way, a clash between human cultures and lifestyles was played out in the owl controversy. The dispute has been seen by some as a battle between the cultures and lifestyles of urban white-collar, affluent environmentalists versus rural, blue-collar, less-well-off logging communities. Another version of this cultural dichotomy viewed owl preservationists as radical anti-growth crazies portrayed against more mainstream pro-growth, middle class Americans. While the reality of the disputing parties was not as clear cut as these dichotomies would suggest, since there are rural, timber-dependent, owl preservationists, and urban, affluent timber advocates, and extreme positions and tactics on all sides, it is true that there were elements of socioeconomic cultural clash inherent in the owl debate.

In interviews, evidence of these perspectives was abundant: According to a timber industry official, the environmentalists:

> wouldn't be happy no matter what we were doing with our national forests, and they probably think that they would be happier in another time in the history of either this country or the Earth. They don't like this time, where we have jet planes and personal computers and high technology. They don't like that and are uncomfortable with it. So they want to preserve their values by making the national forests as one last place that they can preserve from another century. . . . [In their view,] we ought to live in a more holistic way, in sympathy and harmony with nature and natural resources. We shouldn't use them to make our lives better, because our lives would be better if we really lived in a cave and used rocks to make fire, and things like that . . ."

In the words of a staff member of a small Oregon-based environmental group,

> I personally look around at a lot of these loggers, and I feel sorry for them because of their lifestyle. They're uneducated, they're crude, they're not people that I would choose to be around . . . I don't think there's a defensible reason to keep these people doing what they're doing and keep them in their state of ignorance. So I think that we need to find some forms of job retraining, and perhaps ways to educate them. Bring them up so that they can spell, talk, and get along like the rest of us.

While the polarization evident in the owl dispute inherently kept these individuals apart, their images of each other reflected enduring cultural values and lifestyle choices between which it would be hard to find middle ground.

In issues such as these, disputants usually either agree to disagree and find other ways to reach settlement, or they stalemate, as they did in the owl case. Sometimes disputants in value-laden controversies can reach agreement on an allocation scheme that protects differences of values, and, as will be discussed below, there were ways to improve the possibility of better understanding and working around the disparate value orientations in the owl case. But the existence of conflicts over fundamental values and cultural styles contributed to the resistance to resolution of the owl dispute.

A Seemingly Zero-Sum Issue with Few Remaining Options

Making choices among divergent interests is difficult, but not impossible, when one interest can be satisfied at little cost to the other parties, but that was not the widely shared vision of the spotted owl conflict. Repeatedly, the dispute was framed by the media and advocates from all sides as zero sum, that is, the benefits to be achieved by one were obtained only at the expense of the other: owls versus jobs, and environment versus economics. What was on the negotiating table was a pie of fixed size and character, and disputants could only carve it up such that what one got, the other gave up.

Regardless of whether this perception was true, the perception itself has a major effect on our ability to solve problems. The literature on negotiation[21] suggests that negotiators who view the world as a zero sum, fixed-pie place will be extremely competitive in that all they can do is claim the biggest chunk of the pie that they can by whatever means possible, including lying, distorting information, hiding true concerns, etc. But in doing so, they miss opportunities to frame more creative solutions that potentially can meet all parties interests at least minimally. It is in society's and each group's interests to try to look collaboratively for solutions to complex problems, but if each views the dispute as zero sum, they will not do so. The owl dispute was widely viewed as pitting economic against preservation interests in ways that, in the final analysis, would result in one being the winner and the other the loser.

Nobody knows whether the "sloppy" forestry implied in New Forestry is more or less labor-intense, and hence potentially job-creating. Nor is it clear whether or not the parties to the dispute could have agreed on a long-term development strategy, given the realities of the timber supply over the next few decades, examining the full range of private and public lands, and utilizing a complete set of public and private policy strategies, and a diverse set of silvicultural treatments. Given that the dispute was framed as zero sum, it was not in the parties' individual or collective interest to look for such solutions.

The zero sum, outright competitive nature of the dispute was magnified by the fact that it was difficult for most of the disputants to define an absolute bottom line. The outcome that would most clearly minimize threats to the long-term survival of the owl would be one hundred percent old growth protection. The outcome that would maximize raw materials supplies to old growth dependent mills would be one hundred percent old growth utilization. At the same time, both sides would make do with whatever they got. Timber markets would adjust to any quantity of land set aside for timber production. While the owl would clearly die out as the amount of its habitat fell below some threshold, what the threshold was and whether it could be modified through management was not clear, and made the definition of an absolute "bottom line" for preservationists difficult.

Given a zero sum situation and fuzzy bottom lines, the rational approach for disputants is to go for the extremes of what is possible, that is, to argue for all of the pie. In such disputes, interests tend to focus only on supporting their arguments and discrediting those of their opponents, and not to look for creative solutions at the interface. Collaborative problem-solving might be desirable in public decisionmaking, but the framing of the owl controversy as zero sum diminished the possibilities for creative solutions.

It was also the case that by the end of the 1980s, when the issue was politically ripe for settlement, few easy options were left to forge an acceptable compromise among the many different interests at stake. The zero sum perception became a near zero sum reality. In response to an open-ended interview question that asked broadly what the owl controversy was all about, many individuals, from all sides of the issue, highlighted lost options as a major theme, and most of them laid the blame on the doorstep of the Forest Service. Many of these individuals suggested that, had the agency tried harder to resolve the controversy in the early 1980s when the economic recession added slack into the timber base (indeed, timber contractors were given a reprieve from uneconomic contracts through legislation that bought back their obligations) and the politics were less polarized, it would have had a reasonable chance of finding a fairly stable compromise. It is easy, however, to criticize past actions given the benefit of hindsight, and it should be remembered that creating additional land set-asides for environmental reasons flew in the face of the pro-development direction set by the Reagan administration. But at that time, there was a greater set of options—a larger decision space—with which to work when dealing with difficult issues like the spotted owl case.

By the end of the 1980s, trust was also fairly nonexistent. Information from the Forest Service was mistrusted, the Chief's credibility was in question, the Fish and Wildlife Service was seen as unduly politicized, pronouncements from the interest groups were seen as biased and self-serving, and the motives of Congressional decisionmakers were seen as obviously political. Multiparty negotiations run on the engines of interpersonal and interorganizational relationships, and when these relationships are characterized by a complete lack of trust, it makes the negotiations all the more difficult and time-consuming. Trust provides decisionmaking slack in that more creative solutions can be attempted, and implementation concerns do not necessarily doom the potential for agreement. Implementation always requires exercising discretion, and if you do not trust the key actors to act appropriately, your options for settlement are constrained significantly.

The limited decision space and the other characteristics of the issue and response to it combined to create a set of policy choices that were intrinsically difficult. Even with an effective set of decisionmaking institutions, and a lead agency committed to its resolution, the spotted owl controversy would have been a tough nut to crack. We can look back with the benefit of hindsight and criticize our decisionmaking processes and institutions, as the next few chapters do in some detail, but we should remember that the choices implied in any resolution of the owl controversy were difficult to make, involving different patterns of beneficiaries and losers, and many meanings for a much broader array of societal interests and concerns, than simply owls and jobs.

7

Avoiding Tough Choices:
American Decisionmaking Processes

While the spotted owl controversy would have been difficult to resolve under the best of circumstances, our standard decisionmaking processes did not serve us well in molding the conflicting sets of interests, attitudes, and energies into a process of informed choice. Disputes between different interests in society are processed by institutions and procedures with structures and biases of their own. The spotted owl issue moved through various administrative processes at the state and federal levels, saw action in state legislatures and the U.S. Congress, and provided abundant fuel to keep the courtrooms of the Northwest well occupied. Yet, in spite of all these efforts, our standard approaches to making these kinds of societal choices did not do a particularly good job.

In an ideal world, societies would have mechanisms for making collective choices that would generate necessary information, including data about the current issue and its future manifestations, and provide a forum for informed discussion and debate that focused on the real issues of concern. Debate would be substantive and productive, and would consider the merits of alternative arguments as much as their political or judicial import. Government agencies with claims to expertise would base their advice on scientific or technical knowledge and be honest about what they know and what they do not know.

The ideal decisionmaking process would prompt a search for creative solutions that address the real interests of the disputing parties, and would assist in finding a solution that is as good as possible for all stakeholders, and that considers future generations and the needs of nonhuman lifeforms as well. It would produce decisions that are effective at solving the real problem and are durable, in that they are supported by the disputants and can therefore withstand major challenges in their implementation. Through all of this, it would provide the opportunities and incentives for various interests to take the decisionmaking process seriously and participate fully in it. In a time of scarce resources, such processes would be able to encourage and allow decisionmakers to make tough choices, and they would deal with those that incurred significant costs as a result of the decision as compassionately and equitably as possible. But only on the Planet Karma is there an ideal world.

In the real world of societies and governments, institutions have biases, directions, and structures that limit their ability to respond. American resource policymaking is fragmented and uncontrolled, yielding decisions that are slow to appear and often inadequate to deal with the magnitude of the underlying problems. Decisionmaking is generally reactive and crisis-oriented, based on information that is often inadequate. Agencies are not unbiased sources of technical advice, interest groups act adversarially and strategically in ways that conceal accurate information, and elected officials focus on short-term survival in ways that are often counterproductive to the broader public and future public's interest.

All of these participants follow the incentives that are provided to them by the nature of the decisionmaking process; their behavior is rational given the environment in which they function. And the structure of our decisionmaking processes exists for good historical and philosophical reasons. But the net effect of how we do business is to delay dealing with problems, create stalemates, and quite often forge solutions that are not up to the challenges they seek to redress. The next two chapters focus on the nature of these standard decisionmaking processes and their effects on the development and persistence of the spotted owl controversy.

The analysis of the owl controversy must be placed within this broader context, for the failures of decisionmaking are not solely a result of the nature of the issue or the specific set of involved agencies and groups. Rather, they are in part an outgrowth of a broader set of problems in the way we make collective choices. Some of these problems are unavoidable. They are the byproduct of the kind of governance process that we have. As Winston Churchill noted, "Democracy is the worst form of Government except all those other forms that have been tried from time to time." But some of these problems are not inevitable or inherent in our public policy system. Even though all human-created systems are flawed, we clearly can do better, as the spotted owl case demonstrates with flying colors.

Fragmented Policy Processes

Just as landscape fragmentation has reduced the amount of effective habitat for interior-dwelling species in the Pacific Northwest forest, administrative and political fragmentation places constraints on the effectiveness of policy processes to make timely and wise decisions. Unlike the more centralized governments of many other countries, the American resource policy process is characterized by many components with little centralized control and no single group that consistently dominates outcomes.

The most obvious form of administrative fragmentation is seen in the diverse land ownership patterns that are evident across the American landscape. There are literally millions of landowners each of whom influences the long-term future of a piece of ground. For example, the long-term future of the Pacific Northwest forest is determined by the concurrent yet disconnected decisions of federal, state, and county forest managers, private industrial landowners, and private nonindustrial woodlot owners of a variety of types. Even restricting our view to federal lands, checkerboard ownership patterns on BLM's O&C lands, inholdings on national forest lands, and the different objectives and styles associated with FS, BLM, and National Park Service management, each acting with some degree of autonomy, all suggest geographic and administrative fragmentation.

Other aspects of administrative and political fragmentation are important to understand. As anyone who has had high school civics should be able to tell you, power and responsibilities within the federal government are split between branches of government: legislative, executive, and judicial. In theory, Congress passes laws, the President and executive agencies carry them out, and the courts interpret the laws and resolve any disagreements that arise. In fact, all three branches are policymakers, and each provides elements of decision and choice that become public policies.

Within branches of government, power and decisionmaking are also split. For example, among other important divisions within the U. S. Congress, the separation between committees that make substantive law and those that allocate money to carry out programs is important and certainly had a major impact on the owl controversy. Since laws that have no financial support have value only as political symbols, those members of Congress that determine budget allocations have at least as significant a role in setting policy as the members on the ostensibly substantive committees. It is the appropriations committees of the Congress that establish the allowable sale quantity, and fund the annual timber budget of the national forests, and they have as much impact on ground level operations as committees with jurisdiction over the framing of national forest policy legislation.

Functional authority is also split among different administrative units. For example, on FS-managed national forest lands, the U. S. Environmental Protection Agency has the mandate to regulate pesticides, state wildlife agencies have the legal right to wildlife populations, and the FWS enforces endangered species laws. At minimum, functional fragmentation means a lot of hands on deck. In fact, it also means that decisions about forest management are often the amalgam of several different decisionmaking processes. In the spotted owl case, for example, the 1990 Region 6 timber sale program represented concurrent decisionmaking by Congressional appropriations committees and many national forest units carrying out broader policy set by several other Congressional committees, all of which were operating under the review of the FWS as required by the Endangered Species Act.

Decisionmaking authority and the resources to carry out public choices are also split among levels of government (local, state, and national) and public and private sectors. We are a society of some 80,000 units of government, including federal, state, county, municipal, and special districts; and power, resources, and responsibilities are diffused across these units. Similarly, political power is dispersed between governmental and nongovernmental entities in society. It is rare that any one of these parties can get decisions adopted on its own, or implement them without the concurrence and cooperation of the others. Hence, to get policies adopted and/ or implemented, political and administrative fragmentation means that supporters must assemble a coalition of support across bureaucratic units at the same level of government, across levels of government, and between public and private parties.

The reason that our decisionmaking system is so diffuse and complex is partly historical and partly philosophical. The history of the creation of our public lands is not one that exhibits great forethought or planning. The western national forests were by and large laid out by Gifford Pinchot and Teddy Roosevelt sitting on the floor of the Oval Office with the turn-of-the-century equivalent of a magic marker. The national parks were set aside piecemeal as superlative landscapes with value as tourist destinations. The BLM lands in western Oregon had been given to the Oregon and California Railroad to subsidize development of rail access, and only by the good graces of God, and the Depression, did the lands revert to federal control and management. For all these public landscapes, there was no plan, and no strong vision of what the future was to hold for them. The objectives and management practices of each of these land systems were in flux from their beginning, and each of them developed under the influence of different political, geographic, and economic forces. Hence, it is not surprising that their administrative directions and political characteristics are very different from each other.

Much of the political fragmentation is by design. Our political forefathers wanted to create a process that would take account of the interests and concerns of individuals and localities, yet not lose sight of the overall national interest. In addition, having fled a society in which power was highly concentrated, and individual liberties constrained, they feared the abuses of power that could result from centralized government, and at the same time, feared that a government that was too fragmented could not govern and would become impotent or fall apart. Hence, they attempted to craft a process that walked the line between these different extremes.

Given a diverse society with no centralized, controlling sources of authority, another issue arose: how to define the public interest. What evolved over time was an operating definition that was prescribed by notions of pluralism and that goes as follows: If there is a legitimate concern or interest in the population, individuals will get together and press through the political process for a response to their needs. If the interest is ripe for public action, enough pressure will be mounted to force a decision from the system. If the concerns or interests are in conflict with other concerns, the policy processes will provide a forum for adversarial argument, and make a choice that balances the concerns at the table. Given this structure, the role of the government is to manage the conflict inherent in a diverse society, find the balance points, and enforce the decisions that result. This theory of collective choice has been fairly comfortable in that it, in many ways, matches the way that economic markets provide decisions, by aggregating the collective inputs of numerous members of society into a fairly stable outcome.

Piecemeal Solutions, Limited Accountability, and Inconclusive Decisionmaking

Fragmented policy processes, and the theories of government that underlie them, result in a number of outcomes that help us understand the nature and persistence of the spotted owl controversy. First of all, it is hard to get comprehensive solutions to cross-cutting problems, and the owl issue and its component parts required solutions that cut across administrative and political jurisdictions. At the management level, this is obvious. Owls do not stop at administrative boundaries, nor does water, air, or most natural or social forces except red tape, yet BLM, FS, and NPS lands abut each other in the Northwest, with remarkably little coordination between adjacent land managers. In many places in the Northwest, you can easily see the outline of administrative boundaries on the landscape, revealed as the juxtaposition of conflicting land uses, such as clearcut national forest land and unmanaged, old growth national park land.

At the policy level, fragmented policy processes result in conflicting directions being provided to land managers. For forest managers faced with the simultaneous implementation of the Endangered Species Act that mandates protection of critical habitat of sensitive species, the National Forest Management Act that requires deliberate, sustained yield, multiple use land management, and appropriations bills that mandate timber cut levels that are not sustainable and can only be achieved by reducing the amount of owl habitat, there is no chance for success. All are Congressional statements of national policy that are produced by different committees and political consortia, and they are inherently conflicting.

The jigsaw-like pattern of authorities and responsibilities also means that there is a lot of slippage between pieces of the decisionmaking puzzle, and important issues and concerns may simply fall between the cracks. For example, for a significant portion of the chronology of the spotted owl controversy, the owl was not stamped federally approved as a threatened or endangered species. As a result, the Forest Service was under no clear legal obligation to protect it,[1] and the FWS had no obligation to worry about it. For more than ten years after the owl was recognized as a management issue, neither agency worked hard at gathering information to ascertain the current status of the owl's population nor at altering management strategies to provide long-term protection of owl habitat. For most of its life cycle, the owl issue resided in the twilight zone of federal law and responsibilities. The old growth issue still resides in this gray zone. Ecosystem protection has little standing in federal law, and as a result, no one had a clear obligation to deal with the old growth issue. Yet by not dealing effectively with these issues, options were precluded and political crises hastened. The ball was not clearly in anyone's court such that they had to pick it up and run with it.

Fragmentation also results in diminished accountability. If there are so many cooks in the kitchen that it is hard to tell who is in control, it also means that it is hard to figure out who to blame when dinner is burned. The multiplicity of decisionmaking processes and institutions reduces accountability in several ways. Since those who implement public policies, including forest supervisors and district rangers, face conflicting direction from policymakers, they often can do what they want to do and be at least partly within directives.

Fragmentation also gives individuals who do not want to take action a way out, by allowing them to play a set of strategic games:[2]

- "Not My Problem" is a time-honored game played by bureaucrats of all walks of life. Divided jurisdictions and fragmented authority mean that you can always point to the other guy and say he is the one who has to worry about it. In a related way, you can always point to Congress or another level of the agency and say, "It's His Fault, Not Mine," that the situation is what it is.

- "My Hands Are Tied" is a tactical response that can be used when either you want to do something, but cannot, or more often, you do not want to act and avoid taking action by claiming limited discretion.

- In a situation of fragmented authority, individuals and organizations can always bump the problem to a different level or set of agencies by playing the "Surprise! It's Your Problem" game. Congressional acts that set vague policy and shift hard choices back to agencies employ this strategy, as do court decisions that remand decisions back to agencies (though in the latter situation, the judge's decision is generally all that is allowed by judicial norms).

- In a similar way, agencies and legislators can shift decisionmaking responsibility to the future (or to other actors) by using the "We're Dealing with It (When We Are Really Not)" stratagem. Gramm-Rudman budget targets and policies like the National Nuclear Waste Policy Act and the National Forest Management Act shift hard choices to the future, while giving the appearance that the issues are being decided today.

- Finally, all parties except perhaps Congress can employ the "It's Not My Job to Solve Problems" approach. Agencies can claim that they are given specific tasks to carry out that may or may not be problem-solving-oriented. Interest groups can rightly claim that their job is to make the best strategic case for their interests and leave creative problem-solving to someone else.

The administrative and political fragmentation evident in the spotted owl case offered agencies, legislators, courts, and interest groups the opportunity to limit their liability for choices they could make. When faced with demands for altered management, FS and BLM staff could point to Congress and say that the timber bias of on-the-ground management was their fault. Members of Congress on substantive committees could claim that they had done their job: The NFMA provided a framework to resolve these conflicts. It's the appropriations committees and the agencies that mess things up. Appropriations committee members could say that the agencies always said that they could carry out the cut levels with no problems. BLM managers could say that their policies were set by the timber-oriented O&C Act, and that their hands were tied. State and local land managers could claim that the owl issue was a federal lands issue, and the feds were obligated to deal with the impacts of altered timber supplies. Researchers could claim that they had done their job by providing the best information available, and it was the managers' fault. The managers could point to the latent uncertainties in the case, and claim that the researchers were not providing timely, unbiased information.

This litany of examples of finger-pointing and hand-tying can go on and on, but its result is to limit the liability of any one party for making a good or a bad choice. Indeed, looking back at the evolution of the controversy, while there are some individuals and organizations that deserve a more-than-average share of the responsibility for the nature of the response to the dispute, many parties share in the blame. The limited accountability provided by fragmented administrative and policy processes allowed a way to shift the obligation to act, and avoid the blame that comes from inappropriate or inadequate action.

Fragmented policy processes also mean that decisionmaking is slow and often inconclusive. Democratic processes are inherently inefficient, if you look solely at their ability to generate solutions to identified problems. They require the participation by numerous affected parties, and even if they produce outcomes that match the needs of different societal interests better than decisions that are mandated by more authoritarian approaches, they at least take time to do their job. Indeed, the pluralist need to test the legitimacy of various concerns, mandates a slow process

of issue definition and political mobilization that even with the least controversial issues requires a fair bit of time to work out.

The many different decisionmaking processes involved in the spotted
owl case, and the fairly extensive participation of numerous interest
groups, agencies, and political officials in each, guaranteed that any resolution of the issue would be slow to appear. Interest groups sought favorable decisions through state agency processes in Washington and Oregon,
federal administrative processes that involved several levels of the FS, the
U.S. Department of Agriculture, the U.S. Department of the Interior, BLM,
and FWS, federal courtrooms in Portland and Seattle, state legislatures,
and numerous committees in the U.S. Congress. The high level of political participation in each of these arenas, and the difficult nature of this
issue to resolve, meant that any one of them individually would take a
while to produce decisions. The fact that multiple processes existed, most
of which were affected by each other's choices, guaranteed a slow response to the spotted owl issue.

Slowness is not inherently bad, if it implies deliberateness and care in
pursuing an end (as the Hare and the Tortoise fable reminds us). In environmental policy, delay often provides for a careful examination of
choices that will preclude options if we act quickly, and allows for political mobilization that indicates the intensity of public concerns in an issue.
In the spotted owl case, however, delay in dealing with the controversy
meant a shrinking set of options, as suitable habitat was degraded or diminished by harvest activities incurred under existing forest plans. As
time went by, it got tougher to find a solution, not easier.

The multiplicity of access points also meant that decisions, once
reached, rarely held. Those who lost in political competitions could always "appeal" the decision either legally or politically. Hence environmental groups first sought change through decisionmaking processes of
the Forest Service and the BLM in the 1970s. When not satisfied with
the response, they appealed to the Assistant Secretary of Agriculture with
authority over the FS, and the Interior Board of Land Appeals, with authority over the BLM. When not satisfied with that response, appeals were
made through the federal courts, basing arguments on NEPA, NFMA, or
ESA grounds. You can never fully identify all environmental impacts, decide on the one correct way to balance forest uses, or fully resolve all
uncertainties about the status of a species. Challenges can be based both
on unit-level decisions, such as challenging specific timber sales or forest
plans, and on policy-level choices, such as decisions by the FWS on
whether or not to list the owl as a threatened or endangered species.
There is always an ample basis in law to challenge agency decisions, and

this was done repeatedly for substantive and strategic reasons throughout the 1980s history of the owl issue.

Not pleased with the courtroom outcomes, timber interests appealed to state and federal elected officials to deal with the issue. When ad hoc attempts such as the 1989 Oregon Timber Summit failed to produce a viable decision, attention shifted to the U. S. Congress and the activities of both the appropriations and authorizing committees. The decision by the Pacific Northwest regional environmental groups to nationalize the issue represented an awareness that politically they could not win if their argument was carried out in institutions dominated by Pacific Northwest political forces. In essence, then, their strategy of nationalization was a strategy of appealing the choices possible in the politics of the Northwest to a broader national political stage.

The broad set of opportunities to appeal choices made by one agency, branch, or level of government meant that the fragmentation of policy processes resulted in decisions that were inconclusive. Sometimes issues lose currency, or interests simply run out of steam and believe that a choice that has just been made is the best that they are likely to do even though they would prefer another outcome, and a resolution sets in. Such resolutions feel fairly jello-like, but as long as they remain properly refrigerated, they stay in place. But the warmth of the owl issue provided by the nature of the issue itself, and the relatively balanced political power of the two sides, led to a situation of stalemate, where activities continued but no policy solution was allowed to gel.

Finally, efforts to bridge fragmented authorities and activities often meet with limited success. Many administrative studies conclude by noting a need for interagency coordination and cooperation, and organization theory points to the need for so-called boundary-spanning activities, that is, actions that consciously build linkages between units of an organization or between organizations such that activities are not duplicative or counteractive. But experience and theory also suggest that there are powerful incentives that work to preclude cooperative activities. Coordination often implies that one agency has to defer to another, change its style of operation, or at minimum open up its internal decisionmaking processes to external review and criticism, and all of these things are anathema to bureaucratic organizations and political structures. Boundary spanning groups are usually dominated by the parochial concerns of their individual members, dooming efforts to solve crosscutting problems.

In the owl case, there were some early efforts to coordinate habitat management activities which functioned moderately well as fora for exchanging information until they started to challenge agency management practice. With involvement of most of the appropriate state and federal agencies, the Oregon Endangered Species Task Force in the 1970s was the kind of ad hoc group that is often praised as an innovative response to the inevitable problems that fall between administrative cracks. The Task Force provided agency researchers and managers the opportunity to share concerns, which was all fine and good until their advice implied a significant change in management practice by any of the involved organizations. When the Task Force's recommendations were resisted and rejected by FS and BLM decisionmakers, the group's effectiveness as a coordinating body was necessarily diminished, and the agencies lost an opportunity to reduce the magnitude of the coming spotted owl issue.

Groups that span agency boundaries occasionally function well, particularly when the members stand to gain individually from the success of the group. The Interagency Scientific Committee led by Jack Ward Thomas had a fighting chance for success in part because the agencies wanted a legitimized way out of the political controversy, and the Thomas Committee was their best shot at finding the light at the end of the tunnel. In many ways, this result parallels the implications of research into cooperative behavior which suggests that the existence of a superordinate goal, that is, a goal that supersedes day-to-day conflict over political and organizational turf, can build cooperative behavior better than efforts to work through the immediate set of conflicts. In addition, the Thomas Committee also benefited from the visible commitment of agency and political leaders to the effort as well as by the efforts of a strong group leader. Unfortunately, most of these forces do not dominate the functioning of interagency groups, and most cooperative efforts are ineffectual.

A Lack of Problem Focus and Creativity

Policy processes are generally interest-maximizing, not problem-minimizing. They work to maximize political and organizational concurrence, and not necessarily problem resolution. These realities are partly a result of our pluralist underpinnings. If you believe that legitimate concerns will attain representation through an existing interest group or agency, and that the policy process' role is to balance all these conflicting interests, then interest-maximization by default equals problem resolution.

For pure allocation questions, where we are simply trying to figure out which pig at the trough should get the larger share of feed, and where we have a highly mobilized population that brings its concerns fully into decisionmaking processes, interest group concerns may be fairly good

proxies for the problems that society must solve. But in resource management, these conditions rarely hold. First, most forest management choices
involve a technical dimension that influences the character of the problem to be solved and constrains possible solutions. Second, in many policy disputes, not all affected parties, including future generations and
nonhuman lifeforms, are at the negotiating table, thus requiring that
someone else represent their concerns. Finally, policy processes are rarely
neutral balance points among all the different affected interests. Rather,
most processes position some parties over others, and historic patterns of
influence have a major impact on decisionmaking outcomes.

The spotted owl issue was not dealt with by either FS planning or Congressional decisionmaking processes in an unbiased manner. Rather, the
FS's concerns about maintaining timber harvest levels and the ties of key
Senators such as Mark Hatfield to timber interests, guaranteed that analyses and outcomes would reflect these concerns. For example, in the SEIS
analysis, staff members understood that Region 6 leadership would not
allow more than a five percent reduction in the allowable sale quantity
by owl protection. It is not surprising, therefore, that the FS's preferred
alternative resulted in a cut level of 95% of the no action alternative.[3]
Neither administrative nor Congressional choices were based absolutely
on the merits of alternative owl protection strategies.

Judicial checks on administrative and legislative processes are also generally process- and not outcome-oriented: If you go through a good
enough process of decisionmaking, then the outcome is by definition
okay. Did the agency provide an opportunity for adequate review of their
analyses and decisions? Was the public given time and opportunities to
comment on agency choices? Does it appear that the agency looked at
the right kind of things in its decision process? The extreme and most
substantive question is: Did the agency appear to act arbitrarily and capri-
ciously? This latter question is rarely approached anymore in environmental policymaking because agencies have gotten fairly good at following appropriate process, or at least appearing to have done so. But it was
the question asked in the lawsuit challenging the FWS's decision not to
list the spotted owl in 1987, and the court found for the plaintiffs. While
the remedy prescribed by the court was more process, it did have substantive effects. FWS officials were asked to explain how they had reached
their conclusions, and like magic, new information appeared that eventually resulted in the owl being listed.

While process and substance are interconnected, generally our levers
into agency decisionmaking are procedural in nature. Even where there
are grounds for challenging substantive decisionmaking, courts and

legislators have a hard time doing so. Judges can tell if an agency did not describe the impact of a project on an endangered species, and hence violated both NEPA and the ESA, but determining whether the acreage of old growth-forest type in a forest plan is adequate to insure the long-term population viability of the northern spotted owl for the next 150 years is a daunting task for anybody, let alone judges whose knowledge of owl biology is probably based on dissecting earthworms in high school biology thirty years earlier. Federal court judges have neither a lot of slack time nor the appropriate expertise to challenge the substance of agency decisions, even if there are legal grounds for doing so, so they generally refrain from overruling administrative choices. As individual impact statements and forest plans have passed the one ton mark, the task for the courts has become even more weighty. Legislators are on even shakier ground in challenging administrative choices. They may have had experience dissecting lost political campaigns, but generally have little indigenous knowledge of natural resources or environmental science.

One way out of difficult situations where the decisionmaking process constrains action is to employ nontraditional strategies to address a set of concerns. Creative solutions sometimes change the nature of the discussion, excite participants, and mobilize new perspectives on old problems. And occasionally, they provide a way out of tough situations. Unfortunately, many of the mechanisms we employ to make resource policy choices tend to diminish creative thinking, rather than catalyze it. The adversarial, win–lose nature of judicial and administrative appeals promotes strong, one-sided argumentation from each of the affected parties, with very little incentive provided to think of creative solutions that bridge diverse interests. Nor is it generally the task of those who sit in judgment in these proceedings to search for creative solutions, or promote a process of interaction that could lead to new possibilities being exposed. Generally, one side wins, and one side loses, and the rational approach to participating in these processes is to make the strongest case for your side and hope it goes your way.

Administrative agencies have the technical capability and some degree of slack resources to identify and assess creative ideas, but are bound by their own vested interests in particular objectives and ways of doing things. As will be discussed in Chapter 10, bureaucracies develop standard operating procedures (SOPs) that enable them to function day by day, and new ways of doing things usually mean changing SOPs. At minimum, this is time consuming, and generally it is threatening to individuals who are comfortable with the routines and to organizational leaders who fear the potential impacts of a change in routines. Hence an organization like the Forest Service is not just committed to certain objectives, such as timber production, but to certain ways of pursuing those objectives, such as clearcutting in the Northwest. Even if a creative policy proposal is effective at pursuing the organization's objectives, simply because it changes the way the organization carries out its day-to-day tasks means that it would be resisted.

Congressional decisionmaking is more likely to yield compromises that split the stakes, so that all parties benefit somewhat from a decision, but they rarely expand the set of possibilities by creatively seeking new approaches. Members and their staffs are extraordinarily busy, with a very limited attention span for any one issue, and as a result are reliant on others to provide them with definitions of problems and alternative solutions. Generally, this means that interest groups and agencies control most input into Congressional decisionmaking, and since these parties have few incentives to find creative crosscutting solutions to policy problems and often have heavily vested interests in particular kinds of solutions, new ideas are not often at the center of Congressional choices. In addition, the need to build political compromises often guarantees vague, middle-of-the-road policy decisions. Congress may have the theoretical advantage in framing creative solutions, but its institutional realities generally produce lowest common denominator decisions.

The adversarial nature of our decisionmaking processes also tends to reinforce the zero sum preconception that we bring to most disputes, and as was discussed in Chapter 6, leads to a competitive process where disputants are simply trying to claim the biggest share of the pie for themselves. Interestingly, the basic negotiating styles of the major participants in the spotted owl controversy were similar to each other in being highly competitive. The basic approach of many environmental groups is to be as extreme as possible in the hopes that the final compromise will move down the continuum in their direction. If the policy process largely finds balance points between the groups of stakeholders, this is a rational strategic response that has been productive for the groups, but one thing that is lost along the way is creative ways to bridge group interests.

Corporate timber interests were accustomed to a competitive marketplace where the style of negotiation is to win as much as you can. As will be described in more detail below, the Forest Service's leadership approached its interaction with some of the stakeholders in a similar vein. When the Associate Chief says to a lobbyist for the National Wildlife Federation that "we'll take you to the mats on this one," as he did in the heat of the owl debate in the summer of 1989, the tone of decisionmaking is unlikely to yield creative crosscutting solutions. Such solutions may not be available, but it is clear from theory and the history of the owl case, that approaching the issue from a win–lose, outright competitive perspective means that any search for such creative solutions will be stillborn.

Unfortunately, as time goes by, the forces opposing creative solutions get stronger, rather than weaker. Increasingly, conflicts get polarized between positions and sides. Groups get locked into particular solutions, and advocate them as policy positions, sometimes at the expense of their true concerns which may be better met through alternative solutions. Such a dynamic favors polarization and leads to fairly perverse situations, where groups loudly advocate something that seems fairly strange and out of character for them. Watching superintendents of school districts in Oregon and Washington argue that owl and old growth protection should be minimized to protect the values and heritage of future generations of Americans is a case in point. The school directors have legitimate concerns about their revenue base and the health of their student body, but arguing against science and one kind of heritage would seem to miss the point. Having townspeople in northern California try to ban *The Lorax* by Dr. Seuss from their elementary schools because it provides a biased view of loggers (as was attempted in Laytonville in September 1989) seems similarly perverse.

Polarization and locking in on positions are just a couple of reasons that decisionmakers go from bad to worse at creative problem-solving over time. Interestingly, as issues become controversial and needy of creative solutions, the potential for creativity declines despite becoming more possible politically. As issues get more political, they increasingly receive attention from higher levels of the bureaucracy. On one level, this is effective, for most controversial issues are controversial because they demand value-type tradeoffs that only agency leadership should make. Yet often the attention span of leadership is limited and its technical knowledge may be outdated such that agency leaders are not the best source of creative, inventive solutions to pressing problems. Since their primary concerns are protecting their organization and the political chief executives that they serve, focusing on the substance of the problem itself and seeking creative ways out may not be what they do best.

In many ways, the reality of decisionmaking on controversial issues is a Catch 22 situation: you need lower level staff to be inventive, but they cannot legitimize their creations politically. You need leadership to provide political and organizational support, but they are less likely to seek or implement creative solutions. The best problem-solving units have been those that employ both levels of an organization in a way that maximizes the skills of each, but this kind of structure is the exception, rather than the rule, in American resource policymaking.

Short-Term Perspectives and Crisis-Management

In a time of rapid information creation and environmental change, we need institutions that can process and act on information in a way that is forward-looking to avoid crises. Unfortunately, American policy processes are generally short-term-oriented and resistant to change such that policy choices are often incremental and conservative in nature. By limiting your view to the crest of the next hill, you may find the best way to get to that crest only to discover the abyss that lies beyond. Short-range perspectives often doom us to crisis-management, and it was indeed difficult to get decisionmakers to look very far down the trail in the spotted owl case.

Most of our major decisionmaking institutions have a built-in short-range perspective. Even though federal judges have a guaranteed lifetime job, courts are generally backward-looking. They require precedent in order to make a choice, and that precedent is always framed in the social and environmental conditions of the past. Legislators, such as members of the U.S. Congress, are only mildly forward-looking. Congressional cycles operate on a two-year rotation, meaning that choices are primarily evaluated based on their political impact within the next electoral cycle. Even members who want to be forward-thinking are bound by the need to attain short-range political concurrence before they can pass new laws. Presidents and governors, and their administrations, generally operate within a four-year window. In practical terms, their electoral cycle offers a maximum of a couple of years within which to generate a record of accomplishment for the next election contest. Their policies can be forward-looking, but the need to make them politically relevant in the short term, and have them adopted by legislators with even shorter time horizons, tends to make presidential and gubernatorial policy initiatives similarly short-range in orientation.

By virtue of the stability provided by civil service employment, administrative agencies should have the greatest ability to look forward, anticipate future problems, and devise solutions to them. But the stability of their workforce is both a blessing and a curse. Stable workforces mean stable traditions, and agencies look as much to their past for clues to resolve their current problems as to the environmental conditions of the future. In addition, they are also bound by the short-term realities of the broader political process that they depend upon for enabling statutes and operating resources. As a result, even when agency officials look forward and see a need for change, it is hard for them to find the political concurrence necessary to act.

The decisionmaking system that we have is remarkably stable, providing us with a governance process that is basically the same that has been in effect in the country for more than two hundred years: an extraordinary state of affairs when compared with most other modern-day societies. The process was designed to ensure such stability, allowing for transitions of power and personalities in ways that are not disruptive to ongoing social or economic activities. But as a result, the set of decisionmaking institutions yields decisions that are conservative in nature, resist significant change, and generally seek to perpetuate status quo conditions. In addition, the process is protective of itself, that is, it limits changes in the way choices are made.

Even many of the organizations that the pluralist framework relies upon to advance problems into the decisionmaking process seek to protect what they have in the context of current patterns of economic or political power. As a result, the National Education Association lobbies against changes within schools that could well result in a better educational experience, but would cause teachers to have to change their day-to-day actions. Similarly, the Appalachian Mountain Club, a major environmental group in New England, opposed wilderness designation for certain areas in the White Mountain National Forest because they would have had to remove primitive shelters they had created and used in the national forest. The Forest Service opposed the concept of Congressionally designated wilderness in the 1960s because it would take away the agency's ability to make on-the-ground choices, even though Congressional designation was probably appropriate given a declining amount of wilderness lands.

Interestingly, while we have an increasing ability to look forward and anticipate change, our American traditions are generally opposed to planning and planned change. One of the bequests of the Enlightenment period in eighteenth century Europe was a mindset that allowed for rational systematic thought. Planned change as a concept is an element of Western thought, but American traditions and values generally oppose long-range planning. Our economic system is founded on the basis of nonintervention: it is thought that invisible hands will guide the wisdom of the marketplace and generate an efficient and effective set of decisions. Our pluralist political process assumes that similar forces will provide social direction. Planning is seen as overly centralized, potentially socialistic, and subject to the misuses of concentrated power that the governmental system was designed to avoid.

Decisionmaking institutions that are short range- and status quo-oriented, and norms that are suspicious of planned change result in incremental responses to pressing problems, and rarely push the envelope of possibilities out very far. Conservative, incremental decisionmaking is appropriate and effective in an environment that is stable and a society whose character and values are unchanging. But neither of these conditions holds today. As a result, our society seems destined to confront change by crisis-management, since institutions that seek stability in an unstable world are guaranteed frustration.

Crisis management is of course one function of decisionmaking institutions. We will always face external events and unforeseen circumstances that require a short-term, unanticipated response. But the short-range perspective of our policy processes does not work hard at avoiding crises that might be avoidable. In addition, they often enhance the magnitude of unavoidable problems by putting off decisions long enough such that they fester into major sores on the body politic. Options are lost, and the vested interests in the current direction become harder to change. Hence, federal budget decisions that put off inevitable hard choices, simply pass problems to future sets of decisionmakers, whose options will be much more constrained by large debt loads.

In the spotted owl case, short-range decisionmaking by agency officials postponed difficult choices into the future when options were fewer. Agency officials continually avoided the old growth and owl issues in the 1970s and early 1980s by forcing management decisions into the forest planning process, a process that was not designed to resolve such major policy issues. This approach was rational from a short-term political or organizational perspective, but heightened the inevitable controversy when it came home to roost.

The fact that American policy processes are crisis-oriented also means that they are influenced heavily by chance events. Oil spills, earthquakes, and Congressional scandals are not planned, but their occurrence evokes a significant response from the policy process. In the spotted owl case, these dynamics were present occasionally through the chronology. In 1991, rumors were flying around Washington, D. C. that Senator Mark Hatfield had allegedly engaged in questionable financial dealings and might have to resign.[4] Regardless of the truth of the allegations, they at least distracted the Senator and his staff from policy-related matters such as owl-related business, and if proved true, would have had as major an impact on the outcome of the controversy as anything that had occurred through rational, deliberate action.

The crisis-managing, reactive process that we employ to make public choices means that decisionmakers have a hard time putting chance events into perspective; it almost always guarantees a knee-jerk reaction that may or may not be effective in the long term. In addition, it often seems that issues take on a life of their own, rising and falling as events transpire in a now you see them, now you don't cycle of public concern. This type of issue-attention cycle would be functional if it generated a process that resulted in an informed choice on the issue. But often the response is extremely short term and reactive, and has the effect of derailing ongoing processes of analysis and choice that may be more deliberate and thoughtful but not as visible.

Constrained by the Quality of Information

The quality of a policy decision can rarely rise above the quality of the information used to make the choice, and often the information provided to resource policymakers is not up to the needs of the choices at hand. The information problem is not just an issue of uncertainty, as described in Chapter 6. Rather, the information available to decisionmakers is often biased to past needs and methods, and its quality and availability are limited by institutional barriers resulting from fragmented organizations and adversarial choice processes. Inadequate information was a consistent theme in the development of the spotted owl issue, and it remained a theme long after the need for better information was clear, and the resources were available to help out.

Information generation is usually the byproduct of whatever forms of analysis and inquiry are necessary to deal with the specific problem or management process at hand. As a result, the character of available information is usually oriented toward satisfying specific past needs. If your primary organizational concern is in finding and preparing commercially valuable timberland for sale, there is not much purchase in mapping suitable habitat for owls. Nor is there a great need to update maps to show stands that have recently been cut and replanted, since you do not have to worry about them for fifty or more years.

Given limited resources, it is logical to focus information collection on immediate problems and neglect updating the information or looking for bigger picture issues. Up until recently, updating and aggregating information was particularly difficult because most data only existed on paper and was not collected into a computerized geographic information system that now makes information manipulation much easier. Overlaying several variables of information such as tree diameter, number of snags, canopy closure, etc., to indicate effective habitat is a very time-consuming and laborious process when each variable exists on a separate hard copy map. This kind of analysis might at least have been possible if each of the variables had been mapped as raw geophysical information to be combined in different ways. But, since most mapped information corresponded to a particular need, mapped information often represented a conclusion rather than a set of data that could be combined in different ways when new problems and needs developed.

To the extent that conscious processes of inventory and monitoring are carried out, their structures and protocols are often framed within the direction provided by prevailing knowledge and current problems. Hence, if you believe that the primary health problem associated with sulfur dioxide pollution from power plants is local effects around smokestacks (as was our understanding in the 1970s) then a monitoring program would concentrate on ambient levels and health effects within a localized airshed. But such an approach is unlikely to indicate other, unanticipated problems, such as the fact that the larger scale impacts of sulfur dioxide as it is transformed in the atmosphere are now thought to be much more problematic than local effects. Similarly, if you believe that biological diversity is generally associated with the presence of edge between landscape types, and that more homogeneous landscapes like old growth are a biological desert (both of which were conventional wisdom in the 1960s and 1970s) then inventory work would focus on edge and edge-associated species, to the exclusion of interior-dwelling species.

We do not look for what we do not expect to find, and our looking generally seeks to confirm current thinking, rather than seek new understanding. Both of these objectives would be fine, if we did indeed know everything, and what we were looking at did not change over time. But it should give us pause that the early geographers swore that the earth was flat and the astronomer–philosophers of the 1500s and earlier felt confident that the sun and the planets revolved around the earth. Our knowledge base has expanded since then, but by not seeking disconfirming information through deliberate programs of inventory and monitoring, we almost always insure that our current state of information is not up to new challenges and problems.

Part of the disparity between what information generally exists and what information policymakers need is a result of differences in scale. Often the level of aggregation or disaggregation functional for managers is not productive for policymakers, and available information is therefore more likely to exist in the former state. If your primary concern is well-laid-out, legal, and environmentally safeguarded timber sales, you need good quality information at the timber stand level, and minimal quality at the district or forest level. If you are making choices about the long-term future of the spotted owl, however, you need high quality stand information, aggregated up to the regional or multiregional level while preserving a sense of the relationships between stands. At a minimum, this requires numerous administrative units to combine their stand-level information, and do so iteratively over time. Even if they were inclined to do this, the fact that the information, if it exists, resides in hundreds of different drawers at numerous FS district offices, in different forms at varying levels of currency and accuracy, suggests an aggregation task of monumental portions. And rarely are these administrative units inclined to do this.

The quality and character of information available to policymakers is heavily influenced by institutional factors. Not only do staff and financial constraints limit information collection and analysis, but the motives of each of the involved parties influences the information they collect and how they summarize it. In the case of the spotted owl, while there are understandable historical reasons that owl habitat was not mapped in a way suitable for framing by policymakers, it is partly the case that agency officials did not seek information aggressively because they did not want to find out what they did not want to know.

We all do this as individuals when we fear the implications of new knowledge: Not getting that blood test that may indicate high cholesterol levels, meaning dinners will be carrots and beans instead of hamburgers and cake; not taking the car into the shop to find out that the brakes are shot and for $453 plus tax, they can be fixed, and oh by the way, the transmission has problems as well; not asking your child where he was until 4 a. m. last night because you do not want to know the answer. All of these behaviors are satisfying in the short term, but when the heart attack strikes, your brakes fail in downtown Detroit on Saturday night, or you get a late night call from the police that they have picked up your child, you wonder why, why did you not find out more earlier and act on it. Ignorance may be bliss, but only when not acting results in a continued state of bliss, and this is rarely the case.

Ignorance is doubly functional in the short term for agencies. It lets them play the ostrich and keep others from pulling their heads out of the sand. Up until recently, challenges to agency behavior largely relied on information provided by the agencies themselves. It is difficult to muster the time and expertise to develop information parallel to that generated by the agencies. If the agency information base provides both rationale for their choices, and the means to challenge those choices, conflicting information will not be sought and at least occasionally will be suppressed from public view.

Organizational objectives also influence the way that information flows between agencies. Since information is collected for different purposes by different agencies, serious consistency problems between agency data sets generally haunt cooperative efforts. To resolve these inconsistencies requires a sharing of information and mutual sense of purpose that, as was discussed above, is rare. Information is one of an agency's primary sources of influence and power in the policy process, and one of the more tangible products of an agency's existence, and hence it is rare that agency officials want to give it away free. Conflicting information from a sister agency is not embraced because it implies that your agency is doing something wrong, or they are doing something that you should be doing. Since agencies are just as defensive as individuals in justifying their past behavior, contrary information is usually ignored and its sources impugned.

Since individuals and organizations who generate information do so for a self-determined purpose, there sometimes is reason to distrust the messengers of bad news. Information networks have motives. Managers seek to continue their standard management practice, researchers seek the opportunity to do more research, and interest groups seek to prove a point. Every good policymaker considers the objectives and quality of the source of information and discounts the information accordingly. But often, the agency response to contrary information from external sources goes far beyond the nature of the conflicting information. For example, a key item of debate in the spotted owl controversy was the amount and location of old growth owl habitat. In the old growth hearings held by the U. S. Congress in the summer of 1989, the Forest Service officially pegged national forest old growth at 6.2 million acres, while The Wilderness Society, based on a fairly extensive mapping exercise involving FS and volunteer personnel, had concluded that the correct figure was less than 2.4 million acres. Much of the FS staff felt that TWS's data was better than their own, and advised the Chief not to debate the numbers in the Congressional hearing. But he changed his prepared testimony on the way to the hearing to challenge TWS's analysis, much to the shock of his staff.

Information networks have bias, and are exclusionary and self-reinforcing. Who you listen to, and who the research funders pay to generate information, are heavily affected by perceived legitimacy and past relationships. Since so much of science depends on trusting individuals to "do good science," and relying on their judgments to interpret data, who is listened to depends on who is trusted, and that has as much to do with who you know as anything else. The old boy network functions in science, just as it functions in all other aspects of human life; but in science, its effects are perhaps more insidious because we rely on researchers to provide and interpret information in as objective a manner as possible.

In order to understand the history of the spotted owl controversy, it is important to place it within the context of the characteristics of our standard decisionmaking processes. Some of the failings of individual, organizational, and institutional choices made in the owl case represent weaknesses in the American system of government, many of which were adopted for good reasons, and some of which are unavoidable. Fragmented, reactive, unimaginative, and uninformed policy processes cause problems for even simple policy issues, and the owl issue was not simple. Indeed, the interplay of the characteristics of the owl dispute described in Chapter 6 and those of American decisionmaking processes guaranteed that the issue would have a long life.

8

Influencing Tough Choices: Actors in American Decisionmaking Processes

Since decisionmaking processes are not subject to much centralized control and are reactive by nature, outcomes are influenced as much by specific individuals and groups as they are by deliberate analysis and choice. These actors in the policy process include strong personalities, the media, elected officials, interest groups, scientists, and government agencies. Unfortunately, the incentives that influence each of these participants tend to promote behaviors that exacerbate the failings of our decisionmaking processes. This chapter discusses the impact of these actors on the functioning of American resource policy processes. (The behavior of government agencies is discussed in Chapter 10.) What happened in the owl controversy depended heavily on what specific individuals and groups wanted to have happen, and for some, helping to craft a long-term, equitable resolution of the controversy was not perceived to be in their interests.

The Impact of Individual Personalities

In evaluating the evolution of the spotted owl issue, it is hard not to be struck by the impact of specific individuals on the nature of the controversy. Oscar Wilde once said that "It is personalities, not principles, that move the ages," and American resource policy processes bear him out. If a small meteor had landed on top of a key participant in the spotted owl case and removed him or her from the course of history, the issue would not have gone away completely, but the way in which it ultimately played out might have changed significantly.

These keystone personalities included agency, interest group, and Congressional staff members, as well as elected officials of a variety of stripes. Eric Forsman and Howard Wight at Oregon State University forced the owl and old growth issues out of scientific obscurity into the realm of legitimate science with management implications. The fairly high level of autonomy within the line structure of the Forest Service meant that the ground level response to the developing concerns about the owl were largely dependent on the good graces of specific district rangers and forest supervisors, and everyone in the Forest Service agrees that line officer personalities differ greatly and have a major impact on what is actually done on individual FS units. At any point in the controversy, some rangers and forest supervisors were trying to deal with the issue, and others were ignoring it completely.

Given a fairly limited response by the management structure of the FS and the BLM, it was up to nongovernmental organizations to keep the issue moving, and individuals, not organizations, were the key forces in the debate. The major reason that the National Wildlife Federation was a prime player in the mid-1980s was because Andy Stahl worked for them in Portland, and the owl/old growth issue was important to him. NWF dropped out of the controversy, and the Sierra Club Legal Defense Fund got involved in it, when Stahl moved to SCLDF in Seattle. Similarly, individuals like Andy Kerr at the Oregon Natural Resources Council, and Mark Rey at the National Forest Products Association had significant impacts on the course of the controversy, particularly on the tactical character of the debate.

As the controversy increased in size and significance, individuals at the top levels of the agencies began to exert their influence over the course of events, as one would expect, but their actions were as much determined by their own personalities and values as by any other set of forces. For example, FS Wildlife Deputy Director Hal Salwasser had worked on the development of concepts and methods of population viability analysis (PVA), and the choice to use PVA in the FS's spotted owl analysis partly reflected his interests. Having John Crowell as Assistant Agriculture Secretary over the Forest Service insured a pro-development orientation within the agency, and constrained possible FS responses to the developing controversy. The objectives and style of the FWS under Director Frank Dunkle were quite different than those under his successor, John Turner, and had a major influence on the outcome of the FWS status review that determined that the owl did not warrant listing as a threatened or endangered species. Even though he had no direct authority over the FWS, Deputy Assistant Interior Secretary James Cason reinforced Dunkle's inclinations and sought to pressure the agency not to list the owl as a threatened or endangered species. Cason also allegedly suppressed a 1986 BLM report that found that the owl could be endangered by logging.[1]

Similarly, as the controversy wound its way into Congress, key members and staff heavily influenced the course of events. No one can deny the impact of Senator Mark Hatfield on all issues related to timber in the Pacific Northwest. If the meteor landed on Senator Hatfield, the nature of the politics involved in the issue would have changed dramatically, though some would claim that Hatfield's political and spiritual connections would protect him from such intergalactic events.

Even more than the personalities of key members of Congress, aides who happened to be on their staffs had a significant impact on what took place. The major reason that Representative Chester Atkins from Massachusetts got involved in the 1989 appropriations battles over the future of forests in Oregon and Washington was because Nancy Green, a Forest Service employee, was on loan in his office as a Congressional Fellow. Similarly, Indiana Representative Jim Jontz's interest reflected those of staffer Molly Frantz, a former National Audubon Society intern. Vermont Senator Patrick Leahy's interest in the issue reflected the fact that his staff included Tom Tuchmann, who had previously been employed as a resource policy analyst for the Society of American Foresters. House Agriculture Committee staffer Jim Lyons was influential on the issue, and had staffed the SAF's 1983 old growth study that concluded that land reservations were probably the best way to "manage" old growth.

Under any theory of policy formation you could subscribe to, the interests of Congressional staff, who are by and large quite young and not at all reflective of the heterogeneity of interests within the country, should not dictate what is produced as public law. But the reality is that individual personalities with strong interests wherever they lie, can and do have an impact on the outcome of policy controversies, and it is important to understand the personalities involved in any issue to understand its past and future. In many ways, the only way that the conservative, change-resistant nature of the process can be overcome is through the force of personality, and the pluralist need to build political concurrence reinforces this reality. The good news in all this is that individuals can make a difference in determining the course of human events; the bad news is that individual perceptions of the public interest may or may not correspond well with what is needed to respond to a problem, or necessarily how the majority of the citizenry feels.

The Effect of the Media

An informed public is the keystone of a democracy, and while we are blessed with an ideologically unconstrained free press, the operating realities and motives of both print and electronic media limit the effectiveness of media inquiries into complex policy issues. Newspapers only have a few column inches in which they can cover a story, and the trend toward Readers-Digested newspapers à la *USA Today* has magnified this reality. Television news is even more constrained. While a picture is supposed to be worth a thousand words, and talking pictures presumably are worth much more, television news is oriented around twenty-second sound bites that have to summarize a story and be entertaining. A public used to viewing simplified entertainment tends to be comfortable with such news coverage and bored with more elaborate treatments.

Other characteristics of American media tend to reinforce these dynamics. Except in the largest and wealthiest of news organizations, newspaper reporters and television correspondents have to be jacks-of-all-trades, reporting on a range of issues. As a result, their knowledge base about the factors underlying an issue is often limited, and because they are usually operating on short deadlines, they have little time in which to gain a great deal of insight. Their need to act with speed, and appear informed and authoritative on an issue, encourages them to dismiss areas of uncertainty and grayness for themselves and their readers/viewers.

The effect of these operating characteristics is that the media tends to simplify issues into barebones caricatures of the underlying realities at the heart of a controversy. In the owl case that meant portraying the issue as one of owls versus jobs, an easy dichotomy that even small minds can grasp. Other images are equally compelling: the federal government versus the people, the rapacious despoilers of green forest versus benign, slightly anthropomorphic, old-growth–dependent animals, and others. These are simple images, readily conveyed, that convert complex issues into stereotypic conflicts, easy for the public to swallow and digest. As bystanders, we are used to watching simple, two party, win–lose contests take place in football games, presidential campaigns, and soap operas, and the media is very good at forcing complex issues into simple boxes even if a good bit of the issues actually slop over the sides of the box.

Besides, conflict sells well, and the underlying motives of news organizations affects what news is covered, and how it is covered. News organizations are not philanthropic, altruistic entities; rather, they exist to make money, as does any business. In the best of situations, they do this by providing a valued service, though accurate and balanced coverage of controversial issues is sometimes not valued by those who count. Since the bulk of newspaper and television revenues comes from advertising, taking stands on or reporting both sides of issues that may be threatening to potential advertisers can be risky. Since the future of the newspaper or station will probably rise or fall with the economic tides in its reading/viewing area, and because owners are often representative of the economic elite of the local community, news organizations may well be highly sensitive to the economic climate in which they function.[2]

Besides exacerbating the tendency of our public decisionmaking processes to simplify complex issues, media coverage also tends to downplay long-term implications, and avoid continuing coverage of issues that by their very nature will be on the public stage for a long time. Both the public and the media get satiated fairly rapidly on a given topic. The amount of coverage of new mediagenic issues is often large, even if its quality is low, and because individuals cannot process all the sensory input that bombards them daily, they become very good at shutting it out. In addition, with most issues, over a period of time, individuals face a dawning reality that in order to solve the problem at hand, real costs are implied, and such costs are depressing. Newspapers, news magazines, and television both respond to these reactions and reinforce them. As a result, we are faced with crises of the month: the garbage crisis, Earth Day, the AIDS situation, drugs in schools, the old growth crisis, and the like. In almost all situations, it is not the case that the concerns decline over time, rather media attention and our public interest level wanes as satiation occurs, and the next crisis looms on the horizon. Indeed, it is often a disorienting experience to look back at the news magazines of a previous year and wonder what happened to earlier crises that had lost public appeal.

The net effect of these dynamics is that media attention to public issues is often just as reactive and crisis-oriented as that of other public institutions. This is partly a function of the role and organization of the media toward reporting events, as opposed to either interpreting them or providing a forum for debate and the analysis of new ideas and potential solutions. The media is often pulled along by chance events and individual personalities just like the rest of us.

For most news organizations, these organizational factors generally are not reflected in overt bias in news coverage, and like the dynamic we are trying to explore, we can cast it too simply. Actually, over a period of time, some of the media coverage on the owl issue became quite good, particularly that of the regional newspapers such as the Portland *Oregonian*. Nevertheless, it is clear that the operating realities and motives of media organizations tend to diminish their effectiveness as comprehensive analysts of complex, long-range public policy questions.

The Incentives Facing Elected Officials

As issues like the spotted owl dispute become more controversial and rise in visibility, they increasingly demand action from elected officials such as governors, presidents, members of Congress, and state legislators. Since these individuals are nominally the leaders of our governance process, one would hope that they would exercise leadership over public choices and be articulate spokespersons for alternative pathways to the future. Unfortunately the political environment in which they function tends to penalize individuals who take strong stands on controversial issues, and rewards those who are aggressively and benignly mediocre.

Elected officials are no different from any other citizen in that they manifest the same range of motives, interests, personalities, and policy preferences as any other human. But, they also need to build power in order to take action, and be reelected in order to continue doing what they are doing. Since power is fragmented, elected officials have to work at building the political concurrence necessary to allow them to take action, and preserve what influence they have within their operating theater. For example, in the Congress, accumulating power means forming alliances with interest groups, agencies, and Congressional leadership within political parties and committees. Since leadership controls everything from whether action will be scheduled on a member's bill to what office space he or she is assigned, avoiding major conflicts with party and committee leaders is desirable to say the least. And when the leadership has clear stands on issues, as was the case with Speaker of the House Tom Foley (D-WA) on the owl case, the situation can get very dangerous for members who might want to oppose them.

Elected politicians are also faced with the need to support and/or placate their electorate so that they are reelected every two, four, or six years. This electoral reality is true both for those who view the job as a comfortable and interesting life, and those who view it as a means of advocating policy change: If you want to continue your lifestyle or push your policy ideas, you have to be reelected, and that reality influences greatly the behavior of elected officials, even though the truth is that most incumbents will be reelected no matter what.

Electoral stakes result in politicians being enormously concerned with satisfying their constituents and not alienating potential voters. For example, every request made to a Congressional representative by a constituent for information or for help getting their social security check is viewed as a direct connection to a potential vote, and members of Congress use a lot of their staff time to handle so-called casework, that is, dealing with constituent letters and requests. It is less clear how substantive policy work translates to voter behavior, and sometimes young members of Congress only work at casework and at building their power within the Congress. Beyond constituent casework, the other way that politicians reach voters is through on-the-ground campaigning and media advertising, both of which have become inordinately expensive, particularly since campaigns are fought increasingly via television. Hence, elected officials have to spend a fair bit of time on fundraising, and a good bit of what political power they have to affect policy outcomes, on pleasing those who would give them campaign funds.

The incentives that this political environment creates results in a set of decisionmakers who want to do anything but make decisions on difficult issues. Most elected officials have little time to develop expertise that enables them to have great insight into complex issues. On whatever public stands they do take, the limitations of the media requires them to simplify their insights so that their position and rationale can be excerpted into twenty seconds of videotape for the nightly news. The fact that they face a plebiscite on their future every few years means that their time horizon is effectively very short term. Since they need to placate powerful groups, their policy positions generally have to be fairly conservative and status quo oriented to avoid alienating anyone, particularly since the media and numerous interest groups will be watching. Being out on a limb is only a good place to be if there is not a strong wind blowing and the limb is well supported. If the winds of political controversy are strong, or support for your position is from a small, somewhat extreme, subset of the electorate, watch out.

Issues that represent conflicts between seemingly irreconcilable values, that are based in technical arguments that are hard for the electorate to comprehend and the media to convey, and that have their major implications in the future are very difficult for elected politicians to deal with.

The spotted owl/old growth controversy was solidly grounded in this category of issues. Politicians generally seek issues that generate benefits in
the short term and costs in the long term, but the owl issue generated
significant short-term economic costs with the hope that benefits would
result sometime down the line. Politicians embrace issues in which the
opinions of their constituents are clear and fairly united, yet the owl issue
had generated an electorate that was highly fractured and enormously
conflicting, and in the later stages of the story, the balance of power between environmental and timber interests was neither stable nor obvious
to those who had constituents of both types.

Elected officials also avoid issues that affect intensely motivated groups,
and the owl issue presented a set of groups on both sides who were mobilized, angry, and likely to respond to a stance opposed to their interests.
In addition, officials seek issues for which a solution can generate all winners, or at least the stakes can be split among the affected interests. The
owl issue, however, was framed such that it was clear that someone was
going to lose, and for many it was an "all-or-nothing, fight to the death"
kind of situation. The best kinds of issues to support are those that are
framed in terms of so-called "motherhood" issues, that is, those that no
one is against. Many environmental issues fall into this camp, since everyone is for protecting endangered species and against pollution. But the
owl issue was different, because both sides could claim motherhood symbolism, whether it was protecting the future of helpless, marginally charismatic animals, or protecting the future of helpless, marginally charismatic young Americans.

For some members of Congress, such as Bob Smith and Denny Smith,
Republican Representatives from Oregon, the owl issue was easy, because
their constituents and histories were so tied to timber. But for most, the
issue was a no-win situation, and the best anyone could do was to try to
arrange compromises, put pressure on the Forest Service and the Fish and
Wildlife Service, and stand back out of the inevitable firestorm. The history of decisionmaking on the owl case looks a great deal like a group of
school children tossing a hot potato back and forth hoping that either
someone else will hold onto it and burn their hands, someone will find a
technical way out (gloves perhaps), or that it will cool over time before
someone gets hurt.

The Strategic Behavior of Interest Groups

Given the nature of the pluralist system under which we make collective
choices, a public policy represents a temporary balance point between all

the diverse sets of participants. The incentives that this process sets up encourages stakeholders to act as strategically as possible to pull the balance point in their direction. The nature of adversarial argument and the perception that the solutions are of the win–lose variety magnifies this dynamic. Interest groups that succeed in this competition use a full set of tactics, ranging from legal action to mass protest.

Strategic behavior on the part of interest groups is appropriate and understandable, but it also tends to cloud the facts and overstate group interests. Since our institutions do not do a great job of forcing a debate over the technical merits or validity of particular positions, policymakers are often given two or more very different sets of apparent realities from which to make public policy choices. Truth, to the extent that it exists, lies somewhere between the arguments of interest groups. Good decisionmakers work hard to try to get a fix on that truth by triangulating between the positions of different groups. At the same time, since there are lots of opportunities to participate, it is in your interest to keep working the system until your side wins. Encouraged and facilitated by our decisionmaking system, strategic behavior by interest groups tends to make policy choices more confusing, and less enduring.

In the spotted owl case, strategic behavior by interest groups was pervasive. Table 8-1 summarizes some of the sets of strategic behaviors that are commonly used to influence policy debates which were quite evident in the history of the spotted owl controversy. Since our decisionmaking processes provide so many possible places in which action can take place, one set of strategic choices that interest groups face is gaining access to an action channel that is likely to result in a favorable response. Their job is to gain leverage over the outcome either by directly advancing the issue into a forum geared to that kind of issue, or more often, *linking* the current concern to an issue that has already been legitimized politically or legally. For a number of groups, the decision to use the spotted owl as a lever into decisionmaking over forest management was a choice that reflected strategic possibilities as much as real concern about the status of the owl itself. For example, at the Western Public Law Conference in 1988, Andy Stahl outlined the attractiveness of the owl as a legal lever into public decisionmaking, calling it a surrogate for the old growth issue:

The northern spotted owl is the wildlife species of choice to act as a surrogate for old-growth protection, and I've often thought that thank goodness the spotted owl evolved in the Northwest, for if it hadn't, we'd have to genetically engineer it. It's a perfect species for use as a surrogate. First of all, it is unique to old-growth forests and there's no credible scientific dispute on that fact. Second of all, it uses a lot of old-growth. That's convenient because we can use it to protect a lot of old-growth. And third, and this is more a stroke of good fortune in one sense and in another sense, an indication that environmental groups tend to wait too long before they act, it appears that the spotted owl faces an imminent risk of extinction. That's very important, for if it didn't, federal agencies could argue that they could continue to log old-growth and not hurt the spotted owl. It's important that not only it face a risk of extinction but that we haven't gone too far because then federal agencies could argue: Why should we bother to protect old-growth; it's too late already; the spotted owl is doomed. In other words, we have to be right on the edge and by good fortune, it appears that we are in this decade right on that edge.[3]

While Stahl's comments were probably somewhat overstated and many individuals in the environmental community were honestly concerned with the viability of the owl, nevertheless it is clear that using the Endangered Species Act as a lever into FS and BLM decisionmaking was a conscious tactical choice, as it was in the Tellico dam–snail darter case and a number of other frontline environmental controversies.

The ESA is a particularly potent lever because of its absolute mandate to federal agencies to protect endangered species and their habitat, and because it has a fair bit of political support to offset potential backlash. Nevertheless, as is true with any public law, its legal strength is only as good as its political strength, and in the mid-1980s at least, the political power underlying the ESA was an open question. On several occasions, Senator Mark Hatfield pointed out to environmental groups that "we made the law, and we can change the law." As a result, the use of the owl as a judicial lever was hotly debated within the environmental group community. Indeed, the mainstream groups delayed petitioning the FWS to list the owl as an endangered species because of their fears for the ESA and only joined the battle when Green World started the petition process. Once the battle was joined, the more mainstream environmental groups jumped on board because they had much at stake in the outcome.

The way in which an issue is defined will heavily affect everything that follows. *Issue definition* determines who gets involved, their starting power, and the type of decisionmaking arenas that are likely to take action. Was the owl issue about jobs or old growth or endangered wildlife or the Forest Service or schoolchildren? It was about all of these and

TABLE 8-1
Interest Group Strategies in the Spotted Owl Controversy

Gaining Access and Finding a Favorable Arena
    Seeking Leverage by Linking to Other Issues
    Battle for Issue Definition
    Altering the Timing of the Issue
        Uncertainty Delay
        Legal Delay
    Shifting the Site of the Dispute
Making a Case for Your Position
    Battle of the Numbers
    Generating Legitimacy for Your Arguments
    Fait Accompli: Making the Issue Moot
Building Political Concurrence
    Coalition Formation
    Using Influence Networks
    Grassroots Mobilization
    Media Strategies
        Mass Demonstrations
    Working the Symbols
    Compromise

more, but how it was portrayed influenced the reaction of various individuals and institutions and ultimately the way in which it would be resolved. The timber industry repeatedly defined the issue as one of jobs and lifestyles, the environmental groups defined it as a moral and biological diversity issue, and the Forest Service defined it as a forest planning issue.

If the issue was primarily about jobs, it would lead to less old growth protection and more emphasis on community stability measures. If the issue was about biological diversity and our obligations to nonhuman lifeforms, it would engender more old growth protection. If it was about wise forest planning, it would place decisions on old growth protection within the context of unit-level planning processes (processes that, however, were probably not intended to resolve these kinds of value-laden policy questions). How the issue is defined determines the bounds of legitimate response, as well as which Congressional committees, agency actors, and interest groups have a legitimate role in determining a response. Issue definition is critical to issue resolution, and interest groups work hard at defining the issue in their own image.

Timing is also a critical strategic choice, and interest groups try to *shift the timing* of an issue so that policy outcomes are favorable. With a fairly uncontrolled, reactive decisionmaking process that only acts with political concurrence, timing is important in two ways. First, issues are ripe for settlement when adequate information is generated, and enough time has elapsed so that patterns of political support and likely impacts are clear to those who have to make a choice. Since neither available information nor political patterns are absolutes, interest groups will attempt to control the timing of an issue so that information and political support can be generated before a choice has to be made. In addition, since chance events and individual personalities have a major impact on policy possibilities, taking advantage of these opportunities as they arise means that groups need to speed up or slow down decision processes. If you are interested in precluding oil development in the Arctic National Wildlife Refuge, for example, the time to act is in the fairly short window of public and political attention following a major oil spill like the *Exxon Valdez* incident.

Two common mechanisms to shift the timing of an issue are by causing delays based on inadequate information or through judicial intervention. Uncertainty delay works because it appeals to our underlying biases that more information leads to better choices. In the owl case, those benefited by status quo policy argued repeatedly that we did not know enough to take actions that would affect humans and economies in as big a way as was likely in stopping the harvest of old growth. Until the owl issue was settled, forest plans prepared to meet the mandates of NFMA could not be finalized, and since the old forest plans allowed fairly high cut levels, timber interests were benefited by any actions that would stretch out a final decision (as long as injunctions were not placed on timber sales in the interim).

Timber lobbyists constantly highlighted uncertainties about owl biology and economic impacts, and used that as a basis for arguing for delaying any final choice. "Wait a minute. You haven't found owls in habitat other than old growth because you haven't looked elsewhere. We have studies that show that they are in northern California in second growth stands. . . . Do they need that amount of acreage of habitat or simply prefer it? . . . We don't know the magnitude of the economic impacts of this legislation. It is probably huge, and needs to be studied before action can be taken." These and other arguments were made throughout the owl issue, partly reflecting real concerns, but equally reflecting tactical responses to a status quo policy situation that benefited timber interests.

In reflection, the uncertainty delay argument used by timber interests was somewhat strange. The logical next step beyond stating that we do not know enough and need to study things further before taking action, is to protect options so that when you do find out whatever you are going to find out, you have a full range of responses available to you. In this case, it would have meant a moratorium on old growth harvest until it was determined whether owls could exist in second growth or not.

Delays by petitioning the courts or using administrative appeals processes to stop action were another common tactical response. Environmental groups repeatedly sought legal delays, some as an end in and of themselves, but primarily to slow down action and preserve options while they worked the political channels. Court actions largely serve the function of changing the pace of administrative activities so that parallel political strategies can be pursued. While the environmental groups could never have succeeded in changing old growth policy in the late 1970s and early 1980s given the power balance in the Congress, their efforts in the courts provided a braking action while a national political campaign could be developed and implemented.

The nationalization of the owl issue was a conscious tactical choice by regional environmental groups in response to dismal chances for success at the regional level, and much better possibilities for success at the national level. Interest groups often try to *shift the site of the dispute* to find the most favorable decisionmaking arena, a rational tactical choice given the fragmented nature of our decisionmaking institutions. Shifting the locus of action from administrative agencies to administrative appeals processes to the courts to state and federal legislative processes represented a strategic effort to find a decisionmaking arena that would provide a favorable outcome.

Even within one branch of government, interest groups try to pick decisionmakers who are more inclined to resolve an issue in the group's favor. For example, the decision as to where to bring suit will determine which federal district judge will hear a case, and since judges have their own values, biases, and capabilities, which judge is deciding the issue will affect the probability of various outcomes. In the Tellico dam case, for example, the environmentalists tried to bring suit in Washington, D. C., and then Birmingham, Alabama, in order to avoid having their case heard by the federal judge in Knoxville, a courtroom that was located too firmly in the heart of TVA-land. In the spotted owl case, the Ninth Circuit Court of Appeals was seen as pro-environmental, and hence amenable to listening to and acting on pro-owl arguments. Indeed, the Ninth Circuit ruled that the limits on judicial review adopted by the Congress in the 1989 appropriations bill were unconstitutional, yet the Supreme Court overruled them.

In making a case for their positions, one of the more effective strategies in environmental and resource controversies today lies in having numbers that support your argument. Most environmental and resource management decisions are overlaid with a veneer of technical complexity, such that the debate often takes the form of a *battle over numbers*. Regulatory decisions such as how much wood smoke should wood-burning stoves be allowed to emit, and resource allocation decisions such as how much semi-primitive nonmotorized recreation should be allowed on the Huron-Manistee National Forest are supported by reams of paper reflecting the products of numerous environmental impact assessments, technical studies, FORPLAN runs, public comments, and the like. While the analyses no doubt contribute to the GNP by consuming paper and electricity to feed a battalion of copying machines, and keep analysts of a variety of kinds off welfare, it also means that groups seeking to influence the course of a decision must weigh in with their own data and technical experts. In many ways our democratic pluralist system has given way to a system of technical pluralism, where experts battle experts, and studies spawn countervailing studies. It is a process where technical experts and lawyers (since they orchestrate the interactions between experts) do well, and the rest of society foots the bill and watches slightly confused from the sidelines.

Given this system, interest groups work hard to acquire numbers that support their desired solution. In many ways, the environmental groups maintained themselves as an effective lobbying force through the 1980s, in spite of a downturn in overt political or administrative support for their causes, by acquiring good quality technical information and providing it strategically into the political process. Efforts by groups like The Wilderness Society, who built an economics and planning staff in the 1980s, allowed them to counter the technically based arguments of agencies like the FS and groups like the timber lobby.

In the spotted owl case, the numbers game was played and replayed over and over throughout the history of the dispute, with the battle for old growth acreage providing one of the best examples. As was noted above, the amount and location of old growth acreage was a critical question throughout the controversy. By being reluctant (or unable) to provide a definitive, credible answer to this question, the FS opened up a tactical possibility for the environmental groups. Mapping efforts by The Wilderness Society and the National Audubon Society were started in the 1980s to provide their answers to the question, and by the time the Congress started asking the question seriously, the environmental groups' figures were seen as at least as credible as those of the FS, even though they ranged from a FS estimate of 6.2 million acres to a TWS estimate of 2.4 million acres.

The strategic significance of these numbers can be seen in the groups' response as the question changed. If you want to make a case for owl protection, then you want to find the least amount of suitable habitat so that the situation looks particularly dire. At the same time, if you want to advocate the protection of old growth forest lands, you want to be as inclusive as possible in the lands that you target for protection. When the question on Capitol Hill changed from should the owl be protected to what lands should be set aside, the 2.4 million acres figure was a trap for the environmental groups that they then tried to get out of: "So, you believe 2 million acres is necessary for the owl; no problem, we can set that aside." "But wait a minute, it's not really 2 million acres that they need, but 6 million acres . . . "

A similar example can be found in the activities of the California timber industry, whose strategic response to the controversy was to try to prove that the owl was thriving on their lands, but whose efforts were rewarded with a FWS proposal to designate their lands as critical habitat. In 1989 and 1990, the industry funded inventory work on timber industry lands in northern California aimed at locating as many spotted owls as possible in order to prove that the owl did not warrant listing as a federally recognized threatened or endangered species. They found a fair number of owls, many on second growth industry lands, but the strategy failed them when the FWS proposed the areas they identified as critical habitat for the owl. "It just basically backfired on us," said Carl Ehlen, western resource manager for the Georgia-Pacific Corporation, which found about fifty owl nest sites on its lands and had 77,000 acres on the FWS proposed critical habitat list. "We're being penalized now for finding those owls," said Ehlen.[4] Indeed, of the 11.6 million acres proposed as critical habitat, the FWS included 1.4 million acres of private timberland in California, an action that would not have been possible absent the timber company data.

Given that much of the debate hinges on technical argument, it is not only studies that are needed. Studies must also have legitimacy, and interest groups and agencies work hard to *enhance the credibility of their arguments.* One of the easiest ways to do this is to acquire individuals who are perceived credible, and have them make the case for you. The National Audubon Society's efforts in 1985 to organize the Blue Ribbon Panel was an obvious attempt for those concerned with owl preservation to seize the higher ground of technical legitimacy. The Forest Service's recruitment of Jack Ward Thomas as the leader of the Interagency Scientific Committee was their attempt to put their most credible scientist on the line. Similarly the environmental groups flocked to the doorstep of University of Washington and Forest Service researcher Jerry Franklin and supported his efforts to expand his research program because he was seen on Capitol Hill as the guru of forest ecosystem science.

Another way to make a case for your position is to make the issue appear decided. A *fait accompli* strategy calls for groups to keep acting such that by the time a decision is made, its outcome is predetermined. This strategy is a time-honored approach favorable to agencies who want to keep acting the way they have been acting, and interest groups who are supported by status quo action. The Tellico dam–snail darter case provides one of the best examples of this approach, whereby the Tennessee Valley Authority went on a 24-hour construction schedule in the mid-1970s, bulldozing habitat used by the darter, when the darter was proposed for endangered species status. The fait accompli approach is used by interests who sense an inevitable change in direction coming their way, and work hard to maximize the benefits of the status quo before the curtain drops. In the owl case, cutting a lot of old growth timber, and, on some districts, cutting down nest sites in the 1970s or fragmenting owl habitat in the 1980s, could lead to a situation where the environmentalists might win the war, but the spoils/benefits would be nonexistent.

Given the fragmentation of political power inherent in our system, and the incentives facing decisionmakers, it is necessary for proponents of policy change to build political support for their interests, and that means aggregating political power in several ways. *Coalition formation* is a common strategic response that assumes that there is safety and strength in numbers. Forming coalitions between mildly homogeneous interests allows interest groups to pool resources and plot strategy together to increase the efficiency of their efforts, and to gain greater credibility in the political process. The environmental groups became very good at creating formal coalitions starting with the Alaska Coalition in the late 1970s, and continuing through the 1980s.

In the spotted owl case, the Ancient Forest Alliance was organized in December 1988 to represent a collection of 57 groups, focusing for various reasons on old growth forest preservation.[5] The environmental groups had to organize themselves collectively because the nationalization strategy of regional groups could only occur with the explicit involvement of the national groups. Yet even without any strategic motives in joining forces, coalition formation was necessary because once the focus was on Congressional-level negotiation, the groups had to respond to offers collectively. Division within the environmental community would diminish the impact of environmental concerns on the outcomes of the negotiations, and would confuse Congressional decisionmakers.

Coalition formation enables groups to enhance the efficiency and potency of their lobbying efforts by each picking a portion of the campaign to work on. In addition, since members of Congress respond to the appearance of large numbers of citizens backing a proposal, coalition formation allows lobbyists to claim a large number of supporters. The National Wildlife Federation is probably the best example of an organization that claims some six million members via a formalized coalition of affiliated state-level groups. Six million individuals do not pay dues to NWF; rather their membership in the Michigan United Conservation Clubs or the Oregon Wildlife Federation automatically adds them to the NWF body count.

Formal or informal coalitions also allow members to position themselves along an ideological or positional spectrum, such that the center point is legitimized in the direction of the collective groups' interests. Since pluralist decisionmaking seeks balance points in the middle of the spectrum of interests, having EarthFirst!, the Humane Society, or the National Rifle Association in the spectrum of interests pulls the balance point in their direction and legitimizes the more moderate groups. Some would say that such groups force action from an otherwise status-quo seeking system, because mainstream decisionmakers fear the implications of the more extreme groups' positions and hope to placate related groups by more moderate action. Regardless, the net effect of coalition formation is to enhance the impact of the affiliated groups on policy choices beyond the sum of their parts.

Coalition formation also has its liabilities for specific groups, because once you are a member of a coalition, your own ability to act is at least somewhat constrained. Once you are in bed with someone, their feelings and interests have to be taken into account in what you do jointly, or else the relationship is unlikely to last very long. At least, you have to spend a lot of time talking with one another to resolve differences. Members of the Ancient Forest Alliance, for example, spent thousands of dollars on telephone calls and fax charges, trying to resolve internal differences on interests and tactics.

Some of these intracoalition differences were hard to resolve. For example, the stressful negotiations in the appropriations battle of 1989 left the environmental community bitterly divided: the regional groups like

ONRC felt that the national groups had sold them out; and the national groups were split over how much compromise was necessary and appropriate. Individuals like the National Audubon Society's Brock Evans, who had always been viewed by Pacific Northwest environmentalists as capable of walking on water, lost considerable credibility, and had to spend time mending his relationships, so that at least his head was above water.

The fact that coalitions of interest groups are usually collections of diverse objectives, capabilities, and styles sometimes makes it difficult for effective negotiation to take place. One of the possible improvements to our normal decisionmaking process lies in the set of approaches to collaborative problem-solving called alternative dispute resolution. While Chapter 12 argues that such approaches should be utilized more often, issues of representation often arise because of these differences within the interest group community. Just who is an adequate and effective representative to a negotiation process from a set of groups sometimes is not easily determined. In addition, settlements to negotiations only work if the affected parties are bound in some way to the settlement. With diversity within a set of stakeholders, it is often hard to bind the parties to the settlement, and wildcard groups like Green World can upset the negotiations fairly easily.

One of the benefits of working in a political system that has a past is that relationships between various participants already exist prior to the birth of a new issue, and these *influence relationships* can be utilized to affect the destiny of the issue. Relationships exist among interest groups, agencies, and elected officials, and between individuals of all kinds, that provide a pathway through which influence can be placed on policy decisions. Calling an ex-Governor who may be supportive of your interests and asking him to speak on your behalf to a current Senator can be effective because of a past relationship that included the fact that the Governor campaigned for the Senator in his first election campaign. Such interactions are also productive because the ex-Governor's opinions may well serve as an effective proxy for public opinion as perceived by the Senator. Since the voting public's opinion on policy issues often is a mystery to public officials, the opinions of influential acquaintances are likely to be a good indication of the opinions of the subgroup of the public that is already supporting the elected official.

Dendritic patterns of influence flow through the political landscape, and interest groups seek out ways to tap into and influence these flows in order to pull policy decisions in their direction. In the spotted owl case, the longstanding relationship between the regional timber industry and

Senator Mark Hatfield is an example of such an influence relationship. Timber got favorable forest policy via the appropriations process, and the Senator got favorable press, campaign contributions, and the perception that he was doing what he should be doing for his constituents. A similar relationship existed between environmental groups and several influential members of Congress, including Bruce Vento, chairman of the House Interior Subcommittee on National Parks and Public Lands, and Sid Yates, chairman of the powerful House Appropriations Committee. Relationships between individuals inside and outside the government were also used to influence public choices.

While powerful acquaintances of elected officials can have a significant impact on the outcome of policy processes, the opinions of average citizens also have an effect, though indirectly. The nature of pluralist decisionmaking tends to diminish the impact of individuals on the process, and favors the activities of organized groups. Nevertheless, elected officials tend to view the opinions of an individual as those of a potential voter and look at them seriously. Hence mailbag counts that summarize the number of letters for and against an issue are taken seriously, particularly on controversial issues affecting a politician's home district. Hence interest groups work toward *grassroots mobilization* as a direct strategy of influence.

Besides dictating desired policy direction, the mobilization of public support for a position is also needed to allow an elected official to take action, since politicians are likely to be crucified for taking controversial positions that are unsupported back home. One of the major requirements of interest groups that are requesting favors is that support be mobilized before a decisionmaker has to act, and that such support be substantial and demonstrated through letters, civic group action, and local media. Feeling the heat of civic pressure is necessary before a politician can act with immunity from retribution and, just as important, can achieve political benefit in the actions taken.

Interest groups work to mobilize interest on the home front to allow politicians who are inclined to act on the group's interests to take visible action. Environmental and timber groups both worked in this way. Direct mail solicitations were aimed at raising money and awareness. "Already-written" post cards from concerned citizens for the couch potato set were included to increase agency or Congressional mailbag counts. For example, of the more than 40,000 comments received on the Forest Service's draft SEIS on owl management in 1986, some 61% were considered by the agency to be form letters (including clipped coupons) and discounted somewhat as an indication of public sentiment.[6] Indeed, the FS mailbag counts also noted letters that were handwritten but said the same thing in the same words. Some of these letters, by the way, were written by employees of wood products firms who were promised time off or other inducements for writing the letters. According to one such worker,

> I've worked for Roseburg Forest Products for over 20 years and every time the management thinks their profits are going to go down the tube because someone is trying to take their precious timber away from them, they get everyone employed by them to get on a band-wagon and write letters for their benefit, and the sheep just go for it, hook, line and sinker. . . . They have gone so far as to give green stamps to employees totaling $18.00 for writing letters to the Forest Service over this Spotted Owl business. . . . I think it's sad when a company this big has to stoop to bribery to get a letter written.[7]

Besides direct mail solicitations, another common mechanism used by interest groups to influence public opinion and mobilize public action is through a variety of *media strategies*. Such efforts are almost always aimed at influencing decisionmakers by swaying public opinion, directly mobilizing influential citizens, or influencing media editorials that are used by decisionmakers as another proxy of public opinion. In the owl case, interest groups employed media strategies by producing television programs such as that produced by National Audubon Society entitled, "Ancient Forests: Rage Over Trees." They also lobbied the editorial staffs on major newspapers, and seeded magazine reports by providing information and arranging tours of the Northwest that influenced the coverage in articles in *National Geographic,* and other national magazines, and resulted in the owl gracing the cover of *Time* magazine in June, 1990. Public service spots distributed free to both electronic and print media were provided by interest groups on both sides of the controversy. For example, NWF prepared a spot for Saturday morning children's programming that featured Kermit the Frog strolling (or hopping) through the forest and commenting that it was being cut down and should be protected. In Kermit's words, "It's not easy being green." Ribbit. Ribbit.

While the timber industry provided comparable information to media outlets noting that trees are a renewable resource, they also used mass demonstrations as a statement of public concern on the owl/old growth issue. Such demonstrations by and large are aimed at gaining media coverage for a group's position, and since the media is hungry for images of conflict and discontent, such strategies often work well at attracting attention. Logging truck caravans were organized by timber interests to generate media attention to their concerns. Logging trucks circled the U. S. Capitol in August, 1989, and blocked roads in Portland on the Friday evening before Easter weekend, 1990. While television covered this latter event well, it probably generated more negative public response as Portlanders fumed in their cars, and had a clear target to blame for their discomfort. While the loggers wanted that target to be the environmental groups, the mental leap from being stuck behind a logging truck to blaming it on environmental group action on the old growth issue was hard to make for motorists stuck with a car full of screaming kids on a Friday night trying to get to Aunt Edna's house for dinner.

Media strategies are based most effectively in enduring *symbols* of public concern, and interest groups work hard to cast their arguments in symbolic terms. While our society functions largely through economic and political decisions that are based explicitly on self-interest, making an argument for action on a public policy issue that is clearly self-interested is not the best approach. Grounding your arguments on symbols of God and motherhood and baseball and the joy of being free is much more compelling, and media-productive, than saying, "Look, we'll lose a bundle if this policy goes through." As noted above, the media thrives on simple symbols, and the essence of simplicity is to affiliate a cause with something that in the listener's mind is positive.

For the environmentalists, those symbols were furry animals needing protection from greedy developers; untouched, verdant, evergreen-scented forests threatened to become blackened, disrupted clearcuts smelling of smoke and diesel exhaust; and future generations and other lifeforms needing and morally entitled to a decent living environment. For the timber interests, those symbols were schoolchildren entitled to a decent future, neighborhoods whose social fabric would be ripped apart by uncaring urban, overly wealthy preservationists, and future-oriented economic growth through hard work and well-managed natural resources. Both sides sought to cast their arguments on these grounds, for no one is against furry animals, green forests, future generations, schoolchildren, neighborhoods, or a comfortable future standard of living. Both sides could play the symbol game, and while necessary and appropriate given the decisionmaking system under which we operate, such arguments often hide real concerns under the cloak of lofty rhetoric. They

also make decisionmaking difficult, particularly for politicians who will
be on the wrong side of a motherhood argument no matter which way
they go.

In the battle of the symbols, on one level, the environmental groups had
a much easier time of it than the timber groups. Since the base concern of
wood products and logging companies was economic self-interest and
not jobs or community structure, it was hard for them at times to
make motherhood arguments convincing to even mildly knowledgeable
listeners. Since what motivates the environmental groups is less clearly
economic self-interest, though other types of self-interest are most likely
part of their motive, it is easier for them to claim the moral high
ground.

On another level, though, the type of symbolic argument that timber
interests could make was more fundamental to human motivations and
potentially more mobilizing if the argument was made in a way that resulted in a gut level response from various listeners. Threats to jobs,
neighborhood, and lifestyle are threats to basic survival and as such are
lower on the sets of needs that motivate human behavior than the more
lofty images cast by environmentalists. Since human motivation experts
suggest that lower level needs have to be satisfied before humans are motivated by higher order needs, images of threats to basic survival needs can
outweigh more lofty goals. The problem that timber faced in casting its
images effectively, however, is that these needs must be personalized for
the individual listener. And the blue collar, natural resource-based lifestyle that was supposedly threatened by the owl was shared less by an
increasingly urban American population whose opinions counted as the
dispute moved into the national arena.

While groups used a variety of other strategic responses to the political
environment surrounding the owl dispute, a final one we should note is
the use of *compromise*. When it is clear that your side is not going to win
it all, a logical tactical response is to offer a compromise that will potentially bind the other side from further action. If they can get part of what
they want, and the ultimate decision is uncertain, it is possible that they
will buy into your compromise. In many ways, the decision by the Chief
of the Forest Service in his record of decision on the spotted owl SEIS in
December 1988 represented this kind of compromise, as it attempted to
split the difference between the two extreme possibilities promoted by
owl supporters and opponents. This compromise solution failed, however, because neither side (nor agency staff) was happy with it and each
felt that it could get more by acting in other decisionmaking arenas that

were open to them. A similar compromise strategy can be seen in the timber lobby's support of minimal ancient forest legislation in the 102nd Congress. While in previous years, timber's argument was that no additional set-asides were needed given the amount of acreage that had already been restricted from commercial timbering, by 1991, it was clear that additional restrictions would be put into place, and a strategy of supporting such restrictions in a way that would minimize impact on timber was a necessary step to stay in the inevitable political negotiations.

Indeed, casting a proposal as a compromise is also effective strategy, since our political process values compromise. The so-called compromise proposal presented at the 1989 timber summit in Oregon represented such a move. Whether it was a fair compromise or not, timber's support of the proposal ("though it hurts us a great deal"), and the environmental groups opposition to it, shifted power in timber's direction. The environmental groups were cast as extreme and unyielding. Since everyone loves a compromise, casting the environmental groups as extreme and unyielding tended to make them appear greedy and self-interested, regardless of the legitimacy of their position.

The Incentives of Interest Groups

The ways that interest groups participate in the policy process reflect their own needs to maintain themselves as viable organizations. While all organizations have to work at administering themselves, interest groups are unique in that they neither get tax revenues or statutory objectives as do public agencies, nor do they produce the same kind of product that for-profit corporations do. They exist to gain influence over the direction of public policy, and can do so only as long as they continue to be viable organizations. If nothing else, internal maintenance activities can take a lot of time. In addition, the need to raise revenues, set goals, and organize themselves means that internal maintenance issues often get tied up with external influence activities in ways that bias the groups' voice on an issue.

Clearly the choice of issues to champion is influenced by how marketable they are in the market of financial support, and the owl and ancient forest issues were the substance of direct mail fundraising campaigns by a number of groups including Audubon and The Wilderness Society. No matter how legitimate a concern, if an issue cannot be marketed to membership or contributors as worthy of organizational effort, it is unlikely to be a major lobbying priority. On one level, such a dynamic is appropriate, even effective, as it acts to gauge the level of public interest in potential interest group demands. But it also means that interest group activity may be conservative and status quo-oriented, particularly as groups get larger with significant fixed costs to cover. It also suggests that turf battles may be just as hard fought as we assume is the case with bureaucracies. Groups can raise money to support themselves because they have an identity that is partly driven by the issues they cover. If these identities blur in the eyes of potential contributors, the groups have a problem, and as a result, they work hard to differentiate their products by taking on different issues, and different perspectives on the same issues.

The difference in perspective between the regional and national environmental groups illustrates these kinds of concerns. In examining the history of the owl controversy, it is surprising how long it took before the national environmental groups jumped on the old growth/owl bandwagon. The larger groups were not major players in the controversy until 1987 or so. Michael McCloskey, Chairman of the Sierra Club, described the turning point for the national groups as follows: "We had a Group of 10 meeting in Portland about two years ago, and Jerry Franklin presented data on the extent of fragmentation that had already occurred. It became clear that we were moving rapidly from a series of clearcut patches as fragments within a sea of forest to a series of forested patches in a sea of clearcuts."[8] While the regional groups worked hard at nationalizing the issue, they also understood that the issue was not center stage on the national groups' agenda because it was perceived as a regional concern, and because the nationals felt it was not winnable given the strength of the political delegation from Oregon and Washington. The nationals did not want to divert resources from other campaigns, nor did they want to earn the enmity of sometime-supporters from the Pacific Northwest including Representative Les AuCoin and Senator Mark Hatfield.

Since interest groups are always fairly unstable because of uncertainties over resources and fairly high staff turnover, they are also vulnerable to attacks over issues of internal maintenance. For example, one significant resource available to some not-for-profit groups is their ability to provide a destination for tax-deductible charitable contributions. It is clear that the tax-deductibility of financial contributions is important to contributors. Groups organized as 501c3 entities under the federal tax code can provide tax-deductibility to their contributors, but cannot lobby for political action. They can provide "educational materials," however, which is a subtle form of lobbying via molding public perceptions of an issue. Many groups set up two theoretically separate but practically interconnected organizations that gives them the best of both advocacy and the ability to acquire charitable contributions. Hence, the Sierra Club is a 501c4 lobbying organization and the Sierra Club Foundation is a 501c3.

One result of these tax fundamentals is that 501c3 groups are always in danger of crossing the fuzzy lobbying/education line, and have to be careful in the issues that they take on. In the 1970s, for example, the Nixon administration attacked several 501c3 environmental organizations and tried to take away their tax-deductibility. Hitting at internal maintenance issues was also prevalent in the spotted owl case. A prominent example lies in the timber industry's response to the National Audubon Society television special, "Ancient Forests: The Rage Over Trees," that was underwritten by several companies including Stroh's Beer. Lobbyists from the timber industry pointed out to Peter Stroh that loggers and other woodsworkers drink a lot of beer, and should Stroh's underwrite the Audubon special, a beer boycott might result. Stroh's pulled out of the special, as did most of the other corporate sponsors, leaving Audubon holding the bag. The ultimate irony was that the controversy received considerable media attention and probably did more to publicize the show than anything else that Audubon could have done.

Science and Scientists in the Policy Process

Given a set of policy processes that exist largely to satisfy the perceived needs of a variety of human groups, and manage the conflict between them, science and scientists are anomalies. Scientists often see themselves as the purveyors of absolute truth, whose knowledge might be limited by available information, but are unbiased by the kinds of values often expressed in the political process. Political executives, on the other hand, believe that most truths are relative and dependent on the motives of the individuals and groups who bring them forward, and their strategic significance. In their view, the decisionmaking process works via persuasion, influence, and political compromise—activities seen by scientists as illegitimate and unprofessional. As a result, both groups are uncomfortable with the other.

At the same time, science and expertise are political symbols of significant power and influence in decisionmaking processes. This reality was one of the reasons that the environmental groups invested in expertise in the 1980s, enabling them to challenge the traditional sources of expertise, the agencies. In addition, one could almost feel the power of perceived legitimacy surrounding Jack Ward Thomas and Jerry Franklin as they testified on Capitol Hill. The fact that the lack of support of agency biologists for the FS Chief's 1988 decision on the SEIS analysis doomed its ultimate implementation also illustrates the power of expertise, as used by the interest groups.

Since science and scientists can be potent political tools, a dilemma is set up for both scientists and nonscientists. Scientists are pressed into advocacy by interest groups and the media—a role that is neither comfortable nor seen as legitimate by their professional peers. Legislators, political executives, and judges are forced to acquire understanding and make decisions on issues they have neither the training or inclination to work on. They are particularly confused by disagreements between scientists, where neither compromise nor adversarial argument—the basis of most legislative and judicial decisionmaking—is the most effective way to resolve differences between data or theory.

The norms of science also are confusing to decisionmakers: Being conservative about reaching conclusions based on data, the continuing need to undertake one more study to resolve additional areas of uncertainty, the slow deliberateness of the scientific method, and the fragmentation of information caused by the need to own and claim ideas through publication—all frustrate nonscientists, particularly when they are faced with demands for action. All of these norms might be appropriate ways to build scientific understanding, but they cause delays and frustration on the part of decisionmakers who would often be enthralled to have a technically credible solution presented to them, as long as it was within the range of political acceptability.

The set of myths that science and scientists are objective, and uninfluenced by other values they hold, while functional for science, is problematic for social decisionmaking. It often blinds scientists to bias in their analyses and their choice of information and research questions. It also can invalidate the science in the eyes of organizational and political managers, because as they become aware of the value-basis of scientific decisions, they tend to disregard the entire message. Hence, a scientist is branded as an environmentalist, or an industry sell-out, and the value of his or her research is minimized. In actual fact, their messages probably have some element of truth within them, but it is often difficult to sort out truth from bias, and the characteristics of our decisionmaking processes do not help us to do so very much.

The underlying characteristics of American decisionmaking processes and the incentives that they provide to participants help to explain why the owl controversy persisted. Some of these qualities and behaviors are inherent and unavoidable in the decisionmaking system we use. Other factors, though, could have been better managed to mitigate the effects of these processes. Although some areas of forest policy and the structure of the Forest Service itself were designed as ways around the failings of everyday policy processes, the manner in which the policies were implemented and Forest Service norms of behavior, as they evolved over the last eighty years, did as much to create a complex and difficult issue as they did to resolve it. As we shall explore in the next two chapters, neither forest policy nor the organizational characteristics of the FS helped us very much in the evolving controversy.

9

Insufficient Policies
and Misleading Politics

Given the set of political and decisionmaking realities described in the previous two chapters, natural resource managers face a difficult task in trying to carry out land management that is technically valid, socially appropriate, and cognizant of long-term needs and conditions. Yet the Forest Service and many aspects of forest policy were designed to get around some of the vagaries of our standard decisionmaking processes. American resource policy should be a shining star, not a black hole, of decisionmaking undertaken in our collective best interest.

The Forest Service was created in the Progressive era as an apolitical, science-based organization to avoid the perverse effects of politicization that haunted other areas of government activity in the late 1800s. Its task was simultaneously to maximize resource use while protecting the land base, and to do so while looking into the future as best as its staff could. Forest policy in the form of the National Forest Management Act (NFMA) took this orientation one step further by establishing planning processes that were explicitly long-term-oriented, cross-disciplinary, and open to public involvement. In theory at least, planning started with a blank slate on which to inscribe creative solutions in everyone's best interest.

The history of the spotted owl controversy suggests, however, that the Forest Service and federal forest policy have not lived up to these expectations. The concept of multiple use and the implementation of NFMA have been problematic. Both the agency and forest policy are in fact political, reflecting the influences of various Congressional representatives, who prescribe short-term priorities through the appropriations process, and the administration, which constrains the set of possible FS decisions about management of national forest system lands. Endangered species and wildlife policy are primarily reactive and species-oriented, causing problems for the way that other areas of resource policy are carried out. The net effect of these forces was to contribute to the evolution of the owl controversy, rather than facilitate its resolution. This chapter discusses aspects of forest and wildlife policy and administration politics that influenced the evolution of the owl controversy.

The Inherent Problems of Multiple Use

American forest policy in the twentieth century has been aimed squarely at promoting multiple uses of forest resources. *Multiple use,* as defined by the Multiple-Use Sustained-Yield Act of 1960, means "the management of all the various renewable surface resources of the national forests so that they are utilized in the combination that will best meet the needs of the American people."[1] While the set of approved uses has expanded significantly over the past ninety years, from an 1890s set of objectives consisting of timber management and watershed protection to a broad array of purposes in the 1980s, including timber production, grazing, recreation, watershed protection, wilderness, and fish and wildlife management and protection, the concept has remained largely the same. The purpose of the national forests is to maximize human use subject to the constraints of protecting the resource base to insure long-term productivity. Multiple use as a concept is well-ingrained in law, and in the FS as an organization. Scratch the agency and it bleeds multiple-use gray.

Multiple use as a concept of management works well as long as there are enough resources to go around, and uses are not mutually exclusive. But in a time of limited slack resources, when some goals are incompatible, implementing a policy of multiple use becomes difficult. At minimum, it requires that someone make tough choices about the future of specific parcels of land, choices that define winners and losers in the battle for limited forest resources. It also has the potential to create a state of public impotence where decisions do not hold, and the owl case was in exactly this state in the late 1980s and early 1990s.

The national forest system today is close to being appropriated fully to the diverse interests that have legitimized claims on its resources, a remarkable turn of events given the fact that very little demand was placed on the system's 191 million acres prior to the end of the second world war. Up until the 1970s, demands for a variety of national forest "products" were manageable, and could be provided within the constraints of the resource base. Economic demands and political demands were generally in tune with one another, and not so high that those who sought goods and services from the national forests could not be provided with them. Wildlife and timber interests were mildly complementary, since the kind of wildlife sought on the national forests benefited from the openings in the forest canopy created when large trees were cut down. Wilderness interests could be satisfied by setting aside some of the less productive, higher elevation forest lands. There were controversies over the future of the national forests in the 1960s and 1970s, but the demands that they represented did not overwhelm the capacity of the FS to meet them.

In the 1980s, much of the slack in available resources was used up at the same time that increasing demands were being placed on the national forest system. Multiple use means multiple constituencies, and the FS embraced a host of client groups in the 1980s in a noble, but ultimately confounding, process of diversification, as it became unable to meet all of the interests on the table. Multiple constituencies means multiple vested interests, and even when economic demands for forest resources were not increasing rapidly (as was true in the early 1980s recession), political demands for these resources were. Forest resource clients—new and old— brought their claims to an embattled FS and asked for their share of the multiple use contract.

Not only was there a larger set of demands placed on a finite resource, but the type of demands were increasingly incompatible. The newer sets of demands—requests for old growth set-asides, wilderness reservations in mid-elevation forests, and habitat for sensitive wildlife—represented uses of forest lands that were incompatible with timber and deer production. To the extent that ecosystem preservation required the protection of large contiguous chunks of landscape or species needing protection had large habitat requirements, a dilemma was set up for FS managers. They had promised a host of interests that their concerns would be addressed through forest planning, but increasingly appeared unable to satisfy all demands.

Even if the multiple demands on a national forest's resources could be accommodated simultaneously, the pervasive lack of trust in the FS's willingness or ability to implement policy in accord with the interests of stakeholders led groups with demands to seek single purpose set-asides. Environmental groups did not trust the agency to exercise discretion in accord with their interests. New Forestry might be able to provide spotted owls in a managed landscape, but the environmental groups would never trust agency officials enough to give them discretionary control over old growth lands.[2] The environmental groups wanted explicit single-purpose reserves well-defined by boundary lines. In a somewhat similar vein, timber interests would not accept temporary land set-asides while research explored options, because they did not trust the political and administrative processes to ever open them up to timber production. In their view, any land set-asides would become permanently locked-up reserves. Like environmental groups, timber would have been happiest with the designation of dominant-purpose timber production lands, defined by clear boundaries for an indefinite period of time.

Multiple use management might be able to generate "right" answers to allocation problems when the metric used to measure the benefits and costs of alternative products is the same, but when the goods and services sought are incommensurate in value, there are no absolutely correct solutions. Resource economists have tried to quantify all the many demands placed on a national forest so that objective functions can be framed and run to identify optimal allocations of forest resources. Unfortunately, these efforts do not capture the values that are expressed in the political marketplace in which decisions are actually made. As a result, the use of multiple use-based models like FORPLAN to define effective outcomes is doomed to failure.

Some argue that multiple use is not inherently flawed; rather, the FS did not provide a good test of the concept in the way the national forests were managed in the past few decades. Presumably multiple use means maximizing all values produced from national forest lands, which yields an objective function of maximizing all resources subject to the constraints of the land base. Many critics of the agency argue that FS managers maximized timber outputs subject to minimum noncommodity constraints, an equation that inherently biases the resulting allocation. Obviously, evaluating an equation that maximized wildlife habitat and recreation outputs, subject to minimum timber harvest quotas would result in a completely different allocation—one that is equally biased in a different direction. Multiple use may have its problems, but the way that the FS implemented it certainly exacerbated them.

The agency also tended to plan for multiple use production at a fairly small geographic scale, maximizing multiple resource production from each chunk of ground. As a result, uses that exclude other uses tended to be devalued and driven out of forest planning equations. The Multiple-Use Sustained-Yield Act does not require that this happen. It clearly states that multiple use means that "some land will be used for less than all of the resources" and that it does "not necessarily [mean] the combination of uses that will give the greatest dollar return or the greatest unit output."[3] Given the broad set of demands on national forest resources today, it might be the case that true multiple use is possible only when considering the portfolio of forests as a total system, rather than on an individual unit by unit basis. Yet in spite of the national and regional trappings of RPA, decisionmaking on a forest-by-forest basis tends to preclude this type of multiple value planning.

The Unintended Effects of NFMA and Forest Planning

Without judging the merits of the final plans, the existence of NFMA, and the planning process it begat, were not positive forces in the owl controversy. Often, good policy decisions of the past create unanticipated policy problems in the future, and that was the case with NFMA. Since NFMA was enacted partly in response to the clearcutting controversies of the 1960s and early 1970s, the regulations produced to implement the law limited the size of clearcuts and, in the process, promoted forest fragmentation. While forty-acre clearcuts may have been a good way to minimize the visual impacts of timber harvest activities, by creating a patchwork of old growth fragments interspersed with clearcuts and second growth, interior-dwelling, wide-ranging species like the spotted owl lost out as the suitability of their habitat declined. From the perspective of the owls, we would have been better off with large clearcuts leaving larger contiguous areas of old-growth. In this case, solutions to the clearcutting issue in the 1970s created a different sort of problem in the 1980s.

The existence of NFMA planning also let the FS avoid having to make real choices on the spotted owl issue. The consistent response of the agency to demands that it take action on owl habitat protection was that such choices would be made within the NFMA planning process. Some saw the response as tactical: Forest planning was a process that the FS controlled, and putting decisions on owl management into the hopper let the agency maintain control over the character and process of resultant choices. It also let the FS put off making politically difficult choices until planning had run its course, and since thirteen or so forests would be making choices on owl protection, none would be ultimately and individually accountable for the fate of the subspecies. To those most skeptical of FS intent, putting off owl protection while forest planning took place was a way of delaying the ultimate outcome while old-growth was being logged out.

Observers who view the actions of the FS more benignly saw the agency's persistent attempts to place owl decisionmaking within the context of NFMA planning as simply the actions of an agency trying to force a changing situation into normal operating procedures. FS leadership had embarked on forest planning as a way of getting out of endless controversy in a controlled, technically legitimate manner. For some, the goal of getting the plans out became as important as getting the cut out, and the owl issue got in the way. FS staff members who viewed the allowable cut levels as unsustainable and environmentally damaging saw the forest planning process as a way out of the overwhelming dominance of agency politics by timber interests. If you cannot say to Mark Hatfield that "we're overcutting and should not get the cut out," perhaps you could say, "the forest plans, prepared with extensive public involvement in the best democratic tradition, and considering current science and technique, call for a lower cut level."

Regardless of motive, the forest plans became an end in and of themselves. The "Can Do" image of the agency reinforced this pattern of behavior. Congress had mandated forest plans, and they were the way that the FS was going to get forestry out of the courtroom and back to the forests. At times in the planning process, there was the sense that agency officials were saying, "By God, we're going to get these plans done, even if it kills us in the process." Sometimes goals, once established, become the totality of an individual's existence, even if they become counterproductive. It becomes easier to continue to slog sullenly toward the goal than go through the uncomfortable process of reexamination and reevaluation. In the case of forest planning, the FS had a nondiscretionary duty to prepare forest plans, but forest planning as an end tended to override other important choices, including those involving owl protection.

The interplay between the owl issue and forest planning had several major negative effects on forest planning and owl protection. Final forest plans were delayed by the changing requirements and public controversy relating to owl protection. Forest planners had to deal with incredibly obtuse guidance provided by the Regional Guide. The correspondence back and forth between the regional office and the forests described in Chapter 3 suggests the state of confusion at the forest level in response to pseudo-policy direction provided by the regional and Washington offices.

Even when they figured out what guidance was in effect, forest planners had to deal with changing constraints on what they could do in their forest plans. It is hard enough in forest planning to hit a stable target, but moving targets, cloaked in a shroud of politically sensitive fog, become very difficult to discern and aim at. Agency leadership understood some of the effects of changing policy guidelines. For example, in passing along the minimum management requirements in 1982, Al Lampi, Director of Land Management Planning, said: "We realize that many of you are a long way down the road in your planning and that changes in direction will have a heavy impact on you in time and money."[4]

The result was delayed final plans, and the forest plans in Region 6 became some of the most tardy in the national forest system. Delay in and of itself is not necessarily bad. In this case, however, plans that took five or ten years to make meant that, when finally produced, the plans were often based on out-of-date information even though they left a trail of burned-out planners, analysts, and ID team members in their wake. Delayed plans also meant reduced options for owl protection, as extensive timber sale programs ground on, guided by the older generation of unit plans.

In spite of extensive formal public involvement procedures, forest planning also was not very effective at offsetting public criticism of FS decisionmaking or building ownership in the final plans on the part of affected nongovernmental groups,[5] and the way the agency dealt with the owl case exacerbated this situation. To individuals and groups outside the agency, owl decisionmaking looked like, felt like, and smelled like a shell game. Prior to the SEIS process, when groups tried to provide input into FS decisionmaking on owl protection, they were always in the wrong place. Input on the agency's treatment of the OESTF's spotted owl management plan was set aside because owl management was to be analyzed in the Regional Guide. Criticism of the Regional Guide was set aside because the issue was to be analyzed in the forest plans. At the forest plan level, regional guidance on owls was treated as a final constraint. An early letter from the Lane County Audubon Society to the supervisor of the Willamette National Forest indicates the frustration surrounding the owl shell game:

> The Chief of the Forest Service rejected our appeal of the Oregon Interagency Spotted Owl Plan on the grounds that the subject would be thoroughly treated in the Regional Plan. The Draft Regional Plan (at p. 45) kicks the responsibility for number of owl pairs, location and distribution to the Forest Plan level. At the Forest level, the genetic pooling, distribution over historic range and general viability of the owl population cannot be addressed. . . .
>
>   The decision to permit protection only for the minimum 100 pair allotted by the Interagency Endangered Species Task Force . . . makes

it unlikely that the final WNF plan can have an increased number of owl pairs protected. . . . WNF has turned a minimum allotment into a maximum apportionment, without public input, and without (so far as I know) IESTF approval. I could understand your not realizing that you should invite the public to participate from the beginning. Oregonians are accustomed now to State land use planning processes which mandate involving them from the beginning. But, when I called, and specifically asked to be included, you told me that the "shouting was over."[6]

Regardless of the way that NFMA was implemented by the FS and how its implementation affected the evolution of the owl controversy, to some critics of the agency, the fundamental flaw was NFMA itself. According to Andy Kerr, ONRC staff member since the 1970s:

In the Seventies, there was concern about clearcuts, there was concern about the industrialization of the national forests, but Congress was not willing to bite the bullet and say "we can't have everything in our national forests." . . . Up until then, the Forest Service always had back country over the next ridge, and you always had another ridge over here to cut. It started to reach those limits, and Congress, rather than saying, "We've got to decide what the hell the national forests are going to be for," said, "Forest Service, go out and do the most elaborate planning process ever envisioned." And the Forest Service was put in the position of trying to satisfy all these competing interests. The agency needed some basic mandate from the Congress of what the hell it's supposed to be doing.

So the Forest Service's solution was, give us more money and we'll build you more trails, we'll build you more bird boxes, we'll repair more streams that we've messed up by our logging. It just goes on and on and on. So we're reaching those limits now. The owl is but one example of that. I think Congress lateralled—punted—in '76 and didn't bite the bullet and say what our national forests are going to be for. And they handed this thing off to the Forest Service, and the Forest Service did an impossible job, and did it not very well.[7]

The Power of Appropriations

The Forest Service, forest policy, and NFMA were designed to insure that forest management was insulated from politics and driven by scientific analysis harnessed for the public interest, yet politics and political activity

continued. The real power in government is the power of the purse, and
the federal budget process bubbled along quite separately from the implementation of substantive law such as NFMA. The connection between
forest plans and their implementation through the budgeting process was
never clear, but in the spotted owl case, the isolation of the two from each
other was remarkably clear.

The Congressional appropriations committees are the real source of
major direction in the national forests. They establish the Allowable Sale
Quantity in a given fiscal year for the national forests, and the committees'
decisions are based on politics, not rational, administrative analysis. Timber interests were well represented on these committees throughout the
life of the spotted owl controversy, with Mark Hatfield, referred to by
many preservationists as timber's "patron saint," as ranking minority
member on the Senate Appropriations Committee. Since Appropriations
makes things happen on the ground in Congressional districts (dams and
post offices get built; military bases remain open), the Committee's leadership is quite powerful, even in the face of conflicts with committees who
have authorized substantive law. It takes a lot of guts to oppose the leadership of the appropriations committees, and hence it is rarely done. The
economic interests and political leaders in the Northwest wanted a high
level of timber supplied from federal lands, the Congressional delegation
carried these interests into the budgeting process, and the Appropriations
process funded a large timber sale program year after year, regardless of
whether the magnitude of the program conflicted with other law, such as
that contained in NFMA or the ESA.

Nor did the Forest Service complain very much. Large timber sale programs meant large agency budgets, and the agency had significant infrastructure established in timber planning and preparation in the Pacific
Northwest. Besides, even if the FS leadership was concerned about the
magnitude of the Appropriations-defined timber program, it is not an easy
matter for the Chief to explain that concern to a Senate committee bent
on having a large sale program. "What, you cannot do it? Well, we will
get someone who can."

As a result of this separation between ostensible and real decisionmaking processes, environmental and citizen groups were heavily involved in
forest planning activities at the forest level, while the real choices were
being made in Washington. The more cynical observers of forest policy
viewed forest planning as a tool to placate and sidetrack public opposition, while real direction got set elsewhere. Regardless of motive, the
reality of this split led to a lot of frustration within the interest group

community, and to a strong sense (right or wrong) that the agency and its Congressional patrons were not bargaining in good faith.

The Politicization of Scientific Decisionmaking

Just as the owl's biological needs and the management implications of those needs were becoming well defined, the political landscape was changing in major ways. The spotted owl issue came into full bloom at the same time that the so-called Reagan Revolution took place. President Ronald Reagan took office early in 1981, bringing with him a set of policies and administrative appointments that had a significant impact on the course of the owl dispute. The pro-development direction they set, budget cuts they put into place, and their implied definition of what was politically feasible took decisionmaking on the owl case in a direction that did not facilitate its resolution. Indeed, while the owl case would have been difficult no matter what, the executive-level politics of the 1980s had a major influence on the course of events and deserve at least some of the blame for the failures of decisionmaking institutions.

The Reagan administration initiated a set of policies and appointments aimed at cutting the size and cost of government, providing less restrictive environmental regulation, and promoting resource development of all kinds, including the expanded production of natural resource-based commodities. All of these policies spelled trouble for those interested in owl preservation. Reagan administration appointees pursued their agenda with a vengeance, viewing their task as an ideological mission to reform government and foster the growth of the private sector. In their view, a robust private sector unfettered by the constraints of government regulation, yet supported by the resources controlled by government, was the best way to a healthy future. Economic growth would buoy all elements of society.

In the Department of the Interior, an agency that had the conflicting purposes of natural resource development and protection at its core, development interests were well served by the appointment of James Watt as Secretary. Watt, a Colorado attorney who had been president of the Mountain States Legal Foundation, an industry interest group that advocated greater energy exploration on public lands, pursued the Reagan agenda loudly, assisted by his agency heads, Bob Burford at BLM and Frank Dunkle at Fish and Wildlife Service. At the Department of Agriculture, John Crowell, former counsel for Louisiana-Pacific Corporation, one of the largest purchasers of federal timber, was appointed Assistant Secretary over the Forest Service.

While less dogmatic about their agenda, Bush administration appointees generally continued the policy themes of the Reagan years, with a little more concern toward finding middle ground among warring interests. While the appointment of Bill Reilly as EPA Administrator was seen as an attempt to satisfy environmental interests, appointments at Interior and Agriculture were slow to be made, and generally in line with the Reagan tradition. Manuel Lujan, a Congressman from New Mexico, was appointed Secretary of the Interior, and promptly suggested that economic impact be considered before deciding whether to protect endangered species like the spotted owl.[8] At Agriculture, the administration tried to appoint James Cason, a former real estate developer and Reagan appointee in the Interior Department, as Assistant Secretary over the Forest Service, but was rebuffed by opposition in the Congress organized by the environmental groups. The appointment of John Beuter, a timber industry consulting economist, as Deputy Assistant Secretary, was an appointment aimed at promoting timber production on national forest lands.

Administration policies and appointees had a significant effect on the federal land management agencies. When faced with budget cuts, the agencies had to scramble simply to keep up, and the changes in the early 1980s forced forests to cut back services and reallocate staff in a serious way. While some agencies became more creative when faced with budget cuts, most retrenched into core objectives and styles. For the Forest Service, that core was timber, which was also the only resource on the national forests that generated significant revenues for the federal treasury. In constant dollars, the Forest Service overall budget declined from $2.3 billion in 1980 to $1.7 billion in 1985 (the 1985 budget was $2.3 billion unadjusted for inflation.)[9] During this same period of time, the budget for minerals and timber increased ahead of inflation, while funds allocated for recreation, fish and wildlife, range, and soil, water, and air declined. For example, the timber budget rose 7% from 1980 to 1985 (in 1980 dollars), while the fish and wildlife budget declined nine percent.

The Reagan budgets had serious implications for the growing owl issue. Think about it: Here we have FS line officers trying to maintain normal services on a declining budget, while dealing with the burdens of NFMA, and along comes an issue that will complicate their lives. At minimum, the owl issue required labor-intense inventory work and small old growth set-asides. At maximum, it would be a major complication in the lives of everyone on the forest, staff members who were working hard just to tread water. One of the more common means to expand staffing in the FS was to employ seasonal workers, individuals who could be used to do owl and old growth inventory work, but since their positions were the most liquid in the organization, they were the first to lose their jobs given smaller budgets. Coupled with the economic recession in 1982, the budget cutbacks took away slack resources that might otherwise have been available. Organizational risk-taking on issues like owl protection was less likely.

Owl protection was also clearly in opposition to the direction of the Reagan and Bush administration policies, for in no way could the argument be made that setting aside old growth for owls increased regional economic growth or commodity production from the national forests. And the owl issue was seen as a prime example of misguided environmental policy, one that had been foreshadowed by the Tellico Dam case, but was several orders of magnitude worse: "Come on. Are you kidding? Shut down timber mills to save owls?" To political appointees operating with a different value base, it was hard to understand why anyone could be serious about making this kind of trade-off.

The differences between administration policy and owl preservation needs had several effects on what actually happened in the development of the owl controversy. First of all, there was a decided lack of leadership in trying to forge any kind of scientifically credible, economically sensitive compromise in the early 1980s, and a baffling lack of executive leadership on any domestic policy issue early in the Bush administration. 1989 and 1990 were important years in the development of the owl issue, but the Bush administration was unable to put political appointees into place that could facilitate a resolution one way or the other.[10]

When there was leadership from the top on the issue, it led in a direction opposed to owl protection regardless of scientific opinion. There was evidence of undue politicization of administrative processes throughout the 1980s and early 1990s, with political leaders pushing hard against choices that favored owl protection. John Crowell pressed the FS to minimize the economic effects of owl management plans. James Cason and Frank Dunkle pressured FWS leadership to avoid a decision to list the owl as a threatened species, in spite of the ESA's mandate that listing decisions be based on biological evidence only.[11] There are also indications that the White House in 1990 and 1991, led by Chief of Staff John Sununu, was trying to put fuel on the fires surrounding the issue, and limit the effectiveness of agency actions to deal with it. They hoped that the continuing conflagration would engulf the ESA itself, so that when the Act was up for reauthorization, there would be pressure to reduce the amount of protection it provided to entities like the spotted owl.

While politics is always a component of executive-level decisionmaking, the politicization of decisionmaking around the ESA during the Reagan and Bush years was unparalleled. Administration actions reflected the magnitude of the economic and political stakes involved in the outcome of the conflict, and an approach to governance that viewed all choices as short-term-oriented and political, downplaying their significance as statements of rights and future direction. Actions taken to influence the decision of the Endangered Species Committee in 1992 on whether to grant an exemption to the BLM from the provisions of the ESA illustrate the undue politicization of scientific choices during the Bush administration. From several accounts of the deliberations of the God Squad, few members of the committee felt that the criteria for an exemption had been met. It is hard to argue that a handful of BLM timber sales could be considered of regional or national significance, or that there were no reasonable or prudent alternatives. Nor was there the sense that the agency had consulted in good faith, given the speed with which the BLM moved from receiving a jeopardy opinion from the FWS to applying for an exemption.

The committee's final choice expressed the political values of a set of top-level political appointees, acting under clear direction from the White House. Indeed, the key swing vote was held by Dr. John Knauss, Administrator of the National Oceanic and Atmospheric Administration, and it was clear that his job was on the line.[12] His vote was secured at some long-term cost, however. He agreed to support the exemption if a provision was included that the BLM prepare a 10-year timber plan that was based on the recovery plan for the spotted owl. Included in the final exemption, his amendment may ultimately turn out to be a "poison pill" for the BLM, and its political constituents.[13] They got their politically motivated exemption, but at the cost of direction to do better in the future. Nevertheless, the fact that the Bush administration used the God Squad process as a political tool to break the logjam in the Northwest—the first time the process would grant an exemption in the twenty-year history of the Act—demonstrated the level of politicization of scientific decisions that existed in the 1980s and early 1990s.

Even when this activity was motivated by ignorance, rather than by adherence to any grand scheme, such as in the bumbling comments of Secretary of the Interior Manuel Lujan about the Endangered Species Act, the image created at the top was one of bemused lack of concern about the owl, the requirements of the ESA, and the responsibilities of administrative officials in environmental policy issues. The result of the direction set at the top on the spotted owl issue was to squander the opportunity for executive leadership to help us all out of the state of public impotence into which decisionmaking had descended. Given the political forces involved, the only way out of the morass was to generate a scientifically credible, impact-sensitive compromise forged in response to strong incentives to do so, incentives that could only be provided by Congressional or presidential leadership. We were not well served by the administration's response to the issue; ironically, neither were they.

The Reagan and Bush administrations' reactions to the owl issue also had a subtle effect on individuals who were well-intentioned about owl protection, but sensitive to the political realities of the matter. At any point in time, for any issue, there is an envelope of political possibilities. Called "political feasibility" or the "decisionmaking space," the envelope of possibilities is both real and perceived. It is true that, at a given point in time, prevailing political and institutional interests will allow only certain issues to be addressed or decisions to be made. Hence, federal policy to desegregate schools was impossible in the 1950s because the southern-dominated U. S. Senate would not allow it. But no one knows absolutely what is possible. The envelope of possibilities is as much a collective perception as it is anything more tangible. What is thought to be feasible, by definition, tends to be what is feasible, and the political dynamic in Washington, D.C.—of not wanting to be on the losing side of an issue, or worse, of being perceived as politically naive—tends to reinforce the boundaries of this perceived envelope of possibilities.

In the owl case, the politics of the administration tended to define the perceptions of FS leadership as to what was possible, and as a result, limited how aggressively they pushed the owl issue. If the FS Chief had told Senator Hatfield that the Congress was asking for too much timber out of the national forests, no one knows whether he would have been out marking trees in Avery, Idaho the next week. No one truly knows whether a decision to support 1000 acres of old growth for 1000 owl pairs would have held in 1985. No one really knows what would have happened if the agency chose the Audubon Society's proposed plan for owl protection in the SEIS. The only way we know is for an action to be attempted, and response elicited. In Washington, D. C., the normal method is for someone to send up a trial balloon, via leaks or lower level officials taking action. Otherwise, the envelope of political possibilities is only as wide as

potential advocates of change see it, and the FS leadership saw only limited possibilities for a favorable response to the owl issue. Since this perception was consistent with what many of the agency leaders thought
appropriate themselves, the potential for proposals that would push out
the envelope of possibilities was reduced significantly.

The overall effect of the Reagan and Bush administrations' direction
during the 1980s and early 1990s was to mislead the land management
agencies about public values. As described in Chapter 1, there had been
an evolving set of values in public natural resources throughout the twentieth century, and by 1980, this set was extremely diverse and included
a host of values about commodities and noncommodities. The Reagan
Revolution espoused a set of values that moved against this grain, claiming that it had a mandate from the people. While the electorate was concerned about the size and role of government, and their own economic
health, all indications are that they were also concerned about environmental quality and increasingly interested in the use of public lands for
noncommodity purposes. The Reagan Revolution said this was not so,
and they were wrong. But the net effect was to take the country's top level
political and agency leadership out of the loop for facilitating an effective
resolution of the owl controversy. While the FS's conservative response to
the issue partly represented the agency's own enduring characteristics, the
political environment in which it operated constrained effective action
as well.

The Endangered Species Act: Necessary but Not Sufficient

The U. S. Endangered Species Act is a powerful piece of legislation, and
the spotted owl controversy attests to its ability to disrupt the status quo,
but it was intended to be a last chance, stop gap approach to keeping
species from the abyss of extinction, and not a significant means of maintaining ecosystem health. The Act only deals with species that belong in
global intensive care, not declining species or their associated habitat. If
a part of our national objectives is to maintain life support systems by
maintaining environmental quality, the ESA's approach falls short. By the
time a species attains endangered or threatened status, it is often too late
to work backwards and repair the ecosystem damage.

The ESA was framed the way it was because species-level diversity was
what was understood to be important in protecting biological diversity at
the time. In addition, many of the threats that were understood in the
1960s and early 1970s were threats to a specific species, such as overuse

for sporting and commercial purposes, and not threats to a broad ecosystem. Besides, the species-approach probably was necessary in order to regulate inappropriate behavior. It was difficult to define what an ecosystem was, let alone regulate threats to it on a case by case basis. Politically, images of large furry animals or majestic birds drawing their last few breaths were mobilizing, while ecosystems were not.

The effect of this approach has been to make efforts to manage for sensitive species on public lands very reactive. A species does not receive official recognition of its sensitive status until it is at the edge of the abyss. At that point, fewer management options are left, and a lot of cumbersome procedures go into effect. The FWS's involvement brings with it all the normal problems attending interagency coordination and cooperation.

In many ways, the owl case was dealt with better than the average species. Most species are listed when their population is in the hundreds of individuals, not the thousands as was the case with the spotted owl. Most species that have been listed have little habitat left, while the owl still had millions of acres. Indeed, unlike many of the species that have received federal approval as endangered or threatened, the owl is threatened only in a dynamic sense: If we look at the population dynamics today, it is not in terrible shape. Only because the trend line of available habitat is what it is, does the owl warrant federal protection. Finally, an extraordinary amount of effort went into owl research and protection prior to listing, again an unusual situation when compared against other species that have made it onto the list. The fear of listing provides an incentive for an agency like the FS to do what it can to protect a species like the owl, in order to forego a decision to list the species and suffer the inevitable hassle.

While there are some positive aspects of the history of the owl case as it relates to the effectiveness of the ESA, the approach set up in the act is still inadequate. Up until the point where a species is listed in a final rulemaking by the FWS (or NMFS), it has no real protection under the law. Agencies must consult with the FWS under the provisions of the Fish and Wildlife Coordination Act, must take note of state-listed species under the Sikes Act and the Coastal Zone Management Act, and may have a vague mandate to protect fisheries and wildlife in their organic statutes (such as the NFMA). But as seen in the evolution of the owl controversy, when the needs of the species conflicts with the core values of the development agencies, it is hard for them to bite the bullet and forego direction for a not-quite-endangered species. The political environment in which science-based decisions are made reinforces and rewards this pattern of behavior. While the ESA provides incentives to act if the threat of listing is real, it also creates the perverse incentive not to act to learn more about the potential endangeredness of an organism, as long as the agency can get away with it.

Ecosystem care and protection is even more problematic. The ESA was framed with the scientific and political realities of the late 1960s and early 1970s, both of which changed significantly in the late 1970s and 1980s. As more became known about ecosystems, and environmental groups became more familiar with ecosystem concepts, there has been a rising interest in protecting ecosystem dynamics as a means of guarding against the loss of biological diversity. While the purposes section of the ESA references an overall objective of conserving the ecosystems upon which endangered or threatened species depend, the rest of the act is grounded in a species-approach, an important program that will always be reactive.

The only way to have a proactive program for sensitive species management is to focus more completely on the ecosystem level, and Chapter 13 offers some suggestions to this effect. Healthy ecosystems insure the protection of species-level diversity, while healthy species may not insure the protection of other needed aspects of biological diversity. There is so much that we do not know, particularly at the invertebrate level, that only in protecting systems of interacting organisms are we sure to protect what might be there. We can probably grow owls through captive propagation and inter-habitat transfers of genetic stock if that is the primary concern, but in doing so, much of the ultimate objective in having an endangered species law is lost.

Other Problematic Aspects of Forest Policy

The development of the spotted owl controversy was also influenced by the use of several concepts of management that are appropriate and legitimate, but also problematic as they interface with the policy arena. The Management Indicator Species (MIS) concept was one of these. Since most domestic species are threatened largely by habitat loss, their status is a reasonably good indicator of the health of the ecosystem of which they are a part. Faced with the need to act and a lot of uncertainty about ecosystems, the use of MIS's as a proxy for ecosystem understanding seems reasonable for planning purposes, as long as it is remembered that the status of indicator species is to signify (and operationalize) the health of the broader ecosystem. Yet sometimes the underlying reasons and assumptions for making choices are lost in the dust of decisionmaking, and policies are implemented in a way counterproductive to the original reasons for their creation.

Definition of the spotted owl as an MIS focused activities on owl inventory and management, as it should, yet tended to dilute a necessary concern with old growth ecosystem protection. This was particularly true as the issue rose in the hierarchy of the decisionmaking process. The owl was the focus of administrative activities, reinforced by the work on the SEIS, and everyone including the media tended to lose sight of the underlying ecosystem questions. As solutions began to emerge in the form of alternative management plans, they were solutions aimed at owl conservation, which would not necessarily protect the other species of the old growth system for whom the owl was to speak. The cookie-cutter pattern of dispersing owl habitat areas evenly across the landscape separated by 12-mile dispersal distances may have worked for the owl, but was unlikely to work for species unable to move across 12 miles of clearcuts. The indicator species concept tends to focus decisionmaking excessively on solving the problem of the species, not the concerns associated with ecosystem protection. In this instance, industry proposals to transport owls across habitat barriers and breed them in captive populations might have worked to maintain the owl populations, but would have been completely inadequate in protecting other old growth-dependent organisms.

Similarly, the Minimum Viable Population (MVP) concept presented problems in the way it interfaced with the motives and needs of decisionmakers. Basing scientific judgments in analyses that are grounded in concepts of risk and long-term viability and presenting them as a series of possible choices to decisionmakers are laudable improvements over experts emerging from smoke-filled rooms with the answer, like so many moths emerging from the cocoon of expertise.[14] But it runs into problems as it hits the decisionmaking process.

When faced with the need to make a choice, we all seek valid focal points in the sea of possible decisions, and usually choose the alternative that produces a maximum of what we would prefer, yet is minimally legitimate or satisfying to other concerns. For the owl, many decisionmakers believed that the population size and pattern that was minimally viable represented that alternative. As a result, throughout the controversy, the concept of maintaining a viable population was translated into the concept of maintaining a *minimally* viable population.

Since numerous choices had to be made on population parameters, the focus on minima tended to force each choice to the minimum legitimate point. That led researchers to recommend the minimum acreage of old growth that appeared to support owls, and managers to adopt the maximum dispersal distances between these acreages. Both agreed to population numbers that represented the minimum necessary to avoid adverse effects from inbreeding. Even if each of these choices were adequate for owl protection, when all the factors were put together, the ultimate outcome was probably not minimally sufficient for owl protection, let alone optimal to insure survival in the face of random events like fires, windthrow, or volcanic eruptions.

In the real world of diverse motives and objectives, the manner in which alternatives and their effects on survival probabilities are detailed in most minimum viable population studies ("viability analysis") also creates problems for organizations and politicians. As the shroud of absolute scientific judgment is removed, decisionmaking assumptions are revealed that organizations might prefer remain obscure. Elected officials have a similar problem when confronted with a range of choices: Politicians do not want to make choices, for that leaves them vulnerable to attack by those who lose out. In many ways, they prefer singular outcomes from agencies underlain by the force of expertise, in which the agency's recommendation moves in the direction they would like to see it move. Courts are not much better: They are ill-equipped to sort out opposing technical arguments played out on the stage of analytic detail. Choosing whether a very high probability of survival for 150 years at a cost of x jobs is better than a high probability of survival for 150 years at a cost of y jobs is not a task that judges relish or have any great basis in law to figure out. Their hope is that the agencies will make a choice that is at least minimally legitimate on technical grounds, or that the Congress will make the choice that ultimately is its responsibility in our scheme of government.

Part of the problem with viability analysis lies in its basis in concepts of risk. As regulatory decisionmaking has moved to methods such as benefit–cost analysis and risk–benefit analysis, we are getting better at finding ways to portray and evaluate data, but it remains difficult for individuals to understand and evaluate risk and the acceptability of various outcomes. How you might evaluate the worth of a population of owls that has 1000 pairs providing a 0.8 probability of effective survival for 100 years versus other alternatives is not obvious to most people. Problems in understanding the concepts of risk and acceptability are one of the reasons that technically rational administrative processes rarely provide choices that individuals can accept. The political process, therefore, ultimately becomes responsible for doing so. While being intellectually sound and even democratically appropriate, viability analysis unfortunately moves us further into the terrain of technical rationality than we have wandered in the past thirty years, and it makes many individuals and organizations uncomfortable.

A final problem with viability analysis is its data intensity. To provide an accurate assessment of long-term viability, a considerable amount of data are needed to plug into population demography and extinction models, and for many endangered species those data do not exist, and are hard to acquire given the fragile nature of the species in question. Hence, some critics argue that viability analysis does exactly what its proponents argue it guards against: it cloaks value judgments in the veneer of technical credibility.

To many observers of the SEIS analysis done by the FS in 1985–88, the viability analysis was "smoke and mirrors." In their view, assumptions were made that "loaded" the outcomes, and the considerable machinations that went on throughout the analysis and decisionmaking processes were as much oriented toward producing a set of outcomes that were politically correct, as they were in fine-tuning the models to provide more technically accurate choices. Regardless of how you come out on this issue, it is the case that the use of viability analyses set up this dynamic. Indeed, the FS might have been politically and organizationally better off if they had done their job the old-fashioned way: if their staff had simply gone to the mountain and come back with a truth that was minimally supportable in court.

Finally, adaptive management was also problematic given prevailing attitudes about the FS. Adaptive management is a concept of decisionmaking that recognizes the fact that we need to make choices today based on incomplete information. Its guiding principles are: (1) making choices that are seen as experiments, (2) monitoring implementation of the choices, and gathering information to evaluate the long-term appropriateness of the choices, (3) reevaluating the choices at appropriate, guaranteed intervals, and (4) maintaining the ability and commitment to change direction should implementation be ineffective or new information obviate old choices. Adaptive management is a reasonable concept, and no doubt necessary for effective policy and organization choices, but it runs into problems as it moves into the realities of organizational change and interpersonal and interorganizational attitudes.

Individuals and organizations invest in choices and directions, and for a variety of reasons find it difficult to make changes down the line if they find that a past choice is not working out. Setting aside motives and interests, for the FS to have reduced timber harvest in Region 6 in a significant way, they would automatically create a major personnel problem: What do you do with all those timber staff officers in the Pacific Northwest? Even though the economists tell us to write off sunk costs, it is difficult to do so. For example, the garage calls and says that your old car needs $1000 in transmission work. You would not ordinarily pay it, except that you just invested $900 in new tires, brakes, a muffler, and fleece-lined front seat covers. Individuals and organizations also become vested in directions and bound by their behavior in the past. It is hard for them to adapt to new information, and the concept of adaptive management runs head on into this behavioral reality.

In the spotted owl case, adaptive management had greater-than-average problems, because a winning choice required that nongovernmental groups buy in, and no one trusted the agency to change its direction five or ten years down the line. Having seen the agency make decisions that were obviously pro-timber in the past, environmental groups did not believe that the FS would cut its timber program significantly if it found new information that warranted greater owl protection. Besides, the groups had seen new information generated from years of research that had received minimal acknowledgment by agency managers. For example, the 1000 acres of old growth per owl pair that the 1981 Spotted Owl plan proposed was not endorsed by the agencies until long after 95% of the experts felt it legitimate. The agency said an adaptive management strategy was needed in 1988 because additional information was necessary to evaluate long-term owl prognosis. The environmental groups felt that this was just a smokescreen, and that the problem was not uncertainty caused by inadequate data, but a weak stomach on the part of the FS.

Timber was not much more trusting of the FS's ability to change its course downstream. In their view, any commercially valuable forest lands set aside from harvest today were most likely to be permanently set aside. That pattern had been true in the past, and they had no reason to believe that the increasingly powerful regional environmental groups would allow the FS to do anything differently in the future. Old growth set-asides were permanent lock-ups equivalent to wilderness. To timber, then, adaptive management could only mean bad news: If set-aside lands would not be opened to development, change could only come in additional set-asides, which was not an outcome with which timber interests were enamored. In addition, adaptive management introduced an explicit element of uncertainty into the future availability of federal timber. Since certainty was of utmost importance to timber interests so that investment decisions could be made wisely, the possibility of a management redirection in five or ten years was not attractive. In their perspective, FS timber planning in the Northwest had been characterized by a lack of long-term certainty since the beginning of NFMA planning, and enough was enough.

Even if an adaptive management strategy was acceptable to both timber and environmental groups, they could never agree on the starting point. To timber groups, maintaining current cut levels with a slow phase-out (if necessary) would minimize job impacts and more important, allow for adaptive capital investment. In the view of environmental groups, if you truly wanted to implement an adaptive management strategy, you would act very conservatively today by preserving options, gathering more information, and then doing things that reduce options if the new information says to go ahead. That would mean protecting old growth, finding out more about the needs of the owl, and then opening old growth set-asides if it was warranted. In their view, the FS SEIS strategy was a fait accompli approach: cut it now, and make the ultimate decision self-evident. Because neither set of groups trusted the agency to change its behavior down the line, the starting point was doubly important, for in their view, it was the only choice likely to be made, regardless of the adaptive management rhetoric that dressed up the SEIS decision.

Adaptive management, concepts of MIS and MVP, NFMA, forest planning, and the ESA were all well-intended, intellectually compelling concepts of forest and wildlife policy in the 1970s and 1980s, whose characteristics also had the somewhat perverse effect of exacerbating the developing owl controversy. While each intended to remove forestry and wildlife management from the vagaries of political and organizational dynamics, none succeeded, nor was it necessarily possible for them to succeed. Congressional budgetary politics had as much to do with actual day-to-day management of the public lands in the Northwest as did any of these substantive laws. The values expressed by the Reagan and Bush administrations were constraining and misleading to agency officials that might have wanted to adjust their operations in line with shifting public opinion, and executive office politics were pervasive and had a major impact on the evolution of the owl case. While the goal of forest policy was to elevate public land management out of the normal failings of decisionmaking processes, it not only failed to improve on the norm, but contributed mightily to the development of the controversy.

10

Grounded in the Past: Agency Values and Management Approaches

The story of the spotted owl controversy is a story about the Forest Service, and the evolution of its values, traditions, and norms of behavior. The FS was created as something better than the political norm, and for a good portion of its life, it has been. From its origins in the Progressive era through the 1960s or 1970s, the agency was viewed as a model federal agency with an esprit de corps and image of professionalism unmatched by any other. FS leaders saw themselves as the lofty peak standing in a vast wasteland of excessive bureaucracy and politics that characterized most administrative agencies. They took this image seriously and worked to craft an organization that built on Gifford Pinchot's ideals and principles, and responded to their image of national needs.

All organizations and individuals adopt values, norms of behavior, and standard operating procedures to help them carry out day-to-day tasks and cope with demands of the outside world. The FS was unusual only in how deep its traditions and norms went, and how hard it worked to reinforce them in its workforce. Yet these same values and modes of behavior that help us cope with a difficult world also become outmoded as the environment changes around us. For the FS, the legal and political world that they lived in changed dramatically in the 1970s, 1980s, and 1990s, and even though many tried to adapt past behaviors to current conditions, in many cases the effectiveness of their response was hampered by their enduring identity.

The owl issue epitomized these changes, and the agency dealt with it first by downplaying its significance, and then by placing it into normal standard operating procedures. As the debate over owl management became a public controversy, longstanding divisions of interests and values within the FS deepened, and while some groups tried to apply current science and different thinking to the owl problem, others tended to retreat into traditional values and behaviors. Over time, agency leadership seemed to be suspended somewhere between the comfort of their past and the brave new world of their likely future.

Just as adolescence is as hard on parents as on teenagers, those interested in the future of the national forests suffered along with the agency. Neither commodity groups nor environmentalists were happy with the FS in the 1980s. The agency's attempts to find some middle ground between the interests were awkward and unappreciated by either side. Over time, the distance between the sides increased, and the FS, with one foot in each side, was left staring into the widening void beneath them. As previous chapters have outlined, the owl issue was indeed inherently difficult to resolve, and the decisionmaking environment and statutory policy did not help much. Nevertheless, the agency's traditions, values, and norms of behavior also contributed to the developing controversy. An adage of recent years declares that "if you're not part of the solution, you're part of the problem." In many ways, the FS's tradition-bound efforts to solve the problem became problematic themselves. They were both problem and solution, in a mix that was confusing to everyone, not the least of whom was the FS staff and leadership.

Organizational Norms and Standard Operating Procedures

While some norms are culture-specific and some are created by an individual or organization, all serve to provide predictability for their owners. How do American professionals greet each other? By shaking right hands. Of all the possible actions that could take place when two individuals come face to face, we shake right hands. We could kiss cheeks or feet, exchange money, or turn around three times and stand on our heads, but for whatever reason, we shake hands. This norm of behavior provides a great amount of predictability for the two involved individuals, and as a result is efficient for them. They do not have to think very much about what to do, discuss it, or deal with it in any way, except to act rapidly and effectively. Norms are energy-conserving, and they help each of us deal with what would otherwise be a remarkably complex set of human–human and human–environment interactions.

Organizations similarly develop elaborate sets of rules, or standard operating procedures, that govern the behavior of the individuals within them. For example, the Forest Service has written and unwritten norms and standard procedures that determine who gets hired, who gets promoted, what kind of contractors are awarded timber contracts, what kind of safety procedures are used, what kind of objectives are legitimate or considered a priority, and thousands of others. The FS manual, a guidebook that defines policy for many aspects of a FS staffer's job, consists of thousands of pages. The manual is a compilation of rules and standard procedures: an unending definition of appropriate behavior equally applicable to the director of the Wildlife and Fisheries staff in Washington, D. C., and a timber sale planner in an isolated district in Oregon.

For organizations like the FS, these sets of rules and norms are critically important, because their staffs cannot deal with every decision on a case-by-case basis. They have neither the time nor the energy to do so. Nor would they have a very good basis for making decisions without a strong sense of direction based on precedent or generally accepted professional practice, that is, norms of behavior established in the past that presumably were effective in the past. Organizational norms and values provide an overall framework for action that somewhat invisibly coordinates the millions of actions involved in running an organization. By creating incentives that influence the behavior of lots of individual staff members, they help to control a diverse, dispersed workforce, and in the process, ensure overall direction while minimizing the effort leadership must spend on small choices. Indeed, organizations that evidence shared values, translated into norms of behavior, have a much easier time controlling the forces for disruption and chaos that are always present in an agency like the FS, with decentralized operations and thousands of staff members.

These same norms and rules, so important for dealing with day-to-day operations of an agency, can be ineffective or even counterproductive in dealing with nonroutine situations, like the spotted owl case. Standard operating procedures are set up to enable lower level staff members to process certain kinds of situations. Situation A triggers operating procedure Q, and the bureaucracy responds accordingly. But what happens when the situation is somewhat different from those for which operating procedure Q was developed? The tendency is to force the situation into a predefined norm, even if it involves forcing a square peg into a round hole. Since truly nonroutine cases require the attention of agency leadership, and hence consume limited time and energy, the test for "nonroutineness" is how much friction is created when that peg is rammed into the hole. In the best of situations, leadership responds rapidly when even a little heat is generated. But in most cases, the peg is pounded on repeatedly with increasing effort before enough heat is generated for leadership to take it on.

Routine procedures also can become outmoded and dysfunctional over time. The world around an agency is always changing, and new staff members bring new values into organizations. Procedures that have been functional in the past may well become dysfunctional as conditions change, yet organizations and their members tend to cling to them as a comfortable, familiar place in a changeable world.

Changing traditional ways of doing things is difficult because it threatens individuals' sense of security and predictability about their jobs and their lives. Their knowledge may become obsolete or outmoded, and their futures less clear. Even when staff members and leaders of an organization know that change is needed, it takes a lot of energy on their part to bring it about. In a time of limited slack, most individuals are simply trying to keep up rather than make major changes, even when those changes may result in a much better organizational environment in the future. Uncertainty generated by change is bothersome, and as a result, threats to traditional values and operating procedures are resisted.

For those living in the world outside an aging organization, an agency's desire to hold onto familiar ways becomes problematic. At the ground level, elaborate rules become red tape that tends to promote conflict between agency and client, as the reason for the rule becomes less clear or relevant. At the policy level, such organizations take on the character of interest groups seeking a set of self-interested outcomes from the policy process. Biases get further ingrained in agency direction and what it seeks from the policy process. An agency's day-to-day operations become less effective, and its ability to function as an effective articulator of public concerns and needs declines. Increasingly, the agency becomes more a part of the problem than an institutional means to defining a solution.

FS Standard Operating Procedures: Functional for Some Purposes, Problematic for Others

Chapter 1 described the evolution of the FS from its origins at the turn of the century to its style and structure of the 1960s and 1970s. Of all the federal agencies, it was perhaps the most defined by tradition, a sense of mission, and an image of professionalism. Agency leaders worked hard to

transfer these values to their staff by selecting appropriate people, social-
izing them in the right way to think, and rewarding good behavior.

By the mid- to late-1970s, when the spotted owl was becoming an issue
needing agency attention, the FS was solidly grounded in a set of values
and ways of doing things characteristic of a mature bureaucracy. While
some critics would argue that it was more evident of an overmature organization, needing a harvest of old leaders and replanting of new stock, the
set of norms and behaviors were quite functional for the agency and most
of its staff. Budgets were good, the agency's mission was clear, and political support was solid. What they had been doing for thirty years appeared
to be working well.

Substantively, the agency in this period can be described as multiple
use-oriented with a strong pro-timber bias. Pinchot's ideals of maximizing
resource utilization were well established in the agency, generating a production orientation toward natural resources. Of the many possible uses
of national forest resources, timber easily became the domineering
brother of the multiple use family since it was most easily quantified,
translated into political support, and generated tangible dollar revenues.
At times at least, agency leaders were concerned about the demands of
both sides of the interest group community: timber interests, who were
seen as demanding more than good management would likely provide,
and environmentalists, whose demands for reserved, unmanaged lands
were alien to a number of FS values. Both groups' demands were seen as
conflicting with the enduring goal of multiple use management, though
for a variety of reasons, timber tended to drive the show.

The basic method of FS management was landscape manipulation. In
the Pacific Northwest, much of this manipulation involved clearing out
overmature stands through clearcuts so that Douglas fir plantations could
be established and managed on a sixty- or eighty-year rotation. For wildlife purposes, openings could be created in the forest canopy to promote
an understory that was more effective at supporting game species than
was the dark, homogeneous old-growth forest. For recreation, trails could
be cut and campgrounds created. Fisheries enhancement projects required manipulating streambeds and associated habitat.

Since the basic method of most FS activities was hands-on landscape
management, an unmanaged landscape was alien to many of the approaches of FS employees, and less desirable to an agency seeking budgetary resources to carry out projects on the national forests. It also appears
to be the case that many foresters simply like to cut trees. To many FS
employees, forest management is a process of creating a new landscape on

a broad canvas of national forest resources that is exciting and motivating, similar perhaps to the process of creating a new piece of art. As with all tasks in life, there is beauty and a sense of satisfaction in a job well done, and that goes for a good quality clearcut as well as anything else.

The agency also was serious about its dual mission of "Caring for the Land and Serving the People," and viewed itself as having the best idea as to how to do this. One of the core beliefs of the Progressive Conservationists was that technical expertise would provide a better basis for making choices than would the rough and tumble of the political process,[1] and the FS was created and grew in the light of professionalism and expertise. The fact that the methods and values of American forestry derived from the preexisting German forestry profession heightened the FS's sense of professionalism. Agency leaders and staff viewed themselves as experts who had the best basis of anyone in society to make wise long-term management choices for national forest resources. In their view, effective decisions could be made through technical analysis and informed scientific judgment, activities that could only be undertaken in an apolitical, deliberative environment.[2]

Over time, an elaborate set of norms developed that defined organizational styles and prescribed appropriate behavior. In its earliest days, the agency's style was one of technically based, paternal benevolence, reflecting the ideas of Gifford Pinchot, whose concepts of wise use in the public interest reflected his knowledge of the embryonic fields of silviculture and conservation. The style Pinchot and other early agency leaders promoted also reflected their notions of the public good, as reflected in their aristocratic backgrounds, and the political philosophies of the Progressive era.

By the time the owl controversy rolled around, a different basic style had evolved, one that can be described best as "Can Do." The second wave of organizational style can be characterized as quasi-industrial, and militaristic, reflecting the post-War production orientation and the styles of a less aristocratic class of leaders who enlisted in the agency following military service during and after World War II. The style they evolved was one of dependability in action: a kind of "You asked for it; you got it" Get It Done mentality. Eventually, getting it done became as important as figuring out what was most appropriate to be doing. A federal agency that accomplishes a lot of things is not all that common, and Congress rewarded the agency's efforts.

Accompanying the style of the agency as a Can Do, Get It Done organization is a prescription for individual behavior that can be described as the "Good Soldier," a prescription that applies to agency staff and leadership alike. Since the good soldier not only gets it done, but does not question his or her orders, the second wave of organizational style tended to reinforce a downward flow of direction. The good soldier cuts at whatever timber sale level is mandated by Congress, and has a hard time telling his Congressional commanders that he cannot get it done. In this model of an effective organization, obedience is rewarded, even if it is counterproductive in the long run.[3]

While the culture of the FS allows for disagreement and discussion, there is still a strong norm of "Don't Rock the Boat." All organizations tend to jettison dissenting voices, either directly or by not advancing those individuals in the hierarchy of the organization, and the FS was not unusual in this regard. However, the force of its leadership's convictions about the overall direction of the agency tended to reinforce this norm. An organization with less unanimity about its overall direction tends to foster internal factions that provide cover to dissenting voices. But the strong shared sense of mission held by the FS leadership (at least through the 1970s) tended to make a dissenter the odd man out; and dissenters, particularly those who went public with their concerns, were seen as the black sheep of the FS family. The message that went back to individuals who might be vocal about their concerns was to hold back if you want to get ahead. For a Can Do organization, those who take the position of Shouldn't Do are not helpful.

Related to this dynamic was a tendency to "shoot the messenger" of bad news, with the result that information that might flag a developing problem or that is critical of current direction may not be passed along to those who need to act on it. The Good Soldier aura of the organization tends to weaken an upward flow of criticism and the reality-checks that are necessary for agencies to test whether current direction is appropriate, and filtering information upward appears to be a conscious and subconscious dynamic in the FS. This filtering dynamic is common in organizations, since the problems that a subunit faces are seen as partly of its own doing.[4] A squeaky wheel has equivalent chances of getting grease or getting replaced. Hence, information that highlights problems is filtered as it moves upward. The net effect is that problems often are not anticipated or well understood by leaders.

The appropriateness of these norms relating to organizational direction and style were reinforced in the agency's workforce in the 1950s, 1960s, and 1970s through a variety of personnel selection and socialization mechanisms: first, by selecting people from a narrow set of academic disciplines trained in accredited forestry programs; second, by encouraging self-selection at the time that people go into forestry-related careers, in part by projecting an image of the politically correct forester and encouraging only those who fit the mold to apply; third, by on-the-job training and various forms of socialization that define and reinforce appropriate methods, behavior, and values; and fourth, by rewarding those who act appropriately through promotion or other means, and by penalizing those who do not, either directly or more likely, by letting their careers languish on the rocks of an isolated district or research station. Sometimes these approaches for reinforcing correct thinking were deliberate and explicit, and sometimes they were subtle and unintended, but their overall effect was to build a workforce that by and large was committed to the organization and its ways of doing things.

The beauty of these approaches was that while they had the effect of providing stable direction and consistent style, they did not obviously preclude individual discretion in how a staff member approached his or her job. Indeed, the norm of decentralized decisionmaking was also well established in the agency, with district rangers at times having all the powers of a feudal lord. But by the time an individual got into a position of decisionmaking authority in the field, his view of the range of appropriate choices was fairly well bounded by the agency's view. Organizational values played themselves out over and over through vast distances, via a workforce whose values were quite similar, even if individuals had a lot of authority to make choices. From the agency's perspective, this is the best place that they could be: With a workforce that given their druthers would generally do what the agency's leadership wants them to do, the agency does not have to invest in a lot of monitoring. Nor do they have to institute personnel systems that seek control and in doing so, trigger off a lot of resentment and resistance on the part of field-level workers.

One of the mechanisms for building a strong sense of shared identity across a large workforce was the image of the Forest Service as a family. All of us have reference groups, that is, collections of individuals with which we identify and ally. Church groups, professions, neighborhoods, employers, and families are all standard reference groups, exhibiting shared values and relationships that bind their members. In trying to win the hearts and minds of its workforce, the FS tried to be all of these to its staff. By moving staff members every few years, the agency not only builds a staff member's technical capabilities and understanding of the diversity of agency operations, it also builds the staff member's identification with the agency. The common denominator throughout all the different places that a FS staff member could find himself/herself is the agency itself. Particularly when you find yourself in many of the small towns and isolated areas that characterize FS work assignments, the only people you have to be with (or might want to be with) are other FS employees. The agency and its values become a strong force in an individual's life.

The image of the FS as a family is potent in organizational terms, as it built top-down predictability and control by building bottom-up allegiance and support. Like all families, the FS is both supportive and protective of its members somewhat independently of their meritorious characteristics. As a policy matter, the image of the agency as a close family tends to insulate family members from other realities, and for our purposes, a workforce insulated from the variety of interests in society is a problem.

While agency policies and procedures may have encouraged FS staff members to identify with each other, in fact their socioeconomic characteristics were fairly homogeneous anyway. The mechanisms for building shared values, and the set of individuals who were attracted to forestry careers, resulted in a workforce, and particularly a leadership in the 1960s and 1970s, that were quite similar to each other. The average FS employee at that time was a white male from a rural background, trained in forestry, who viewed himself as fairly conservative, and liked to hunt and fish. Women and racial minorities, urbanites, social liberals, environmental preservationists, and individuals from nontraditional academic backgrounds were neither sought by the pre-1980s agency nor rewarded within it. Even when such individuals were hired into the agency, the pervasive set of values and norms evolved and reinforced by the old boys who populated the agency made it difficult for nontraditional newcomers to feel comfortable and prosper.

All this homogeneity within the organization is not to suggest that divisions in values and interests were not present in the FS, for like all collections of more than two individuals, cohorts and groupings were evident. Differences in style and interests between line officers and staff members, researchers/scientists and managers, and Washington, Regional, and District offices were and continue to be prevalent. For example, differences in regional styles exist, with strong biases between the two west coast regions: Region 5 being viewed by Region 6 as California loonies, and Region 6 being viewed by Region 5 as timber beasts. These kinds of distinctions become problematic when a problem crosses the boundaries between region or type of worker, and needs a common response, as was the case with the owl controversy. Negotiating the way through these differences is difficult and time-consuming, and confusing to extra-agency parties.

One of the most interesting distinctions in this case is the difference between timber and wildlife staff. In an agency that was aggressively pursuing a large scale timber program, wildlife staff were okay as long as they did not make too many demands, and were largely concerned with growing game animals that benefited from active timber management. To the extent that the interests or values of the wildlife staff diverged, wildlife lost out in the organizational pecking order. Wildlife was a small portion of the agency's budget, and a small component of the agency's expertise base. As a result, the wildlife staff was chronically underempowered. They were used to getting the table scraps from the timber program, and did not see the point in complaining too loudly. Indeed, the optimization formulas for determining national forest objectives generally maximized timber outputs subject to minimum wildlife outputs. Timber was an output; wildlife was a constraint. The image of maximum timber and minimum wildlife was reproduced in many ways in the FS in the 1960s and 1970s, not the least of which was the agency's early response to the owl controversy.

Underempowerment is generally self-reinforcing, as individuals who see themselves as having little influence or control over outcomes generally do not try hard to exert influence. As a result, they have little effect on major outcomes. It is a circular and continuous process, that leaves the powerful powerful, and the weak weak. Weaker interests tend to downplay their concerns and needs, and until the owl issue increased the significance and power of wildlife interests in the agency, wildlife staff tended to understand their subservient role in the agency, and acted accordingly.

The set of organizational styles and norms that were well-established in the FS extended to their handling of policy issues. Under all circumstances, it was important to maintain control. Organizations and individuals seek autonomy as both a means of carrying out their everyday tasks, and an end in and of itself. The ability to control one's own life is a powerful motivator of human behavior, and individuals translate their needs into organizational needs. To be out of control means a loss of predictability and stability, and requires the expenditure of a lot of energy to regain balance and momentum. Organizations that are out of control rapidly become inefficient and ineffective, and highly stressful places to be.

For the FS, control was not just an individual and organizational prerogative, but a professional and political one as well. Foresters must believe that they can control the destiny of landscapes for tens or hundreds of years. In addition, since the objective of the FS was technically based management of the national forests, it was important to maintain control over management decisions so that uninformed, ill-advised politicians and citizens would not take over and make bad choices. One of the worst possible outcomes was for courts to take control of the direction of national forest management. Judges have no scientific basis for making management decisions. If the agency lost control of an issue such that decisions were being superimposed on them by the outside world, bad management choices would be made: bad in terms of what science dictated, and bad in that the choices would more than likely conflict with the values of the agency.

A related organizational norm was to avoid politicization of forest management decisions. One characteristic of FS leadership that the staff points to with pride is that there are no political appointees in the agency. It is remarkable how many of those interviewed for this research noted the real and symbolic value of this organizational feature. The fact that the Chief is not a political appointee is highly valued by the agency staff, and they contrast their agency and its historic success with that of the Department of Interior agencies, which are seen as much more yielding to political pressure.

While the agency has no political appointees, it is clear that its leadership plays politics. It is also evident that everyone fears that if the agency's leadership alienates political leaders, a change might occur that will provide a politically appointed Chief in the future. This fear of future politicization leads the agency to be more political in the short term. It is an open question whether the fear of politics leads the agency to be more political than it has to be. Some staff members suggested that the agency's leadership avoid pushing the political envelope as far out as it might go, because of their fear of the ultimate horror: a political Chief. That is, they play politics to avoid politics, and may make bad policy choices along the way. The jury is still out on this argument, though it is clear that the behavior of the agency's leadership throughout the owl case was quite conservative and cognizant of political forces. We will never know whether agency leaders could have changed the political forces operating on the owl decision without paying the ultimate price.

One way that the agency maintains control and avoids politicization is to study issues to death. Since their ultimate legitimacy is their technical wisdom, agency officials need to ground their choices in technical analysis. This is both a historical birthright and political reality. Having seen what happens in court to agencies who have not provided adequate technical justification for what they intend to do, the "Study It" approach was also seen as a judicial necessity. It is also a good organizational strategy for staying in control over the definition of issues, and determining the rules of the game for how they will be resolved. For example, forest planning debates were conducted squarely in the FS's court up until appeals were filed. Draft forest plans were massive documents that were difficult for nongovernmental parties to work with in spite of the agency's efforts to address a lay audience. Their approach, and the terms of debate that they fostered, guaranteed the FS an upper hand in the negotiations that ultimately took place.

A norm of analyzing issues before taking action can hardly be criticized, but at times the agency's use of studies as a means of controlling the issue might have outweighed their value as technical inputs to tough policy choices. Often agency action appears to follow the thought process that if we just study this long enough, we will find an answer that will survive public scrutiny. For some forest management questions, though, there may not be right answers. Instead, their solutions represent temporary balance points between all the interests at stake, and may require changes in agency values and methods, neither of which can be discerned through technical analysis.

As the 1960s and 1970s ground on, public involvement in administrative decisionmaking became a political and legal necessity of agency life, and the FS developed elaborate public participation processes, but their approaches tended to view public involvement as a linear process, not a dynamic, interactive one. The agency's approach contained four steps: let the public express their concerns; go back to the office and figure out what is best for them, given the agency's understanding of its responsibilities and the capabilities of the resource base; produce a draft plan and provide a period for public comment; and make a decision. This approach reflected the agency's view of itself as the repository of technical expertise, its understanding of its statutory obligations, and its interest in controlling the flow of the decision at hand. What it did not provide was an opportunity for interchange between affected interests, a chance for creative solutions to emerge (that were not within the normal vision of agency personnel), or much ownership of the resulting decision on the part of affected interests.

As the 1980s began to prove ineffective this model of public involvement, agency officials increasingly began to view their role as the unbiased arbiter of conflicting interests: the FS as societal balance point. They came to view their role as trying to find a balance point between the different points of view, with one measure of success being whether everyone was angry with the agency. According to one FS staff member, "We must be doing something right if everybody's yelling at us."

While responsive to a simple notion of the pluralist ideal, this approach unfortunately leaves something to be desired. The sum of the interest group parts misses a lot about the future and about other agency mandates. While right answers may not be attainable for many issues, there are still better and worse ones based on an understanding of technical realities, and an image of the needs of the future. Even if the agency sees itself as a neutral facilitator of compromises between interest groups, it is not a sign of good process to have everyone angry at the end. The point of balance that the agency found would have been one in form only, as the aggrieved groups would insure that decisions were tied up forever in appeals, the courts, and the political process.

The values, norms, and standard operating procedures that were created and reinforced by the FS were appropriate and functional for most of the organizational dilemmas the agency faced up through the 1970s. They worked, at least for the agency, and in many ways for a society that consumed greater and greater amounts of wood products, and began to demand an increasing set of other products from the national forests. As public demands changed, the amount of unmanaged forest declined, scientific knowledge developed, and the character of forest policy controversies shifted, these same sets of values and norms became a trap for the FS.

Norms, Values, and Traditions versus the Owl Controversy

For an agency steeped in multiple use management, biased toward timber production, and oriented toward active manipulation of forest landscapes, the owl issue was anathema. The owl appeared to require single purpose land set-asides on which little or no management could take place. The old-growth lands that were to be set-aside were traditionally viewed as inefficient and wasteful, as the overmature timber on them was long past the point of efficient wood fiber production. The preservation approach to land management that seemed to be required for owl protection was not within the "normal" professional paradigm of foresters or wildlife managers. For a Can-Do organization, preservation meant "Can't Do," and that automatically conflicted with organizational norms, as well as political and economic realities.

The size of the set-asides was also problematic. Single purpose set-asides were not new to the FS of recent years, even though they appeared to conflict with the notion of multiple use. But the size of the owl reservations and their value as commercial timberland set them apart from previous designations. As one more interest group demanding goods from the national forests, the owl appeared to be not only needy, but greedy. A few scattered owl protection areas were okay, just like a set of scattered wilderness areas, research natural areas, and areas for nonmotorized primitive recreation were not a problem. But the notion of six or eleven million acres of unmanaged, single-purpose land designations to protect an owl was hard for many FS staff members and leaders to understand, let alone accept, given their traditional ways of looking at the world.

Because the needs of the owl were so incompatible with traditional FS norms and behaviors, the policy controversy that developed around owls and old growth was threatening to the core values of the agency and many of its staff. Less old-growth harvest meant fewer (or smaller) timber sales, which had serious implications for agency budgets and size of staff. Just as important, the new directions implied by owl protection plans questioned the legitimacy of the expertise base of the agency, and required at least some retooling of individual skills. Constituency relationships with timber and other interest groups were disrupted, and political ties with traditional supporters were affected.

As the issue developed into a national policy debate, it increasingly became threatening to the agency's sense of autonomy and control, and challenged its concept of professionalism. It also tended to empower groups like the wildlife staff within the agency, and as a result, promoted divisiveness within the FS family. All in all, the owl issue had the potential to upset what had become for many staff members and leaders, the normal way of doing business, a comfortable and predictable day-to-day existence.

Nor did the culture of the organization facilitate the internal criticism that would have promoted a more aggressive response to the budding issue. While the agency held a diversity of viewpoints within its boundaries, longstanding processes of socialization, the norm of being the good soldier, and the image of the FS as a family tended to suppress divisive criticism. As employees filtered the information they sent upward, problems and perhaps an appropriate sense of urgency were not conveyed to top agency leadership until somewhat late in the chronology. Wildlife interests within the agency were accustomed to having little influence, and decisionmakers were accustomed to downgrading wildlife's concerns, and both wildlife's timidity and leadership's callousness were reflected in the evolution of the controversy.

The inherent clash between FS norms and the nature of the owl issue foreshadowed problems in how the agency would respond, because it was likely that when threatened, agency leaders would rely on traditional ways of doing things. The standard response of organizations when faced with a threatening situation is to resist change, passively and actively, and the first response by the FS to the owl issue was to try to avoid dealing with it at all. For the first five years of the controversy, the FS largely ignored the developing issue, and while they sent a staff member to participate in the activities of the Oregon Endangered Species Task Force, not much else took place. The owl was a nonissue for the agency. It was neither threatening nor demanding, because all of the experience of the leadership suggested that the owl could be dealt with in a minimal fashion with no resulting organizational or political fall-out. What owl preservation implied was so far out of standard practice that a significant response was not considered legitimate, nor necessary, given the pressures on the agency at the time.

As it increasingly became necessary for agency officials to respond in some way to the owl issue, their response was that of anyone dealing with a new situation: place it into normal standard operating procedures; that is, deal with it in a time-honored way that is comfortable and familiar. For much of the history of the spotted owl controversy, the failings of the FS in dealing with the issue was not that it was not doing anything, but rather that it was responding in time-honored ways that were inappropriate to the owl case or the times.

Indeed, one can argue that the agency misread the significance of the issue throughout much of its lifecycle. For example, when given a major opportunity to redeem themselves as technically credible analysts through the SEIS, one of the first agency actions was to appoint a disgraced forest supervisor to head up the effort. It was widely known in the region that the head of the project had been moved to the SEIS effort as a means of reprimanding him, by removing him from a position of line authority, because he had been caught illegally cutting firewood on his forest. Even if this individual did a great job as project leader, the message it sent was of an agency unconcerned about the quality or credibility of the SEIS effort.

Many of the other responses to the owl issue reflected the agency's normal way of doing things, and failed in the end to adequately deal with the evolving issue. The second order response of the agency was to direct forest supervisors to establish small set-aside areas for owl protection. Owl management areas were consistently set at the absolute minimum level. Three hundred acres, then a thousand acres, were the absolute minimum that any owl seemed to use. It was not the mean or median value, which would have been a more legitimate standard for wildlife management.

Timber remained the primary concern of much of the agency leadership throughout the issue, partly due to the politics involved, and partly due to their sense of agency priorities and values. A number of the staff members who had been involved in the SEIS effort indicated that a five percent reduction in the allowable sale quantity of timber was what agency leadership defined as the allowable impact of the owl issue on the agency's timber program. Owl planning in the SEIS was constrained by this specified size of economic impact. According to one observer of the agency, the five percent figure was the smallest number seen as credible, and it had appeared before in FS responses to other protection issues. Five percent was big enough to seem significant, but small enough that it had a fairly insignificant impact on timber.

Perhaps the most consistent response of the agency to the issue was to pursue a "Study It" strategy. The Spotted Owl Subcommittee work, the Regional Plan/Guide work, the Old Growth RD&A, the MMR analysis, the Forest Plans, the SEIS effort, the Spotted Owl RD&A, and the Interagency Scientific Committee report were all attempts to further define the issues and provide a seemingly legitimate basis for making a choice. There was a sense that if we just got the analysis down right, that the politics would follow. In addition, continuing to study the issue ensured that any ultimate action would be delayed. The most cynical observers felt that the delays were orchestrated to enable old-growth to be cut in the interim. While some in the agency might have been so motivated, probably the delays were reflective of how complicated and politically hot an issue this really was. Certainly the amount of time it took to move from the draft SEIS to the final SEIS and decision by the Chief was reflective of the political sensitivity of the ultimate choice.

In many ways, the FS got trapped in its traditions of using technical studies to justify political choices. Whether or not they realized it, agency leaders had been making resource allocation choices that were reflective of political and organizational biases for years, even though they were dressed in a veneer of science and technical opinion. They worked hard to cultivate an image of themselves as sage experts, and the image was a potent barrier to those critical of agency direction. The way that they dealt with the owl issue was consistent with this traditional approach: they tended to make choices whose veneer of technical analysis was compelling while hiding the value choices that had been made. The SEIS illustrates this approach in that it was grounded solidly and voluminously in technical analysis, but the final decision was aimed at responding to political realities. While some agency staffers tried to make the distinction between the SEIS analysis and the Chief's decision clear, the overall effect was to generate a decisionmaking analysis and conclusion that appeared to be based on extensive scientific grounds. The subtleties were lost on the media and the Congress and were confusing to many.

As the science base changed and nongovernmental parties acquired the ability to make scientific arguments, the effectiveness of the FS approach declined. Increasingly, the agency's science could be (and was) challenged, and new ideas were brought to the policy arena by nongovernmental interest groups. In addition, internal advocates of change were empowered by the developing science and the enhanced responsiveness of the political environment outside their agency, and could more credibly lobby for different choices both from within and outside the agency. While courts could not evaluate how good the new ideas were, they could delay final actions and suggest that the FS think about its decisions again.

Studying various aspects of the issue endlessly was also a way to maintain control over the course of events related to the issue, and FS efforts to stay in control of the timing of events and the magnitude and character of response were pervasive. The Spotted Owl RD&A, the Interagency Memorandum of Understanding signed in late 1988, and even the Interagency Scientific Committee (the Jack Ward Thomas group) were all mechanisms to ensure FS control over direction of the issue. (Even though the ISC was only partly FS researchers, creation of the group with the specific set of individuals that were named as its members ensured the outcome at the outset. Agency leaders must have known this, and have been willing to be bound by the report's recommendations, as a way out of the controversy.)

What motivated any response to the issue at all was largely a fear of losing control over it, and with it, losing control over the direction of national forest management. Listing the owl as a federally recognized threatened or endangered species was consistently seen as the ultimate horror. While some who viewed it this way may have been concerned about the meaning of a pro-listing choice for the owl itself, the fear of listing was partly a fear of having the FWS involved in FS decisionmaking, which at minimum meant more red tape in national forest management.

Just as concerning were the new levers that a listing decision would give to nongovernmental groups, enabling them to use the courts to contest management actions. The desire to maintain control provided an incentive for the FS to act, but because the goal was keeping the subspecies off the endangered list and not protection of the owl, the direction became how to take minimal action to avoid listing. As it turned out, minimal action was not enough, at least to satisfy owl supporters.

The FS also continued to deal with public involvement as a linear process, though they did explore the potential for using other approaches. The first response to the concerns about public participation was to deem early choices on owl management as non-decisions, not needing public review. Decisions of the Oregon Endangered Species Task Force and the Spotted Owl Subcommittee fell in an administrative never-never land: The interagency groups were not decisionmaking bodies in a statutory sense—no statute gave them the authority to make a choice on owl management—but nor were they a child of the FS, BLM, or the FWS. They may have looked, smelled, and tasted like an administrative decisionmaking process, but because they resided in the interstitial spaces of formal institutions, their obligations for administrative review and public involvement were not clear.

The SEIS analysis in 1985–86 was the first time the public was involved in a meaningful way in owl management decisions, and the agency did a bang-up job of carrying out a linear participation process. They solicited comments on the draft report, and received over 40,000 pieces of mail which they diligently coded and analyzed by computer. They also convened several working sessions of interest groups to keep them informed of developments related to owl management. And they even provided a special comment period after the final SEIS was produced, before the Chief produced his Record of Decision.

What they did not try to do was to convene a multiparty working group to attempt to either find some middle ground, or more likely, to negotiate their way through some of the data questions at the heart of the dispute. To the agency's credit, they did hire a consultant to explore the possibility of a mediated settlement, and he concluded that mediation was unlikely to make much progress given the polarization of the parties, and the types of values at stake. His conclusions were probably correct. Since interest groups had many avenues to pursue their interests, the incentives facing the stakeholders did not favor the success of a mediated, collaborative process. Nevertheless, it is unfortunate that such a process was not attempted since it might have focused on some of the data questions that were so confounding to decisionmakers and used strategically by many of the involved parties. The agency's traditional ways of doing things did not favor this kind of multiparty decisionmaking, and agency leaders were comfortable with the consultant's conclusions. It meant they did not have to engage in a process that they would not control and that they perceived was counter to agency traditions of informed expert choice.

With time, the agency's definition of an appropriate role in society became schizophrenic, leaving them somewhere between acting as the expert and playing the role of arbitrating the neutral balance point between the various interests in society. The SEIS provides a great example of this bimodal behavior on the part of the agency. While the analysis undertaken in the SEIS had weaknesses, it was reasonably good given the constraints of data and time that the planning team was faced with, and many of the experts involved in the case supported it. But the choice was not satisfying to anyone. The analysis followed the role of the agency as expert, and employed pretty good science. The decision followed the role of the agency as balance point between all the interests at the FS table. The seeming inconsistency between the two was confusing to many, and fatally doomed the analysis as a way out of the public controversy.

The third component of the agency's response to the evolving issue was to act strategically on its own behalf. All individuals and institutions do this; to assume otherwise is to negate a large component of human behavior that is self-interest-driven and competitive. Even the most altruistic of individuals or organizations will subconsciously bias decisionmaking in their own direction, by evaluating information in certain ways or choosing a course of action that is judged most effective when evaluated through a lens of institutional bias. The FS's spin on the owl issue was both subconscious and tactical, incorporating moves that strategically advanced their interests.

Most of the interest group strategies described in Chapter 8 were evident in the behavior of the involved agencies. Through much of the debate, the FS's consistent remonstrations that what looked like decisions were not really decisions was tactical behavior designed to minimize the accountability of the agency, and to eliminate the red tape and possible appeals that accompany an agency decision. The OESTF's spotted owl plan and the Regional Plan/Guide were claimed not to be choices, since the ultimate choices would be made in forest plans. The adaptive management strategy utilized in the SEIS was seen as the logical successor to the game of "Not a Decision," since it claimed that choices made today were not really important choices: The really important choice would be made ten or fifteen years down the line, when we have better information. Many people inside and outside the agency viewed the SEIS's use of adaptive management as "smoke and mirrors," whether intentional or not.

Agency officials also acted strategically in the way they revealed and concealed information throughout the evolution of the owl controversy. The persistent lack of agreement on a map of existing old growth, and/or owl habitat, could have been resolved much earlier if agency leaders wanted it to be. Indeed, choices were made on owl management within the agency from the mid-1970s onward that relied on mapped information that was as accurate as the information used to make other more routine choices. But the consistent message played to the outside world was that we lacked a good inventory of old-growth or owl habitat information, and had to get it before long-term management choices could be made.

Another example of tactical information management was in the way that agency leadership portrayed the support for the conclusions of the SEIS. Associate Chief George Leonard reportedly told several groups that half of the agency biologists supported it, and half did not, when in fact, only a handful of agency biologists were willing to take the stand in court in support of the SEIS's conclusions. It was clear to agency leaders that they had a problem with their staff biologists and researchers if the SEIS got into court, so they deliberately overstated the scientific support for the analysis so as not to discredit the Chief's decision.

All of these agency responses to the owl controversy were individually and organizationally rational, given traditional perspectives, but they were not terribly productive from the point of view of the rest of society. Information that should have been generated and legitimized was not. Creative, new approaches to multiple value management were not explored. A new political consensus on national forest management was not assembled, and its possible parts were made to fit even less well. The time had come for a different kind of response to the evolving owl issue, and agency officials were unable to carry it out.[5] Even those who tried were bound excessively by the character of the agency as it had evolved and been perfected over time.

Failures of Leadership, Problems of Cognitive Bias

Some of the problems were the result of the activities and styles of the leaders of the federal agencies. While the worst of this behavior was evident in the patently political moves of FWS and DOI leadership, FS leaders deserve their share of the blame for the failure of the policy process to act more aggressively and conclusively. At times, agency leaders appeared paralyzed by events and frustrated by their obvious ineffectiveness at finding a way out of the spotted owl maze. Sometimes it appeared that the more they struggled, the deeper and more entangled they became, much like an animal caught in quicksand.

There is no doubt that FS leaders were in a difficult position since the issue itself was intrinsically difficult to solve, and no one else in the policy process wanted to let them off the hook. But it was also the case that many of the actions taken by agency leaders were counterproductive to the long-term resolution of the issue. Indeed, some observers of agency behavior argue that FS leadership never truly believed that they would have to deal conclusively with the issue as their political supporters on Capitol Hill would take care of it for them.

Effective leadership involves several activities: First and foremost, it requires leaders to articulate a vision for their organization and to mobilize support for that vision within and outside the agency. Relevant to their image of the organization is their role of facilitating the aspirations of the agency's staff. There is wisdom in staff attitudes and objectives, and it would be foolish not to listen closely and find ways to enable staff members to do their jobs with creativity and enthusiasm. A third role for agency leadership is that of change agent. Since the agency exists because actions are needed to deal with current problems or likely future states, effective agencies constantly seek change in the way that they are doing things. While sometimes agency leaders can control their external environment and succeed by creating an unchanging organization, rarely can they do so forever. Sooner or later, they will need to foster organizational change responsive to changing circumstances.

For agency leaders, the role of change agent means promoting innovation and responsiveness inside their organization while simultaneously promoting understanding for the policy change outside the organization. They must work at gaining the resources needed to implement change, and the political and organizational support to allow it to happen. Building political concurrence is a required activity of leadership, particularly as the power to act has become increasingly dispersed across organizational units and political entities. Indeed, even the power of the president has been described largely as the power to build coalitions of support for desired actions.[6] Those who have failed to do this have failed to accomplish their policy objectives.[7]

A fourth role for organizational leaders is to act as a translator and communication channel between the internal operations of the agency and the external world. Those outside the agency need a clear and honest understanding of the abilities and inclinations of organizational staff, and those inside the agency need a clear and honest assessment of the policy environment in which they are operating. This role of translator is difficult in part because there is a strategic dimension to the actions of leaders in the negotiations that occur within and outside their organizations, and in part because they perceive some of their power coming from their role as the gatekeeper of organizational information. This dimension is overplayed by leaders, however, and they would be better off conveying honest assessments in both directions, particularly since the days of absolute control over information are long past. Leaders who are not true to the knowledge base of their organizations can be exposed and their credibility challenged.

The final two roles for agency leaders relate to the image that they set as it is perceived within and outside the agency. Agency leaders must model good behavior if they expect it of their staff. The most hardworking organizations are those whose leaders are equally and visibly hardworking. Whether they want it to or not, their behavior becomes a role model for aspiring leaders in their organizations, and that makes it doubly important, for not only do their actions have impact on today's issues and choices, but they have a longer term effect on the styles of management carried out throughout their organizations.

Agency leaders are often the most visible representatives of the agency to the external world. They represent the public persona of their agency, and hence their role as "image of the organization" becomes very important. If they are charismatic and obviously knowledgeable, the organization is viewed similarly. If they are seen as dull and uninformed, the agency loses some of its luster as well. For most people in the policy world, their primary experience of the FS is with agency leadership, and hence they translate the attitudes, styles, and behaviors of the leaders to the organization as a whole.

The experience of the spotted owl case suggests that agency leaders did not live up to their obligations defined by these six roles. The long-term vision of the FS and the national forest system was at best murky; they often acted in ways that were not understood by their staffs and sometimes counter to staff hopes and wishes; and communication between the agency and the external world was filtered considerably. In addition, agency leaders were seen less as agents for needed long-term change than as part of the problem that needed solving. The staff was demoralized by their behavior, and the image of the agency as a professional, technically based source of credible choices declined.

Part of the reason that FS leaders were ineffective in dealing with the owl issue was the result of underlying biases in the ways that humans think. The cognitive psychologists tell us that our powers of judgment are biased systematically, and that to be effective, decisionmakers need to be aware of these biases.[8] In the owl case, the effects of cognitive bias were evident in the behaviors of many of the involved parties. Some of these biases can be characterized as follows:

- We tend to generalize from our own personal experiences into situations that might warrant a closer look. Hence if my limited experience with environmentalists is that they are loud and abrasive, I will tend to believe that all environmentalists are loud and abrasive and act toward them accordingly.

- We tend to perceive things selectively by filtering and evaluating the information that we take note of. Of the millions of stimuli that surround us daily, it is necessary to select those we recognize and deal with, and this filtering process is influenced by our past and the maps we use implicitly to make sense of the world around us.

- Selective perception is biased in a direction in which we would like reality to be. For example, we tend to hear what we want or expect to hear. That is, we tend to selectively make sense of information conveyed to us in a way that confirms our prior sense of what the information should say.[9]

- Perception also tends to be unduly influenced by how information is framed. For example, in one interesting experiment, one group of individuals was asked to estimate the product of a set of numbers arranged in ascending order ($1 \times 2 \times 3 \times 4 \times 5 \times 6 \times 7 \times 8$), and an equivalent group was asked to estimate the product of the identical set of numbers arranged in descending order. While obviously the answer is the same, the estimates are vastly different with the first group providing much lower estimates. The existence of the first few numbers tended to frame individual judgments.

- We tend to act on our selective perceptions in ways that make them self-fulfilling prophecies. If I believe that someone is acting in a certain way because he or she is a jerk, and I act with hostility toward them, not surprisingly they act back the same way and confirm my prior hypothesis.

- Individuals also tend to be poor at perspective-taking, that is, understanding the substantive and psychological viewpoint of others such that they have a good sense of what is motivating the others' behavior. If you do not truly understand another group's interests, it is hard to develop effective responses to them.

- Decisionmakers also tend to be overconfident in their own abilities to make wise choices, and to discount uncertainties in the information on which their choices were based. Further, they tend to seek information that confirms their prior choices, and tend not to collect or acknowledge information that might call into question the direction that they chose.

- Individuals tend to get socialized fairly rapidly into the value structures that surround them, meaning that what you believe is heavily influenced by those with whom you associate. Decisionmakers rely on these networks of associates for good quality information and right thinking, and as a result, tend to have their prior judgments reinforced.

- We all tend to get entrapped by prior decisions and courses of action, and act aggressively to maintain direction, in part to minimize the energy required in changing direction, and in part to rationalize prior choices.

All of these normal cognitive biases were evident in the development of the spotted owl case, both on the part of agency leaders and many others involved in the case. For example, some level of misjudgment on the part of the FS Chief and leadership comes from the tendency to generalize from one's own experience. When they were line officers on districts fifteen years previously, they were able to simultaneously satisfy the sets of demands placed on them by constituents and Congress, not the least of which was getting the cut out. The timber was there, and other demands were not so excessive, and the tendency is to generalize from this personal experience into the present. "I was able to get it done, why can't they?" Just like it is hard to transfer the lessons of your own experience to your children (since they rarely believe you until they have experienced problems on their own) it is difficult to substitute information transferred from lower levels in an organization for personal experience at the top. The cognitive tendency is to fall back on what the leader truly knows because he/she experienced it, and that experience may well be out of date.

The situation also showed a tendency on the part of the involved agencies to get entrapped in courses of action established in the past. The overriding concern with getting the forest plans out, no matter what, overshadowed the developing owl and old growth issues. Continuing to pursue the traditional behaviors described earlier in this chapter even after some became dysfunctional similarly represents the trap of tried-and-true behavior. The knee-jerk reaction of the FS leadership to the environmentalists' counterproposal in the appropriations battle of 1989 also illustrates the polarizing and entrapping effects of defending a prior position. Even the Chief's dispute of The Wilderness Society's old growth data in the Congressional hearings in the summer of 1989 can be seen as entrapment, since his role as agency defender promoted his defensiveness. Rather than begin to address new issues and proposals on their merits, we tend to dig in deeper as the going gets tougher.

The effects of framing can be seen in a number of places in the history of the dispute. The establishment of the fairly arbitrary five percent ASQ drawdown as the maximum allowed impact of the SEIS owl plan became a focal point that defined a line in the sand that the FS leadership was unwilling to cross. The magic 500 number for species viability, the easily remembered 1000-acre number for old growth habitat areas, and others were arbitrary distinctions in the universe of possible numbers that became real determinants of wildlife policy.

The visual effects of overflights in the Pacific Northwest were also dramatic mobilizers of Congressional action, as the view from the air revealed an image of a set of scattered fragments of forest in a set of clearcuts. Congressman Jim Jontz and staff members from Sid Yates' and Bruce Vento's offices, among others, were compelled by what they saw. The involvement of the national environmental groups was motivated similarly by the personal experiences of the groups' leaders at a meeting held in the Northwest in 1987. It is the imagery of the case that has had as much impact on public opinion as anything else, and whether the situation is portrayed as a battle of owl crazies against school children, or greedy timber profiteers against helpless furry creatures, it has had a major impact on the response of viewers.

Selective perception was evident throughout the case. Indeed, it was amazing to carry out interviews that on some days moved from representatives of one side of the issue to those of another. By and large, timber interests misperceived the concerns and character of the environmentalists, and environmental activists misunderstood the interests and character of the timber interests. No one effectively understood the Forest

Service. And what was just as amazing was that there was very little dialogue among the groups, at least in the region, that would let them test their hypotheses about each other.

There was a marked tendency on the part of all involved parties to do three things: Simplify their image of what motivated the different sides; overstate the magnitude of the demands and tactical approaches of their opponents; and ascribe malignant intent to their opponents' actions where no intent or incompetence was more likely to be the case. There was very little understanding of the complexity of motivation, the diversity of representatives within a "side," or any potential for finding common ground. Timber tended to overstate the image of environmentalists as against all economic progress, and environmentalists tended to overstate the image of timber interests as crude, dirty profiteers. Part of this overstatement was tactical, no doubt, but another part was misunderstanding born of ignorance.

It was also clear from the interviews that intragroup socialization had as much to do with the framing of the values held by the representatives of the different sides, as anything else that could characterize the involved individuals. We all tend to adopt part of the community of values of the individuals with whom we associate, and the individuals representing the timber industry, and those representing the environmental groups, are not very different in terms of socioeconomic or academic backgrounds. Rather, their near-term experience has framed their values and what they are willing to defend, and if we had shuffled individuals between experiences some time in the past, the same conflict might be playing itself out with the roles reversed. This argument can be overstated, and there is no way to test it, but clearly value socialization that comes from community and job association is a strong determinant of where an individual comes down on an issue like the spotted owl controversy.

It is also clear that familiarity with a set of individuals and their values, and our competitive nature, makes people defend their associates long after they have been discredited. You see this in the behavior of presidents who defend problematic appointees long after they should, and you see this in the behavior of individuals involved in policy disputes. They see it as their role to defend their colleagues, partly to rationalize their own relationships, and tend to get entrapped in points of argument.

These types of cognitive biases influence our behavior in everything we do. There is nothing to suggest that we check them at the door of a public policy dispute, and indeed, the history of the owl case suggests that decisionmaker bias has a significant effect on the courses of action decisionmakers choose. In addition, since often these kinds of biases become more evident as individuals get stressed and overtired, they help to explain why we tend to get worse at resolving issues with time, rather than better.

Organizational norms, traditions, and ways of doing things, and the cognitive biases associated with individuals, have a major effect on the course of policy events. Whether you view them as organizational and individual personality, or the behavioral factors that appear to "muck up" well-intended policy, their quirks and characteristics often determine how individuals and agencies respond to the demands for action that are placed upon them. Oddly enough, these kinds of characteristics are generally not well accounted for in estimates of rational policy analysis, yet their effects are unquestionable and probably undeniable. As will be discussed in the final section of this book, there are ways that one can guard against bias in individual judgment, and approaches to test for excessive bureaucratization in how an agency responds to an evolving controversy. We can do better in spite of ourselves in dealing with controversies like the spotted owl case.

THREE

Policy Implications for the 1990s and Beyond

11

The Context for Change

Previous chapters of this book have argued that American resource policy processes are fragmented and often impotent, short-term-oriented and reactive, conservative and status-quo seeking, and not particularly creative or good at seeking and dealing honestly with information needed to make wise choices. Interest groups, elected officials, and the media tend to exacerbate these tendencies. Individuals tend to exhibit cognitive bias and patterns of value-formation that similarly reinforce these inclinations. While administrative agencies have the potential to rise above these problematic characteristics, in fact they tend to look backward and seek to deal with new problems in time-honored ways, whether or not they are functional or effective. Finally, forest and wildlife policy, while sometimes aimed directly at rising above the short-term nature of the policy process, has failed to escape the limitations of the policy process, in part because the executive-level politics of the 1980s pushed in a direction that was in conflict with a broader set of evolving public values in American natural resources.

While these behaviors sometimes took place for understandable historical, ideological, and functional reasons, the diagnosis is not very encouraging, partly because it is not limited to the spotted owl issue. Just as the owl has become a surrogate for the larger old growth issue in the Pacific Northwest, the owl controversy and the characteristics of the decisionmaking processes and institutions that contributed to it are good indicators of the state of many current natural resource and environmental systems, policies, and agencies. They describe natural systems under stress, and management institutions and decisionmaking processes that are ineffective at managing that stress or at crafting an appropriate future.

Stressed systems and ineffective management institutions are evident at all levels of government, and in most areas of the United States and the world. The problems indicative of such stresses range from endangered species–development controversies that seem to pop into public view at regular intervals to toxic contamination problems at federally owned facilities. They include global scale issues such as the problems caused by climate change and the transshipment of hazardous materials, and local-level conflicts including numerous battles over the siting of noxious facilities such as solid waste landfills. These diverse sets of problems have commonalties not just in the fact that they all influence our collective future, but they also thrive on the medium that nurtured the owl controversy. And the degree to which we will succeed at crafting a viable future depends on whether we can overcome these problems.

It is interesting that the problems evident at the end of the twentieth century have parallels in the state of affairs at the end of nineteenth century America that led to major reforms in resource management policy. Decades of overexploitation of natural resources with inadequate and ineffective management institutions, and a political structure that turned a blind eye to resource and environmental management, produced a set of serious resource and environmental conditions. These conditions resulted in the decline of many wildlife species, in high levels of environmental contaminants in and around urban areas, and in conflicts over the appropriate use of federally owned lands.

From these problems came a twenty-five-year reform period that swept many areas of social policy, including that of natural resource and environmental management. The Progressive Conservation movement created the norms of multiple use that were supposed to be responsive to diverse human needs while husbanding resources, and management direction set by technical elites, who could craft effective future states through the benign wisdom of science. It was a well-needed and important change in government and the organization of collective action. At the same time, while the problems evident in the 1890s were serious, their solution was relatively obvious, and they could be instituted by fairly simple mechanisms. A major change for the better in the 1890s could come from simply piping sewage downstream and screening it coarsely, stopping the hunting of plumage birds such as snowy egrets and roseate spoonbills, and creating forest reserves out of the vast public domain.

The 1990s must see a comparable set of reforms in the way we approach resource and environmental management. The problems that are evident today are much more subtle, and the nature of the institutional challenges much greater. Understanding global meteorological patterns and atmospheric chemistry, forecasting patterns of genetic depletion of species like the Pacific salmon, devising global-scale institutions such as new treaties and international bodies, and crafting management direction out of a fragmented political system that allows the involvement of a much more diverse set of interests than was extant in the late 1890s— all are challenging tasks needing assistance from more basic changes in resource policies and management institutions. Indeed, some of the necessary changes are aimed precisely at updating the reforms of the Progressive movement for the conditions of the twenty-first century.

This chapter describes the context for change, and why it will be a difficult but not impossible task. The social and political context into which new natural resource policies and management approaches will be born is characterized by *heightened complexity* caused by:

- expanding and conflicting public values,
- ambiguous and conflicting norms of collective choice, and
- inherently complicated future environmental issues;

and by a *reduced capacity* to meet the demands created by these values due to the existence of:

- declining slack in the natural resource base,
- declining slack in the ability of government to act proactively due to discouraging fiscal realities,
- unstable coalitions due to fragmented power and multiple interests, and
- limited vision and guidance from elected and appointed officials, and the management institutions that they control.

While many of the challenges facing citizens and decisionmakers are daunting, there are reasons to be optimistic that necessary change can come about. The last section of this chapter describes these reasons for optimism as a stepping stone to the prescriptions for change that are contained in the last two chapters.

Expanding and Conflicting Public Values

Throughout American history, public values in natural resources and environmental quality have continually evolved. Chapter 1 described the evolution of American public values in natural resources up through the

1970s. By all indications, these trends continued through the 1980s and into the 1990s, even as the national political leadership changed. While the general public in the 1980s might not have been convinced that the instrument of environmental progress necessarily must be government, their values were consistent with trends of the preceding decades. They continued to value environmental quality high on their list of priorities for social action. For example, a 1984 Gallup poll reported that 61% of the American population felt that environmental protection should be given priority, even at the risk of curbing economic growth. Only 28% of those surveyed felt that economic growth should have precedence over environmental protection.[1]

In the 1990s, the environment still ranks high on the list of public concerns. A 1991 Gallup poll reported that 71% of those surveyed agreed that environmental protection should be given priority even at the risk of curbing economic growth, while only 1 in 5 felt the converse.[2] More than half (57%) favored taking "immediate, drastic action" toward environmental protection, while only 8% felt we should "continue as now." Seventy-eight percent considered themselves to be environmentalists, and half (51%) had contributed money to an environmental cause. A 1990 New York Times/CBS News poll reported similar findings.[3] Seventy-one percent of a national sample agreed that "we must protect the environment, even if it means increased government spending and higher taxes." Even when placed in a more local context, so that individual costs might be more tangible, 56% agreed that "we must protect the environment, even if it means jobs in the local community are lost." Thirty-six percent disagreed with the statement.

The polling data on public support for endangered species protection is consistent with the more global public concern about environmental protection. A national poll conducted by the Greenberg/Lake and Tarrance Groups in December 1991 under contract to The Nature Conservancy and The National Audubon Society found that two-thirds of voters across the country supported the Endangered Species Act, and 40% supported it strongly.[4] Nearly three-quarters of those surveyed said that a political candidate's stand in protecting endangered species was an important reason to support a candidate. The poll found that these results were consistent across all regions of the country, and were independent of political party identification. Only different income groups showed much difference toward ESA support, with lower-income voters slightly less likely to support the Act than upper- and middle-income voters, though even among voters earning under $10,000 a year, a majority (52%) supported the act. The polling results are also important because the poll was conducted at the peak of the early 1990s economic recession and at a time when endangered species issues had been well aired in the popular media. The spotted owl was a household word across much of America in 1991, and the White House and commodity interests were aggressively trying to discredit the ESA.

There is no reason to believe that the overall concern with environmental and endangered species protection will change in the future. The environment will continue to be a good media issue, continuing problems from a legacy of toxic pollution and resource mismanagement ensure that new crises will emerge, and well-organized interest groups will continue to press environmental issues onto the public agenda. Surveys of school children, the citizenry of the future, also support the notion that a strong set of environmental values will persist over time. For example, in a survey of 86,000 school children conducted by *Weekly Reader* in November 1992, cleaning up the environment was viewed overwhelmingly as the number one priority for the new president and administration. According to Adam Kudamik of Choconut Valley Elementary School in Friendsville, Pennsylvania, "We fifth-graders in Mrs. Barnhart's class think cleaning the earth is most important . . . Come on! We can do it!"[5]

The changing demographics of the United States also favor the continuing evolution of environmental values as a central public concern as the interests, styles, and values of the aging Baby-Boomers dominate the American policy agenda, perhaps for the next 25 or 30 years.[6] While there is considerable uncertainty about how the future will shape those values, there is a literature that argues that basic values are set down in the first twenty years of life.[7] The first two decades of the "Boomers'" lives were largely the 1950s, 1960s, and 1970s, and the concerns and issues on the public agenda during this period were much different than were those of the preceding generation. The race for outer space and the associated development of science education, a growing petrochemical industry and a rising awareness of environmental pollution, suburbanization and a parallel concern about the loss of open space, the social activism spawned by the Great Society of the 1960s and the war on poverty and the opposition to the Vietnam war, and a growing cynicism about mainstream institutions—government, industry, organized religion, etc.—combined to create a generation whose values were quite different from its parents.

The relative affluence of the times also created a generation less concerned with the need to strive for continued material progress, ironically even as they were consuming unprecedented amounts of natural resources. The Boomers provided a stark contrast to the preceding generation, whose most prominent memory during their formative period was the Great Depression, highlighting the need to strive for material comforts. Many of these forces supported the development of the environment as an important theme of the Boomer's values, and Earth Day and the environmental movement of the 1960s and 1970s were smack dab in the formative years of this set of individuals.

An illustration of these differing values lies in the different response to the endangered species poll described above, as a function of age. Table 11-1 cross-tabulates the response of individuals to the following question: "When you hear a news story about how some local industry is being hurt by laws protecting a bird, fish, plant, or animal, do you find that your sympathies are usually more with protecting the wildlife or more with protecting the local businesses and jobs?"[8] The Boomers, represented in age classes 40–54 and 30–39, are considerably more sympathetic to wildlife concerns than their predecessors, who are either neutral or more sympathetic to business. It is, of course, possible that as the Boomers age, they will become more concerned with economic matters (that is, age is the most influential variable and not year of birth), and some measure of this has already taken place. But it is more likely that some measure of the higher level of environmental values held by the Boomers will remain as they age.

None of this is to argue that economic well-being will not be a major concern of the public in the future. While the changing demographics of the country will reinforce the evolving trend toward environmental values, Americans will certainly be concerned simultaneously about their own economic status as well as the position of our country in relation to global markets. Indeed, the legacy of public and private debt, and the shrinking segment of the population at work (as the Boomers move into retirement) will bring economic concerns to center stage. In fact, concerns about the economy have been the top-ranked problem perceived by Americans since the end of the Vietnam War reflecting a return to the Depression-era concern with the state of the economy.[9]

When thinking about future concerns over economic and environmental status, it is important to remember that the public wants "well-being." This includes both economic prosperity and a decent living environment. These two sets of values are basic human motivations, and they exist simultaneously in individuals. While a fragmented political process rarely forces individuals to make choices between these values, when forced to make a choice, short-term economic prosperity is likely to win out. But

TABLE 11-1

Percentage of Population Expressing Sympathy for Wildlife or Business,
By Age[8]

	Age (years)				
	18–29	30–39	40–54	55–64	>64
Wildlife	61	58	51	37	31
Business/jobs	19	23	28	36	41
Margin	+42	+35	+23	+ 1	−10

the public (and their elected officials) do not want to make that Faustian bargain, either for themselves or their children. They understand intuitively that sustainable economic prosperity will not come from a policy that trashes the environment, nor is a high level of environmental quality likely to result from a policy that promotes environmental protection at any cost. The environment is thus an economic issue partly because impoverished societies are less likely to invest in long-term environmental protection. The obligation of our public choice processes and our resource management agencies is to find credible and creative ways out of the dilemma, and to help craft public understanding and political concurrence for policy direction.

Ambiguous and Conflicting Norms

The public policy dilemma created by a public that simultaneously values environmental protection and economic prosperity is made more complex by other ambiguities and uncertainties in American norms of collective choice. Just as the owl controversy reflected the conflict between an aggressive timber program and an activist endangered species policy, other conflicts have emerged over what is valued and who is trusted in the way we make and implement choices in our collective interest. These include ambiguities about:

- the role of government, and how much it should be trusted to act in our best interest;
- the role of experts versus that of individual citizens;
- the role of interest groups;
- the ways in which policy processes measure value (science, economics, politics, or ethics).

It is easier to make a binding decision when the values that underlie that decision are clear and consistent, but that is not the case in American society in the early 1990s.

Throughout American history, there has been a love–hate *relationship with government,* and it is unclear where the balance in that continuum lies today. Most people would agree that the government-created and -controlled programs of the Great Society of the 1960s were fraught with problems, and the boom-and-bust patterns of hands-off government in the 1980s were equally problematic. What remains today is a great deal of cynicism about the effectiveness of either government or corporate decisionmaking and a great deal of uncertainty about the appropriate role of each.

Surveys of the American public demonstrate declines in public faith in government. According to the University of Michigan's Center for Political Studies, the public's faith in government has been dropping steadily for several decades. In 1964, some 63% of adults said they could trust the government most of the time. By 1988, that figure had dropped to 37%. In 1964, some 70% said government was run for the benefit of all. By 1988, some 67% said government was run for the benefit of a few special interest groups.[10]

Patterns in American voting behavior also confirm these trends in American attitudes toward government. With the exception of the turnout for the 1992 presidential elections, Americans increasingly have stayed away from the voting booths. For example, in the 1960 presidential election, approximately two-thirds of those eligible to vote did so.[11] In 1988, that number had dropped to roughly half of eligible voters.[12] Only a third of eligible voters vote in Congressional elections.[13] The lack of participation among young people is even worse: Only a third of eligible 18- to 20-year olds voted in the 1988 elections,[14] and 20% vote in Congressional elections.[15] These low voting levels, particularly among young adults, have resulted in the United States having the one of the worst voting records worldwide. Indeed, a Congressional Research Service study examined 143 elections in 28 democratic nations between 1969 and 1986 and found that the U. S. placed dead last.[16]

Public alienation is near an all-time high.[17] More than six in ten American adults feel a sense of powerlessness and disenchantment with the institutions that influence much of their lives. For example, the Gallup poll reported declining confidence in the U. S. Congress from 42% of respondents in 1973 to 32% in 1989.[18] Indeed, less than half of the respondents expressed significant levels of confidence in many U. S. institutions including the Supreme Court, banks, public schools, television, organized labor, and big business.

The declining attitudes of the American public toward government, elected officials, and societal institutions create very real dilemmas in the ways that policy issues play themselves out. They add a set of uncertainties about what government should do, and reduce the trust often necessary for elected or appointed officials to craft solutions to public problems. Chapter 6 described some of these factors in the context of the spotted owl controversy, including uncertainties about the role of government in dealing with regional economic transformations, the obligations of the federal government as a dominant landowner in many regions, and the appropriate division of governmental responsibility among the three levels of government (local, state, and national).

Many of these same ambiguities exist today across other areas of resource and environmental policy. Regulation as a means of commanding appropriate behavior to insure adequate environmental protection was questioned during the twelve years of Reagan and Bush control, and while many people today would support regulatory control beyond that sought during the 1980s, how much regulation, and to what end, is uncertain. For example, at what point does regulation of development on private lands become a taking under the U. S. Constitution? This question is a very real one being played out in the courts today. It is an issue of great significance to those who seek to limit development on private lands because of its impact on endangered species, coastal processes, and open space esthetics.

Incentives and subsidies are another dominant set of tools that allow federal influence on state, local, and private decisionmaking. Whether or not they are effective and appropriate is (like regulation) also in question. Both drain the federal treasury and as a result must be on the chopping block when federal officials seek to deal with the federal deficit. Unfortunately, such programs also tend to create vested interests and result in mutual dependency relationships between government and private individuals as well as the continuation of such programs long after the public purpose for their creation has been exhausted. Federal agriculture, forestry, mineral, and rangeland programs often fall into this category. As a result, the objectives of federal land management can get subverted by these dependency relationships, leading to mishandling of such issues as incompatible uses on national wildlife refuges, the direction of fire policy on national forests and parks, or the perceived rights of concessionaires on all public lands.

Part of the ambiguity about the appropriate role of government lies in a question about *the role of experts* versus that of individual citizens in setting policy direction. While democratic norms suggest that each individual has the right and obligation to be represented in collective choices, as issues have become more technically complex, we have tended to elevate the role of the technical expert. Scientists, decisionmakers, and the general public have been willing co-conspirators in this deification of expertise. If science can provide the answers, then elected officials and the general public do not have to make difficult choices. Numerous technically based agencies have been established to deal with what are effectively social problems. Hence we rely on the Federal Aviation Administration and the National Transportation Safety Board to worry about hazards when we fly, the Consumer Product Safety Commission to regulate hazards in the products we purchase, and the USFS, the FWS, the BLM, and the EPA to take care of the quality of our environment. In many ways, the responsibility of each citizen to participate has been lateralled to agencies, who presumably take care of us in a benevolent, right-seeking way.

Many resource policies are framed in this manner. The 1973 ESA authorizes biologists to determine what species need federal assistance based solely on the best scientific judgment. The 1976 National Forest Management Act mandates the USFS to plan the future of each national forest unit by applying fairly elaborate technical analysis and inviting the input of the public. Primary air and water quality standards are set through risk–benefit analysis, assessing the potential risks of various pollutant levels to human health. The theory is that through benevolent science and its inherent technical wisdom, answers would be provided and society would move forward.

It has not worked out that way, and the history of the late 1970s and 1980s tended to shatter any remaining myths. The 1979 accident at the Three Mile Island nuclear power plant called into question the effectiveness of Nuclear Regulatory Commission oversight. Scandals over the conduct of the EPA's hazardous waste clean-up program in the early 1980s fed public cynicism about the ethics and quality of technical agency decisionmaking. The image of the Challenger space shuttle vaporizing into a white cloud in a brilliant blue sky cast doubt upon one of the most cherished accomplishments of American science, the ability to move humans through space to the Moon and back safely.

Public, mediagenic controversies such as over the significance and direction of global climate change have created at least confusion in the general public, if not a sense of hopelessness about whether there are ever correct answers to these large-scale questions. The obvious and well-publicized politicization of science that occurred in the FWS and FS decisions on the spotted owl undermined the public's willingness to trust government agencies, and an administration that opposed the direction suggested by good science continued to feed the public's mistrust. All of these events suggest that not only is scientific and technical judgment limited by our state of knowledge, but that scientists and technical experts are subject to human frailties. It is clear that the inherent uncertainty in technical judgments requires that experts exercise discretion. Historically, however, this judgment has been subject to bias, corruption, and politics.

What remains, and is an important component of the context for policy change in the 1990s, is a lingering distrust of government agencies and technical experts, along with a parallel fear that individual citizens can never fully comprehend the important issues of our day. Are we satisfied with biologists making value-laden choices for the rest of society, as could be the case with the implementation of the ESA? Can we expect politicians and the general public to be cognizant of the dynamics of complex issues, and able to rise above their tendency to seek short-term, tangible objectives at the expense of goals more effective in the long term? Neither extreme is likely, and the ultimate effect of this duality is to require that both sides work harder: scientists, at communicating and lending credence to their wisdom and recognizing its limits; politicians and citizens, at becoming informed, and communicating their interests to scientists.

A third ambiguity about our social choice mechanisms lies in the appropriate *role of interest groups*. As political parties have declined, and other social institutions such as organized religion have diminished in significance in shaping public opinion, interest groups have risen as a dominant boundary-spanning force. They act as a two-way channel of information and influence, whereby electorally significant public concerns are communicated to elected officials, and the related actions of the elected officials are communicated back to the public. As issues have become more complex, interest groups have played an important role in educating the public and informing decisionmakers, and they are one of the ways that the general public can be represented in public policy decisionmaking. Interest groups also play important roles as "fixers" of implementation, that is, as monitors and watchdogs of administrative agencies as the agencies carry out public policies. Indeed, the importance of having interest group watchdogs was highlighted through the 1980s, as agency capacity and quality declined, and as scientific decisionmaking became more political. Finally, as political campaigns have become increasingly expensive, interest groups have risen as a major electoral force. All of these roles—communication, citizen representation, fixers of implementation, and political fundraising—have ensured that interest groups are major forces in public policy.

While interest groups are important elements of our collective choice processes, their dominance is problematic in several ways: Since they are primarily single-interest oriented, they promote the fragmentation of values and public purposes so threatening to our ability to articulate dominant sets of social values that help agencies and decisionmakers craft effective policy direction. In addition, since their principal role is to influence the course of policy direction, their biased involvement in information-provision is often misleading to decisionmakers and the public at large. They also become vested in certain directions, deserving of the pejorative term "special interest group," as they lobby for policy provisions that benefit their members at a cost to others, and oppose changes in policy direction that might well be warranted. And the proliferation of interest groups often has led to impasses, as few compromises can be arranged to meet all groups' interests. At times we seem to face a situation of public impotence, as disputes arise, fester, and fail to scab over. Healing seems impossible.

Multiple and conflicting attitudes toward the heightened involvement of interest groups in public policy processes lead to some very real policy dilemmas about how policies and policy processes are structured. The right of the citizenry to challenge government direction is exercised largely by interest groups, and at times, endless courtroom appeals have damaged our ability to take action. Sometimes delay is fine, and it often works to the advantage of environmental objectives—stopping or slowing a proposed development, so that information and opposition can be mustered. But at other times, we would be better served by action, and the excessive use of judicial channels works against the public good, and leads to a very real question about how much access interest groups should have to judicial review. This issue was clearly on the table in the old growth and owl debates of the early 1990s, and was discussed in the debate over revising the regulations governing national forest planning.

Issues over the appropriate roles of government, interest groups, and scientists suggest another even more fundamental ambiguity that exists in our collective choice processes today: *whether we measure value through science, economics, politics, or ethics.* All decisionmaking processes need to assess the magnitude of different values, and the way in which such measurements are weighted has significance for what actions are ultimately taken. For example, in the case of endangered species protection, if we weigh potential courses of action based on scientific value, keystone species would receive top priority for protection. We would focus our protection efforts on saving tropical rainforest habitat where much more diversity exists, and write off Pacific Northwest forests, and genetic basket-cases like the California condor.

An approach that relies on economics as a valuation process would focus on those species with a tangible monetary value, either as a source of a good (like the cancer-fighting properties of taxol extracted from the bark of the Pacific Yew) or as a result of calculating nonmarket value based on a species' existence or option values.[19] This approach would extend protection primarily to those plants and animals that either were known to have commercial value, or were understood and popular with the general public. Few species would survive the clash between economic interests such as those in the Pacific Northwest.

Politics provides a different construct for valuing public choices. Political processes value lifeforms with a human face, that is, those that attract the support of an interest group, agency, or elected official with enough influence to win in the political bargaining contest. Political processes are swayed by the values and arguments of science, economics, and ethics, but which species are protected depends on who wins, and winning has more to do with the characteristics of the human guardians than those of the species in question. Do they have enough political influence to get action on a proposed action? Is the species symbolic enough to attract general public attention and support? An endangered species protection program based solely on political values would tend to protect the warm fuzzies: popular, mediagenic species.

The value structure provided by a perceived sense of ethical responsibilities sometimes reflects all of the preceding values, and sometimes prescribes a completely different set of values. Ethics define rights and obligations, and hence structure the nature of the relationship between different elements of a society. However, because ethics are defined by humans even if they are divinely inspired, they usually reflect other human values. An approach to endangered species protection based on ethics generally takes an all-or-nothing tack. If our ethical responsibilities include the obligation to not consciously let a species or subspecies go extinct, then the protection policy must be extended to all species or subspecies. We must spend our time protecting beetles, clams, and rats as well as eagles, condors, and spotted owls. Indeed, since programs generally have a budget constraint, this approach of necessity bogs down into protection for the most critically endangered, even if triage would protect more biological diversity over time.

American policy processes consist of an amalgam of all four of these valuation approaches. Which is most influential on decisionmaking has varied at different stages of the country's history, reflecting different levels of knowledge, affluence, and popular opinion. What is unique about today's situation is that we have unprecedented levels of scientific understanding of issues like the loss of biodiversity, at the same time that we have a declining economic situation and a fragmented political system. How far people are willing to go to extend survival rights to other life-forms is less clear today than at any time in the preceding twenty years.

These ambiguities add complexity to the task facing policymakers and agency leaders because while they are dealing with substantive issues, they simultaneously resolve these more fundamental governance issues and articulate the prevailing set of social values. Hence, when the Senate Appropriations Committee defines limits to judicial review to provide some certainty to a funded timber sale program, as was attempted in the 1990 appropriations bill, it implicitly establishes guidelines on the role of interest groups in resource policy. When the U. S. Congress debates whether or not a greater level of balancing of interests should take place within the ESA, it is implicitly defining a set of roles for experts and an operative set of ethics. The underlying uncertainty about where the public is on many of these fundamental questions adds considerable complexity to choices that are already difficult.

The Character of Future Issues

The nature of future environmental and resource issues will add an additional measure of complexity to the ability of individuals and organizations to craft effective policies and management strategies. For example, many future issues will demand cross-boundary solutions, since ecosystems do not respect political boundaries, nor do the air, water, nutrients, organisms, or processes that define them. Yet most issues that we will face in the future cut across political boundaries both nationally and internationally. Solutions to these problems necessarily must be cross-interest, as the array of interests with stakes in environmental matters grows in size and complexity. If nothing else, simply the increasing size of the world's population, coupled with heightened expectations in all regions of the world, will force interconnections between nations and regions.

Increasingly, decisionmaking will require a better understanding of complex physical, natural, and social processes, and their implications for effective resource management. Implementing plans to slow down processes of global climate change requires a comprehensive understanding of atmospheric chemistry and meteorology, and the human and nonhuman processes that create greenhouse gases. Managing forests for a broad array of landscape values requires much more understanding of terrestrial system processes and social system characteristics than has ever been the case in either public or private sector forestry.

Contrasting the situation facing a FS land planner in the Pacific Northwest in the early 1960s with his counterpart in the 1990s is illustrative: In the 1960s, the planner had a fairly straightforward task that did not require a lot of creativity or new knowledge. He faced one dominant objective, harvesting timber on a moderately sustainable basis. Old-growth was decadent timber, beyond the point where it would add a significant amount of wood fiber to the system, and hence, it was appropriate to harvest it and replant with a Douglas fir monoculture. Clearcutting was efficient and effective for regeneration. Few other human demands had to be dealt with.

The 1990s FS land planner is faced with a much more complex job. Most clearcutting is out, so she has to choose from a much more creative set of silvicultural strategies. She has to simultaneously consider the biological requirements of ecosystems and species of plants and animals all the way down to fungi, with the needs of rarer organisms trumping those of more abundant and commercially valuable species. Her work is defined by the legal requirements of NEPA, the ESA, the Freedom of Information Act, and the NFMA, and her job environment is influenced by occupational health and safety rules, and antidiscrimination rulings. Her decisions must be influenced by numerous human concerns and demands, including those of resource-dependent communities, industry, recreationists, and preservationists. Whereas once individuals who were inept at dealing with people could find a satisfying career in forestry, the 1990s ecosystem manager must be as good with people as she is with trees. Forest ecosystem management in the future will be even more information-intensive and complex, with regional scale concerns about wildlife corridors, environmental degradation, and international trade overlaying a much more complicated situation at the site-level.

The benefit of these trends is that they provide creative people with a large set of interesting and challenging problems. At the same time, the technical complexity of future resource and environmental issues requires a better informed public, and professionals who possess much more effective skills at organizing and understanding a wide array of information. This in turn means that agencies and decisionmaking institutions must be effective at acquiring needed information and acting on it.

Declining Slack: Stressed Natural Systems

Sometimes when decisionmakers are faced with the dilemmas caused by an expanded and conflicting set of demands, and ambiguous direction on how to choose among them, they can find a way out by giving all sides what they want. But that is only possible in situations where there are slack resources, and increasingly that is not the case in American natural resource and environmental policy. Evidence of stressed natural systems and overtaxed physical infrastructure exists at all levels of government and in most regions of the country. These stresses range from incompatible uses on national wildlife refuges and adjacent lands problems at national parks, to overtaxed facilities at state parks and limited solid waste landfill space at the local level. Analysts have claimed that pollution and popularity are choking the national parks,[20] excessive and improper logging has eroded the ability of many national forests to grow trees,[21] and overgrazing has damaged the capacity of BLM and FS range lands.[22] The Grand Canyon and Yosemite Valley have air pollution problems, and the Everglades have seen a 90% reduction in wading birds since the 1930s caused by major changes in the use and distribution of water in south Florida.[23] Independence Hall, the cradle of American liberty, has a leaky roof and adjacent buildings are closed to public use because they are unsafe,[24] and logging in the Olympic National Forest has resulted in so much siltation of streambeds that salmon and other fish populations have been decimated.[25]

While some of these claims tend toward rhetorical excess, in fact there is considerable evidence that most public land systems today are threatened by overuse or inappropriate use on-site, activities on adjacent lands that threaten the integrity of public resources, and limited staff and monetary resources. The national forests provide a good example of declining slack due to stresses caused by improper on-site management. Indeed, what is amazing is the extent to which criticism of the agency's management direction has become public and is generally assumed to be true. For example, Forest supervisors in Region 1 (Montana, Idaho, and the Dakotas) wrote to the Chief late in 1989 expressing their concerns with FS direction:

These are troubling times for many of us. The values of our public and our employees have been rapidly changing and have become increasingly divergent, increasing the level of controversy surrounding the management of National Forests. We are seeing a drastic increase in the number of challenges to our land and resource management activities, challenges which are not easily overcome by throwing more money at them or working harder to educate our publics or increasing the amount of documentation. Many people, internally as well as externally, believe the current emphasis of National Forest Programs does not reflect the land stewardship values embodied in our forest plans. Congressional emphasis and our traditional methods and practices continue to focus on commodity resources. We are not meeting the quality land management expectations of our public and our employees. We are not being viewed as the "conservation leaders" Gifford Pinchot would have had us become . . . We have become a dysfunctional Forest Service family. . . . The stress in our organization is serious. A "can do" attitude will not save us this time. We are spread too thin. It is time that we start dealing with our internal problems, before we crack apart at the seams.[26]

A similar message was sent to the Chief from the forest supervisors in Regions 1, 2, 3, and 4 following a 1989 meeting that is generally referred to as the Sunbird conference. Representing roughly half of the national forest system, these supervisors blamed the state of affairs on intraorganizational factors, constraints placed on the agency by the administration and Congress, and the changing values of the public: "During the first half of this century, we operated in an environment of rural values. We are now operating in an environment where about 5% of the population relates to a rural setting."[27] Their prescription was fairly straightforward: Work more with outside publics and elected officials, recognize changing values and respond to them, use the NEPA process honestly as a means of making good choices, and change internal management processes to shift behavior.[28]

Numerous recent studies have indicated problems in the National Park System as well, related to on-site management issues and the effects of activities adjacent to national park units. For example, the report of a blue ribbon panel on the state of the parks on their 75th year anniversary noted the threats facing the parks from adjacent lands, and highlighted administrative and management problems within the NPS, including a decline in professionalism, poor training, politicized decisionmaking, poor communication, an eroding budget, and inadequate management and research capabilities.[29] The report concluded that, "the ability of the National Park Service to achieve the most fundamental aspects of its mission have been compromised."[30]

National wildlife refuge lands are also stressed by on-site problems, adjacent lands issues, and limited administrative resources. A 1991 FWS survey of national wildlife refuge managers confirmed the earlier conclusions of a 1989 GAO report: that a significant portion of the refuges were experiencing secondary uses that were harmful to the fish and wildlife that the refuges were supposed to protect. "Incompatible" secondary uses were found on 301 of 478 refuge units surveyed, including military air exercises, oil and gas drilling, and a variety of recreational activities including offroad vehicles, airboats, jet skis, and waterskiing.[31]

Deleterious off-site activities are even more of a problem for national wildlife refuges (NWRs) than for national parks, because most refuges are fairly small islands in a sea of urban or agricultural lands, and there are many more refuges than national parks. One of the more visible controversies over adjacent lands problems in the wildlife refuge system came in 1985, when the contamination of the Kesterson National Wildlife Refuge in southern California by high concentrations of selenium in irrigation water flowing through the refuge became public. Government scientists had known for quite some time that selenium, a naturally occurring mineral in the San Joaquim Valley, was accumulating in increasing concentrations in the refuge, but the public knew little about the contamination until the FWS linked widespread death and deformities in refuge birds to the high selenium concentrations in refuge waters.[32] Today, Kesterson is listed as a hazardous waste site and its wetlands areas have been filled in to keep waterfowl from landing there.[33]

Kesterson is in fact only the tip of the iceberg of contamination issues threatening wildlife refuges. A refuge-wide survey in 1983 found that toxic chemicals were causing water-quality problems at 121 refuges and other FWS sites.[34] As with other federal lands with huge toxic contamination problems, including military bases and DOE national laboratories, the magnitude of the NWR toxics problem is just beginning to be understood, and because much of the problem comes from adjacent, privately owned lands, its solution is much more difficult. The magnitude of the long-term cost for cleaning up the refuges could be massive, with estimates for Kesterson's clean-up running as high as $195 million.[35]

Evidence of limited slack in stressed systems comes from controversies other than public lands. Many wildlife species of commercial and recreational value have declined in recent decades, despite active management to build their populations. These species include the population of mallards, which are now at 55% of their 1955 population, even though a majority of the FWS management dollars are spent on waterfowl habitat enhancement.[36] Similarly, populations of Pacific northwest salmonids, perhaps the single most important recreational and commercial fishery in the United States are at record low numbers. A comparable picture exists for striped bass populations on the east coast, and lake trout populations in the Great Lakes. Oyster landings are roughly a third of what they were a hundred years ago, due to siltation-induced habitat loss, and changes in the freshwater–saltwater balances in many estuarine areas. Populations of nongame birds such as the black tern, loggerhead shrike, grasshopper sparrow, lark sparrow, and lark bunting are down, caused primarily by a loss of wetlands, forest, and grassland habitats.

Many of the growing numbers of large-scale endangered species issues are another indicator of natural systems that are fully appropriated to a set of public and private land uses. The endangered species that pokes its ugly (or lovely) head into a development pattern is more than likely a mere indicator of ecosystem loss or damage. In many ways, it is unfair to blame the Pacific salmon for conflicts over water use and quality in the Columbia River basin, the California gnatcatcher for questions about the level of development of the California coastal scrub, or the northern spotted owl for battles over the future of the old growth Douglas fir forests of the Pacific Northwest. Issues over the management and development of these systems have been extant for many years. The fact that species are endangered is simply a further indicator of systemic stress. For some time, all of these areas have needed a process of deliberate planning to avoid and resolve conflicts among users, and to avoid and mitigate damage to natural processes. It is unfortunate that endangered species laws have become the regulatory lever of last resort to force this kind of planning. We should not have to wait for a bird, or a fish, to force us to act.

Through all of these examples, it is not that we lack land and natural resources. Indeed, the American landscape is a huge, diverse, and magnificent estate. Rather, limited slack and hence difficult choices, are a result of: what we have done with and to our environment, rising and conflicting demands placed on these resources facilitated by statute and

changing political organization, a public choice process that does not succeed at promoting long-term rational courses of action by offsetting the
incentives that encourage most individuals and organizations to seek
short-term rational behavior, and the fragmentation of political power
and public purposes so evident in our policy and politics.

Declining slack is also a function of improvements in how we understand the world. We have a more mature sense of the interconnections of
activities across time and space. Who would have thought that wildlife
managers who were maximizing species diversity (and deer populations)
by creating edge, through numerous small openings in the forest, would
simultaneously doom a set of interior-dwelling species, such as the spotted owl, northern goshawk, and marten. Ignorance may well have been
bliss, as decisionmakers now discover that the ecological maxim—everything is connected to everything else—means that no decisions are easy.
At times, when faced with the complexity of many resource management
decisions, one gets the sense that many well-intended decisionmakers
simply want to pull the covers over their heads and go back to sleep.
Their problem is limited slack in part due to stressed systems, and it is a
problem that we will have to continue to cope with for quite some time.

The prevalence of stressed natural systems does not generate an absolute constraint on future development. Rather, future resource development requires more creativity, deliberation, and forethought than has
been needed in the past. Sustainable management of a finite set of resources is necessary and possible. Some of the concepts of forest management that are codified in federal law are appropriate and can be effective,
if they are implemented in accordance with the spirit of the law and the
realities of the times. In other areas, such as federal mining, the laws
themselves need to be updated to reflect the current state of natural and
environmental resources. In the United States, we are finally to a stage
of development where a new generation of resource and environmental
policies is necessary and appropriate: a set of policies based on an understanding that resources are finite and limited, human and natural processes are interconnected across vast geographic areas, and our success
is equally dependent on dealing with current and future human needs
and wants.

Declining Slack: Fiscal Realities

One key component of the context for management is a fiscal situation
that has contributed to the stresses evident over the past decade, and will

continue to make the job of resource management highly challenging. The financial resources available to resource managers have been declining or at best constant in the face of increasing demands and needs. For example, if the budgets of the Agriculture and Interior Departments and the Environmental Protection Agency are taken as a proxy for spending, it is clear that the revenues available to these agencies have declined as a percentage of the federal budget, and the agencies have lost purchasing power over the 1980s. The budget for these three units of federal government declined from 7.6% of the budget to 4.6% from 1980 to 1990.[37] While the total outlays for these departments increased from $44.8 billion to $56.9 billion over the decade, their purchasing power nevertheless declined.[38]

A quick look at the budgets of the major public land management agencies reinforces this overall trend of declining budgets in real terms. For example, a 1988 analysis of the budgets of the FS, BLM, NPS, and FWS indicated that the FS, BLM, and NPS had lost purchasing power over the decade of the 1980s.[39] While the FWS showed a 22% increase over the period, much of this was a result of legislation in 1984 that expanded the Dingell-Johnson program, whereby taxes on fishing equipment are used to provide grants to states for sport-fisheries projects.[40]

Not surprisingly, some programs were hit harder than others by declining resources. For example, while FS fish and wildlife habitat management and research showed modest gains in real terms over the 1981–87 period, other programs took direct hits. Recreation management declined 3%; soil, water, and air management declined 15%; and range management declined 18%.[41] In the BLM, most renewable resource programs took major cuts: recreation down 9%; wildlife habitat management down 21%; grazing management down 32%; soil, water, and air management down 42%; and range improvements down 45%.[42] A 1991 GAO report concluded that FWS law enforcement agents were not able to conduct necessary fieldwork, and as a result, were only selectively enforcing wildlife laws. Lack of funding precluded effective enforcement with operating funds decreasing from $24,100 per agent in 1984 to $11,800 in 1990.[43]

Public acquisition of lands for conservation, open space, and recreation purposes was also down considerably through the 1980s. Endowed by depletion taxes (outer continental shelf oil royalties and motor boat fuel taxes), the Land and Water Conservation Fund has been a major source of revenues for acquisition purposes for all levels of government. Between 1965 and 1986, some 3.2 million acres of recreation lands were purchased by federal agencies with $3.6 billion of LWCF funds, and 32,000 outdoor recreation projects at the state and local levels were funded, leveraging $3.2 billion in state and local funds. By the end of FY 1986, the NPS had spent roughly $2.1 billion in LWCF money to acquire 1.6 million acres of national park lands, and the FWS had acquired some 390,000 acres of refuge lands.[44]

But the LWCF, undoubtedly one of the most successful public mechanisms for acquiring land for conservation and recreation purposes, ran head on into the Reagan administration's philosophy of "less government is better." Land acquisition became a low priority for the administration, and Presidential budgets sought to zero out appropriations for these purposes. While Congress did not follow the President's lead, in part because of the political benefits of LWCF funding at the local level, appropriations were considerably less than during the Carter administration. In 1978, appropriations for LWCF-funded acquisition by the federal resource management agencies were $490.1 million.[45] In 1981, these dropped to $114.9 million, and throughout the 1980s, levels of appropriations were roughly in the $100–200 million range. As the federal budget deficit grew, the LWCF funds became a target of the Reagan and Bush administrations for deficit-reduction purposes. Since the Land and Water Conservation Fund was not a dedicated trust fund, its receipts only existed on paper, and as such, the almost $1 billion in the fund could be used to offset a portion of the deficit. Not acquiring new lands became a priority for the Bush administration less for ideological reasons than because of the increasing need to demonstrate deficit reduction in the smoke and mirrors process called federal budgeting.

Declining funds for natural resource management purposes translated into smaller workforces. Overall, the number of civilian federal employees has stayed fairly constant since 1970, rising slightly from roughly 3.0 million in 1970 to 3.2 million in 1990.[46] Over this time, however, the national labor force expanded, and the nation's population increased by about 22%. Federal civilian employees as a percentage of the American workforce actually declined from some 3.7 to 2.7% of the labor force.[47]

For some agencies and programs, these trends translated into only slightly diminished capacity; for others they meant drastic reductions. For example, while FS staffing levels declined somewhat from 38,524 FTEs in 1985 to 36,958 three years later, some forests and seasonal workers were hit particularly hard.[48] While the overall level of BLM employment stayed the same, those elements of the agency regarded as the periphery of the BLM's mission saw significant reductions. For example, wildlife personnel declined from 363 FTEs in 1980 to 259 in 1989, representing a decrease of 47% in fish biologists and 34% of wildlife biologists.[49] BLM range management staff over the decade was trimmed from 551 to 413 FTEs.[50]

While state and local employment generally increased in this period, in part to make up for the shifts in federal programs, state-level tax revolts in the late 1970s and early 1980s and economic recessions in the early 1980s and the early 1990s put major constraints on what state and local government could do in a time of diminished federal revenues. Natural resource and environmental management was seen as one of the more liquid components of state and local budgets, and hence more easily cut to accommodate growth in social welfare programs. For example, in Connecticut, state park officials proposed closing 13 of Connecticut's 92 state parks in 1990 because of budget reductions.[51] In Michigan, the budget of the state Department of Natural Resources decreased roughly 13% in real terms from 1986 to 1991.[52]

Declines in resources might not be so devastating if job demands facing public employees had stayed constant, but through the 1970s and 1980s, the tasks facing land managers and pollution regulators grew more difficult, the demands facing public natural resources increased, and our understanding of environmental and resource problems expanded. Each of these changes added burdens to staffs that simultaneously were being reduced in size. For example, as the FS and BLM resource base became more stressed, the challenges to agency biologists and botanists in helping to design timber sales and range management strategies grew much more difficult, even without considering the growing public and media attention to their work. BLM wildlife biologists, who in 1989 were responsible for roughly 1–2 million acres of rangeland apiece, had their hands more than full with NEPA compliance, resource management and activity plans, wilderness study reviews, and endangered species issues. The few BLM botanists (15 nationwide in 1989) were overwhelmed with work as the number of rare plant species increased. For example, Utah had one botanist to manage over 100 rare plants, many of which are candidates for endangered status.[53]

The FWS endangered species program showed similar growth in tasks, with limited growth in staffing.[54] For example, the number of listed species grew from 748 to 927 between 1981 and 1987, while funding for the endangered species program stayed fairly constant in real terms over the same time period.[55] Currently there are 1258 listed species, including 728 domestic species.[56] Approximately half lack recovery plans and more than four thousand species await review as candidate species. Conflicts and litigation over the spotted owl and other controversial species exacerbated the problems caused by historically limited administrative resources, as money and staff were siphoned off from other work, forcing lots of people to become owl biologists whether they wanted to or not.

The pattern of rising demands on natural resources at a time of diminished capacity occurs throughout many areas of resource and environmental management, and the expanded power of numerous interests in the policy process made resolution of the conflict between the two difficult. The deteriorating management situation cried out for more work, more thought, and more creativity, but did so at a time of declining capacity for effective change. Even though funding levels turned around somewhat in the Bush administration budgets—some of the resource management programs received increases in real dollars in 1989–1992—the legacy left by the spending pattern in the 1980s has overwhelmed our ability to spend our way out of the problems we face.

The sources of resource management problems in the 1990s are different from those of a decade ago. The fiscal constraints have less to do with ideology, and are primarily the result of the legacy of mismanagement and misallocation of the 1980s. Rising social service spending, paying off the excesses of the past decade such as federally insured bank failures, and the mammoth federal debt will continue to constrain spending on discretionary domestic programs such as natural resource management and environmental protection. For example, in 1990, mandatory spending on entitlements (social security, Medicare, Medicaid, etc.) and interest on the federal debt accounted for roughly two-thirds of the entire budget.[57] While defense spending may continue to decline as a percentage of the overall budget, a large portion of the defense budget is fixed. This means the primary source of flexibility in the budget is in domestic programs, which account for roughly 15% of the budget. The fact that 17% of the budget is deficit spending suggests how tight the situation truly is. If the primary strategy to deal with the federal deficit is to cut spending, there is not a lot of slack with which to do so. Entitlements will no doubt come on the chopping block, but an aging population will increase the political pressure to avoid major cuts in entitlements, while simultaneously driving up their costs. A growing economy is another way out, but forecasts do not project a major growth period in the near future.[58]

Hence, resource managers and political decisionmakers will find that the job of resolving conflicts due to rising demands on public resources at a time of diminished fiscal and administrative capacity will prove to be extremely difficult. As will be discussed in the final two chapters, creativity, innovative financing mechanisms, and public–private partnerships will be needed to get around the fiscal realities of the next decade. The alternative—to continue to overconsume public resources—will cause irreparable harm to natural systems already stressed by a decade of overindulgence. It is time to put future generations on a sound footing, economically and environmentally, and the legacy of the 1980s has made this task extraordinarily difficult.

Fragmented Power and Conflicting Interests

Chapter 7 described the fragmented state of American resource policymaking processes, and its impacts on our ability to make effective policy choices. While fragmented power is intentional in our decisionmaking system, the inherent fragmentation has become worse over the past thirty years, partly as a result of an increase in interest group politics. While it is difficult to determine exactly how many interest groups exist, they number at least in the tens of thousands, and several studies suggest that more than half of the groups were formed in the last thirty years.[59] Upon his return to Congress in 1987 (after a twelve-year absence), Congressman Wayne Owens (D-UT) noted that "in those twelve years I was gone, basically every group you can think of has developed a Washington office or a national association aimed at presenting their case to Congress."[60] The growth rate of citizen and public interest groups was high during these years, and that includes groups organized around environmental and natural resources issues.

No doubt part of the reason for the growth in the numbers of groups during this period was a spin-off of the social action movements of the 1960s and 1970s. The maturation of direct mail fundraising technology allowed groups to target segments of the population that might support their causes, and raise funds and mobilize constituent influence, in ways (and to groups of people) not possible in earlier times. Increasingly, interest groups were a more efficient way for citizens to have their concerns represented in the decisionmaking process. If an individual is concerned about pesticides or population issues, he/she can give money to the Environmental Defense Fund or Planned Parenthood, and respond to action alerts accordingly. The alternative—supporting political parties and specific elected officials—is necessarily more diffuse: Both have to craft platforms and agendas across a broad set of issues, and in order to be successful, usually have to shy away from controversial issues. Interest groups allow an individual to choose a concern from Column A and one from

Column B, ensuring that his or her specific concerns will get carried into the political process.

One of the institutional responses to the rise of interest group action has been to create more opportunities for groups to take action, insuring that a broader set of concerns could be brought into the decisionmaking process, and promoting the fragmentation of decisionmaking responsibility and power evident today. In the environmental realm, provisions for citizen's suits contained in laws such as the Endangered Species Act allow groups who formerly lacked standing in court to take action. Clear performance standards contained in laws such as the 1970 Clean Air Act, the 1970 National Environmental Policy Act, the 1972 Water Pollution Control Act Amendments, and the 1976 National Forest Management Act gave intervenors a basis for challenging administration actions. The reporting requirements contained in many of these laws gave nongovernmental parties the information with which to understand and question agency direction, and the requirements for public involvement gave individuals and groups a forum in which to be heard.

The expanding set of channels for action created more opportunities for interest groups to act, and to demonstrate to a potential membership that they had acted. It also fragmented decisionmaking authority, as groups who lost in one arena moved their concerns, and chance for success, to another. The spotted owl case amply demonstrated this game of musical decisionmaking chairs, as administration decisions were challenged first via administrative appeals then by judicial appeals, then by appeals to Congress, and first by NFMA, then by NEPA, and then by the ESA.

The increasingly technical nature of many issues, including those in the environmental and natural resources realm, also created a new demand for information, and interest groups took on the role of providing expertise to decisionmakers. Groups like The Wilderness Society and the Environmental Defense Fund developed considerable expertise that could be used to challenge agency assertions. While this was an important strategic decision on their parts, it also contributed to a fragmentation of information in decisionmaking arenas, such that technical pluralism has become the norm in legislative debates and judicial contests.

The rise in numbers of groups and a deepening of their influence is significant because citizens and decisionmakers increasingly are reliant upon them. The information intensity of many current issues forces both citizens and decisionmakers to seek expertise, and interest groups can be used for this purpose. Indeed, while the growth in numbers of groups in the environmental and natural resources area tended to level off in the 1980s, their importance as a source of information, and a means of representation, increased. Ironically, the administration of Ronald Reagan and James Watt did more to build the capacity of the larger national environmental organizations, through increased membership, contributions, and legitimacy, than any previous political force.

Changes in campaign finance rules in 1974, the rising cost of political campaigns, and the expansion of interest groups into electoral politics also deepened politicians' reliance on interest groups. Political action committees (PACs) have become a major source of funding for electoral campaigns,[61] and groups interested in natural resource and environmental management issues from both sides of the development–preservation spectrum use PACs as a means of gaining influence over policy decisions. There is a FishPAC and a FurPAC. The Realtors PAC and BuildPAC, representing the National Association of Home Builders are among the largest contributors to election campaigns. While considerably smaller, the environmental PACs are growing. These groups include the Friends of the Earth PAC, the League of Conservation Voters PAC, and the Sierra Club PAC. In the 1992 elections, for example, Sierra Club endorsed 226 candidates for Congress and raised a record $1.25 million for campaign contributions.[62]

PACs are important because in a time when campaigns can run into the millions of dollars, candidates need PAC support, but in accepting it, they lock themselves into positions and make themselves beholden to single interest groups. Mickey Kantor, a lawyer active in California and national Democratic party politics, and a Clinton advisor during the 1992 presidential campaign, notes the effect of PAC-driven politics: "It tends to make [the political process] money-driven. It takes the ideology out of it. It stifles creativity and strong stands, all of which is not good for the country."[63] Senate majority leader George Mitchell (D-ME) says the flow of campaign money "tends to keep you locked in to your views."[64] The net effect of this pattern of campaign financing is to reinforce a tendency toward political fragmentation and to stymie political leadership.

Just as important to the rise of interest group politics is the decline of other so-called boundary-spanning institutions in American society, that is, forces that cut across a broad spectrum of interests to create more holistic images of desired direction. Historically, political parties, strong well-defined leadership, community scale networks such as churches and

Rotary clubs, and generally shared public values acted to cut across the range of specific interests at stake in public policymaking, and to craft broad direction. But all of these forces are in decline in the 1990s.

Even while America was becoming more and more dependent on the interest groups, the fragmentation of interests and dispersion of power that the groups represent was having effects on public policy decisionmaking. As the groups matured, they tended to define narrow, single issue niches, which while making citizen representation easier, also made effective problem-solving more difficult. Trade-offs almost always have to be made in public choices, and that is more likely when the stakeholders have diverse interests rather than indivisible, narrow interests. For example, in crafting an effective energy policy for the future, no doubt some environmental and economic tradeoffs will have to be made. But they are unlikely in a political system where groups are organized to oppose (and support) every specific energy technology at every specific location. There is no way for all groups to win, and with the empowerment that has come from the fragmentation of decisionmaking arenas, the result is that no binding decisions are made. Instead, huge conflicts are created and impasses produced, leading to seemingly unending choice processes.

The spotted owl case provides a good illustration of this type of impasse. By the end of 1992, all the involved parties had a pretty good idea of what the ultimate solution would be: some component of ancient forest set-asides, a smaller component of lands designated and available for timber harvest, compensation and retraining for displaced workers, economic development assistance and compensation for timber-dependent communities, and additional constraints on log exports to allow mills to adapt to the changing supply picture. What was lacking at the end of this twenty-year long saga was not an idea of the solution, but rather the political will to buy into the solution. Part of the reason that politicians could not act was due to the fragmentation of power between the major stakeholders.

While strong leaders might emerge to guide us out of the current state of affairs, or strong shared public values might crystallize, there are other forces that will tend to promote rather than diminish fragmentation of interests and political will. The rise of the retirees as a major political force,[65] the empowerment of more ethnic minorities and women,[66] and a move toward term limitations for state and federal elected officials[67] will tend to add to the dispersion of political power within the United States. The fact that the solutions to future issues will increasingly involve other nations adds additional complexity to the political forces involved in public policy choices, as problems such as the loss of biological diversity, global climate change, and the movement of toxics and hazards around the globe require action.

In the area of American natural resources and environmental management, more players on a field of dispersed power will tend to promote fragmentation of interests and continue to generate public policy impasses. For example, one of the responses to the rising power of the national environmental groups has been an effort by commodities industries and dependent communities to organize at the local and national levels. The so-called "wise use" movement, tactically named to pick up on Gifford Pinchot's classic conservation statement, aims to counter environmental group pressures by mobilizing grass roots concerns. Clearly, timber and other commodity groups fell behind the environmental groups in supporting their case to decisionmakers with an appearance of broad public support. In 1989, they found themselves playing catch-up to a well-organized and diverse community of environmental groups. The American Forest Resources Alliance was one response to the owl controversy, and a broader wise-use movement responded to a broader set of concerns that commodity interests, and the communities they support, were losing in the battle to set and implement resource policy.

By joining such diverse groups as farmers, loggers, ranchers, miners, and ORV recreationists, and coordinating the advocacy of their interests with those of the more traditional lobbyists on Capitol Hill, such as the American Farm Bureau Association and the National Rifle Association, the wise use movement has considerable potential for counterbalancing the force of environmental interests.[68] And, given the financial resources that a coordinated wise use movement represents and the power of money in election campaigns, the groups could have significant impact on both electoral contests and policy disputes.

At the same time, while the environmental movement might not see as much of an increase in resources available to them in the near future as was the case in the early 1980s, schisms within the environmental community might lead toward more fragmentation of advocacy roles. Ironically, the increased effectiveness of the larger national environmental groups, and an increased sense that groups like Audubon and National Wildlife Federation are as much a part of the "establishment" as the Farm Bureau, the National Association of Manufacturers, and the National Forest Products Association, has lead to resentment at the grassroots level.[69] As a result, a number of state- and local-level groups are working to craft a more strident message, separate from that of the large national groups.

Certainly, the professionalization of the environmental groups, their increased willingness to make the trades necessary to secure federal policy, and the symbolism created by magnificent and expensive offices in D. C. and New York yield a dilemma for groups that historically have been dependent on being perceived as fighting the Washington system. While their emergence as a normal component of the policymaking process has secured sizable victories in environmental and resource management policy, it threatens their grassroots and "anti-establishment" bases. The net result might well be the development of other groups at the grassroots level who appear more in the image of the environmental movement of the late 1960s and early 1970s, and less like the well-heeled lobbying groups of the late 1980s and early 1990s.

The overall effect of these developments most likely will be more fragmentation across the political landscape. While there are opportunities in a situation of dispersed power, as will be discussed in the next chapter, in the short run, heightened fragmentation yields impasses and chaos in policy arenas. All in all, a situation of fragmented power and conflicting interests makes us dependent on effective boundary-spanning institutions, and unfortunately, there are not a lot of good integrative role-models to emulate.

Limited Vision and Insufficient Guidance

Strong, articulate leadership can sometimes provide a way out of the problems caused by fragmentation of power and interests, but the context for change in the 1990s makes the exercise of leadership difficult. Chapter 10 discussed the problems of leadership during the 1980s. Within the resource management agencies, the Old Guard tried to hold onto objectives and modes of action framed in the post-World War II period, and failed to re-create their organizations in a style appropriate to the changing times. The land base suffered, resource and organizational slack was squandered that could have been used to craft a different future, employee morale suffered, and the agencies were no longer regarded as credible sources of wisdom and insight.

A part of the reason agency leaders failed was caused by broader failures of national political leadership, as the Reagan and Bush administrations led in a direction contrary to public values on environmental and natural resources issues. Leaders can rely on several validating forces, including strong, well-articulated values, technical knowledge, and a public/political consensus. While Ronald Reagan was very effective at articulating an overall sense of direction, the merging of politics and science that occurred during his administration tended to undermine his ability to govern by weakening the administrative agencies.

While less dogmatic, the Bush administration was ineffective at drawing support for its actions by any of the three forces. It neither articulated a clear sense of public direction and values, relied on the wisdom of the agencies' technical skills and knowledge, nor worked very hard at crafting public support (and a political consensus) for its actions. Nor did it support its staffs when vision emerged from technical analysis, such as that provided by the Jack Ward Thomas interagency panel report or the FWS critical habitat study. Indeed, not only was the Bush administration generally antigovernment, it was hostile to the idea of being visionary. President Bush joked about the "vision thing," preferring short-term crisis management as a mode of government action.

A visionless presidency is subject to control by the worst kind of political pressures, and efforts to influence and redefine federal environmental law were evident throughout the Bush administration at the highest levels of government. Many of the efforts of the White House Council on Competitiveness, headed by Vice-President Dan Quayle, sought to restrict environmental regulation.[70] For example, a major controversy occurred when the Council led the way in rewriting a federal manual for delineating wetlands in the United States, a move that would eliminate some 30 to 80% of currently regulated wetlands from federal regulatory jurisdiction. Field testers indicated that the proposed hydrology standard was "totally arbitrary," representing a "departure from the scientific understanding of wetlands ecology."[71]

The lack of vision in the Bush administration also reflected the context for exercising leadership, and this context has important implications for the success of future presidents and agency leaders. While Bill Clinton and his appointees in the environmental and resource management agencies have a number of advantages over their predecessors, the context for change will also affect their ability to craft effective visions, and guide the American public to a decent future. The context for leadership includes a state of the country and world that is more ambiguous about necessary and desired direction than any time since the 1920s. The 1990s leave us with a fundamental redefinition of world power and national problems, both of which yield great opportunities, but also significant uncertainty about what is appropriate, effective, and likely to be supported by the American public.

The conditions of limited slack described earlier also have implications for the ability of leaders to take action. Less slack means that difficult choices have to be made between conflicting demands. While John Kennedy and Lyndon Johnson had the ability to spend their way out of public problems, Bill Clinton and his successors are constrained from doing so. The management of fiscal constraints may have spawned such homilies as "doing more with less" and "excellence through reprogramming," but the fact is that managing programs with declining resources is not a lot of fun. The fact that stressed natural systems will require restoration and protection from excessive short-term demands means that conscious choices will have to be made that favor long-term sustainable development over short-term exploitation. In essence, the citizen-consumers of the next two decades will have to take the hit in order to build a sustainable economy and environment for the long term, and a message of sacrifice is bad politics for those who seek political leadership positions.[72]

The misinformation of the last twelve years does not help. The Forest Service and its political overseers did little to moderate the expectations of dependent interests in the Pacific Northwest about the timber shortage that would inevitably occur, nor were political leaders effective at educating the public about the costs of the gluttony of the 1980s. Throughout the period, political leaders continued to act as if there was slack in our resource and financial base, neglecting the realities of resource depletion and deficit spending. Ironically, while there were short-term gains to be had, the long-term legacy of recent political leadership is one of environmental and human degradation.

The contrast between the historical American perception of unlimited resources, compounded by a number of years of political blinders, and a future that requires leaders to articulate a more moderated view of global production capabilities, puts political leaders into an incredible bind. Vision is needed, but it will not be easy to sell it to the electorate. The fragmentation of power, interests, and avenues for intervention in the political process makes this task even more difficult. Assembling coalitions across diverse interests is tough work, particularly when you know that difficult choices will be challenged through a variety of political, legislative, administrative, and judicial means.

The ability to lead is also influenced by the news media. The ubiquitousness of media ensures that the actions of leaders will be viewed under the glare of media attention and public visibility, and the intensity of the spotlight is not necessarily the most productive environment for the deliberation, consultation, and creativity often necessary to deal with difficult problems. In addition, if the leader's message is at all complex, as are most effective representations of necessary courses of future action, the media will tend to simplify it, or may overlook it completely. By caricaturing complexity and valuing simple images of conflict, the media tends to polarize the players in the decisionmaking environment, and if nothing else, ensure that any strong stances will be opposed. Fragmentation of single-issue interest groups makes this dynamic worse. The net result is that no matter what a leader tries to do, it is likely that he or she will be attacked. Only a few people truly enjoy pain, and the pain and suffering that results from most strong stances tend to make people want to avoid them. Hence, politicians frame their campaigns around motherhood issues and negative attacks on the other guy's character, and the media and the public eat it up, while bemoaning the deterioration of public debate.

The importance of money in reelection contests, the growth of the opinion polling industry, and the ability to manipulate voter opinion through the media and computer-guided, direct mail campaigns pose considerable risks for elected officials who choose to follow a vision of what they believe is right, if it disagrees with constituent demands and/or the interests of an intensely motivated interest group. For example, the 1992 defeat of Indiana Congressman Jim Jontz was evidence that coordinated opposition to environmental agendas could cost a politician his/her job, even if their district was far removed from the area of environmental concern. In Jontz's case, timber and mining interests contributed significantly to his Republican opponent in a rural, Republican district, in direct response to his sponsorship of ancient forest legislation for the Pacific Northwest. Indeed, advertisements in the Portland *Oregonian* sought contributions to oppose Jontz's reelection. The message sent to other environmental champions was clear: watch your back. While some measure of informed reflection is always appropriate, constantly looking over one's shoulder strains the body politic and causes well-meaning individuals to stumble.

None of this is to say that leadership will not be important. Indeed, strong, visionary leaders who can articulate a future direction and craft the political will to get there are one of the best integrative, boundary-spanning forces we can muster. But the context for leadership will be discouraging to some who would otherwise get involved, and it will be frustrating for many who marshall the courage to run for office and accept appointments to leadership positions in agencies. Just when we need strong, compelling, well-thought-out visions of the future and how to get there, the ability of elected officials and their appointees to lead is constrained by the very conditions that demand leadership.

But There Are Reasons to Be Optimistic

While the environmental and resource problems facing our society are challenging, and our basic decisionmaking processes are problematic, there are reasons to be optimistic that necessary change can come about. This is particularly true when thinking about the agencies and policies specifically involved in the spotted owl controversy, and still true—though less so—of the broader set of societal institutions that need to deal with the global-scale environmental challenges of the future. Four key factors define the potential for important change:

- Conflict creates windows of opportunity for change, forcing societies to adapt and innovate to find ways out of the conflict.

- Agency and political leaders are evidencing a generational change, with new leaders drawn from the post-World War II generation.

- While there are many ways to improve resource and environmental policy, some aspects of federal land management policy have worked fairly well, and can serve as the basis for continued federal actions if they are updated for the times.

- The "idea" base underlying resource and environmental management decisions is developing rapidly, including advances in science, communications, and information processing technology, and ways to organize the human activity involved in making and implementing collective choices. Together these advances can provide the infrastructure for more effective decisionmaking needed in the 1990s and beyond.

While our society tends to view conflict as a negative force that is inefficient and needs to be managed to resolution, conflict is in fact the driving force for social change and is inherent both in pluralist bargaining and free market competition, central themes in American political and economic systems. Conflict results from the interaction between the divergent human interests inherent in a diverse and heterogeneous society, and from the interaction between the interests of those who benefit by present day conditions, and those who seek a change for the future. Conflict mobilizes human energy and creativity, and forces human institutions to generate information and clarify and change policy direction. More fundamentally, conflict forces us to ratify and adapt values in response to changing human interests and environmental conditions. Since we live in a changing world, and in a state of incomplete information where our current understanding may be totally in error, forces that encourage thought and response to changing conditions and information are critical to future success. Indeed, complacent societies fail over time, even if their members die happy.

While the conflict evident in the spotted owl controversy could have been managed so that it produced a better and more timely outcome, there are in fact positive aspects of the huge amount of conflict it generated. It did result in a public that was more informed and educated about public lands, wildlife, and the difficult choices inherent in federal land management, as the media and interest groups increased their coverage of these issues. It also produced a set of political representatives interested in public lands issues that went beyond those traditionally involved in determining forest policy or budget decisions.

In our society, citizen empowerment is both a desirable end and a sought-after means to producing better choices, and the owl issue mobilized individuals and organizations across the spectrum of political interests in public lands issues. It is important for the long-term efficacy and wisdom of our policy process that both community activists in Forks, Washington, and urban environmentalists in New York City were motivated to take action, if only to read an article or write a letter. An informed and involved public is a requirement of an effective democracy, yet many social forces work in opposition to this goal. Institutionalization, centralization of authority, the increased technical complexity of issues, and mental overload caused by information saturation from media and other sources all work to disempower and distance individuals from the institutions and arenas in which choices that affect them are made. To the extent that these processes result in under-representation of some interests and concerns, and a lack of ownership in the decisions that are made, policy formation and implementation will be ineffective and possibly counterproductive. The inordinate amount of conflict evident in the history of the owl case resulted in political participation by a variety of previously uninvolved citizens, and this outcome must be viewed as positive, even if it had no impact on the disposition of the owl issue itself.

At the same time, the level of involvement in the owl controversy signified something real about the amount of interest in the direction of public resource management, and the extent of this concern is another reason to view the situation with optimism. Only when threatened do we as individuals and members of societies tend to clarify and reaffirm our values and commitments. If there is any benefit in war, for example, it is largely that it forces a clarification and affirmation of national purpose and values, and that is true in battles over policy as well as those over ground. Except for a handful of individuals and organizations, most want the Forest Service to succeed; only a few have written off the agency as unreformable. Staff members of administrative agencies across government understand the implications of the Forest Service not succeeding: If the least structurally political agency with the most resources cannot succeed, what then for agencies like the Fish and Wildlife Service that function in an explicitly political environment with meager resources? Few of these officials want agency decisionmaking to be carried out in the courtroom, and even fewer want Congress to dictate on-the-ground management direction.

At the same time, most members of Congress also want the agency to succeed. They want the Forest Service to live up to its earliest expectations as a source of thoughtful decisions that are based on the best science available, and that hopefully also have political support. Politicians do not want to make these choices on their own, and usually appreciate agencies that are honest providers of good quality technical information. While it might make their political lives difficult on an issue-by-issue basis, elected officials hope that technically correct and politically feasible solutions to problems do exist, and prefer not to stick their necks out to find them if they do not have to.

The vast majority of individuals within the interest group community also want the Forest Service to be a success; they seek to redirect the agency, not to destroy it. Many groups understand the history of the public lands management agencies and, as described above, would prefer that the Forest Service be seen as a success. The alternative is complete reliance on the broader political process to provide direction, and most groups understand that the winds of political compromise can blow against you as often as they blow with you. Having administrative agencies that are effective is in most groups' interests, and they understand this reality, at least most of the time. In their view, of course, being effective partly means that the FS is doing what the groups would like them to do, but most of the groups are savvy enough to understand that being effective also means that the agency is making choices honestly and credibly regardless of the outcomes. While clearly there are organizational advantages to casting the relationship between a group and the Forest Service as adversarial—"we're going to make them do things right"—and these advantages help in fundraising and defining the mission of the group, there are lots of opportunities for interest groups to sell themselves as the (contribution-deserving) fighter of evil forces. Unending battles over the same ground are not the core objective of these groups.

The fact that there is a lot of care and concern about the FS and the direction of resource policy provides a window of opportunity for trying out new perspectives. No matter how strong the internal agency forces for change, if there is no political will that can be mobilized externally, change will be suppressed and will only occur subversively. The owl issue and the broader debate over public forest policy have produced lots of new political possibilities. Opportunities exist to link forest policy to emerging public concerns over global change, the loss of biological diversity, and the need for science education, among others. There is an opportunity today for resource managers to forge a broader coalition of support for public forest management than at any prior time. The envelope of politically feasible policy options has widened significantly, and the FS and other resource agencies have not taken full advantage of what is possible.

There are other reasons to view the current state of affairs with hope. Sometimes organizations needing change are so stressed by their situation that the energy needed for creative change does not exist. Staff and leadership burnout, low morale, and a brain drain that looks like so many rats leaving a sinking ship can result in an organization incapable of making the very changes needed for renewal and revitalization. Like trying to start a car with a dying battery on a cold January night, no matter what you do, it is hard to overcome the limited energy reserves.

While the federal land management agencies certainly have their share of tired, overly stressed workers and leaders, many of whom deserve a rest, the good news is that there are a lot of indigenous forces for change within the agencies. Many individuals within the Forest Service, Bureau of Land Management, and the Fish and Wildlife Service still believe in the potential of their organizations, and are ready and willing to try out new behaviors. Many of the policies and practices that exacerbated the spotted owl situation have been reviewed and are in transition, at least in agency rhetoric. Moves to broaden the measures used to evaluate the performance of Forest Service line officers, approaches that work more effectively at involving a broader set of public groups, and activities that diversify the value base of the agency are all positive steps as long as they produce real and lasting change in organizational behavior and performance. Indeed, the development of grassroots, employee-based organizations, such as the Association of Forest Service Employees for Environmental Ethics, should be viewed similarly as a positive step that can mobilize creative energies and foster productive change.

The demographics of the agencies also offer hope. The generation of agency leaders who were influenced by the styles and attitudes of the Depression and World War II, who brought a Can Do, top-down, militaristic leadership style back from the War and into the USFS, the USFWS, and countless state resource agencies, will be replaced by a different generation of leaders, whose values and norms are more in line with the general public's concern with environmental quality. The new guard never worked in a time when environmental concerns were not part of law, and environmental impact statements, public involvement, the National Forest Management Act, the Endangered Species Act, and the like are an ordinary part of their administrative lives. As the new set of leaders emerges, they will bring with them a different set of disciplinary backgrounds and a broader set of values about the objectives of forest management.

If nothing else, expanded workforce diversity,[73] if it succeeds, will bring into the agencies segments of society that have not been well represented before, and that hold at least somewhat different values. An expanded value base will move the agencies closer to the diversity of values in the American public, and provide them with a broader set of ingredients with which to make forest management choices. In-house socialization and the conservative nature of the forestry profession can possibly defeat some of the positive aspects of this demographic change, but the potential is there. The new guard will see itself as less bound by the polarized positions of the past.

In interviewing numerous individuals inside and outside the federal agencies for this research, whether or not the spotted owl case was seen as a watershed event for the Forest Service, or the final nail in the agency's coffin, depended on whether the respondent felt that the agency had the potential to change. Clearly the jury is out on that question, but the energy and desire of many FS staff members to do things differently is there; it is unmistakable and ripe with promise. If nothing else, the conclusions from this research—that individuals, and individual actions, do make a difference in the course of historical events—suggests that change will occur at least in scattered districts and potentially more broadly.

In many ways, demographic changes in the agencies are indicative of a broader set of generational changes in decisionmaking structures within the United States and elsewhere in the world. The generation of leaders whose values were formed primarily during the Great Depression, World War II, and the Cold War are retiring and moving out of leadership positions. The shift in power in the United States from George Bush to Bill Clinton is an important symbol of this generational change. The Boomer generation is moving into leadership positions, bringing with it a broader view of desirable economic development, and less of a sense of a bipolar, win–lose, good-and-evil world. This broader, more ambiguous view may help us frame a more complex understanding of the intentions and aspirations of other countries, perhaps making it more likely to work across national boundaries toward common purposes. If nothing else, changes in political leadership will at least make it more likely that necessary reforms in public sector agencies will come about.

While much of the discussion in this book has viewed the current resource and environmental management situation as a glass that is mostly empty, it is important to recognize that it is at least partly full: some aspects of current federal resource management policy are effective. In viewing the drama of the owl issue, it is true that on one level at least, the policy process worked as it should have because it forced a great deal of public attention on a set of choices that clearly could only be made at the highest levels of the policy process. As the issue became more controversial and more obviously value-based, it was bumped to different and higher levels of the policy process, moving from regional level administrative agency action to national level agency decisionmaking, then to administrative appeals processes and to the courts, and finally to the Congress and Executive Office. Major policy choices should receive attention by high level agency and elected officials, and it is clear that the Chief of the Forest Service, the leadership of the U. S. Congress, and the Executive Office of the President gave the issue a great deal of attention.

In addition, some aspects of federal policy worked fairly well, even if they worked in unintended ways. While largely a reactive policy, the Endangered Species Act created a set of incentives that encouraged most parties to try to avoid a listing decision. Forest Service and BLM officials feared the limitations that might be placed on their management choices if the owl was listed. Indeed, much of the FS's efforts in the spotted owl SEIS analysis, and associated interagency dynamics, was designed to avoid a listing decision because of this fear. Fish and Wildlife Service officials did not look forward to the time and energy that would be needed to oversee implementation of a listing decision, since it would involve them in countless FS and BLM decisions. Environmental groups and a number of agency and Congressional staff members feared the potential for a political backlash against the Endangered Species Act itself, if listing the owl was perceived to involve significant economic impacts.

Since the ultimate goal of federal endangered species policy should be to avoid the need to list species, policies that create incentives for appropriate action are at least as effective, if not more so, as policies that attempt to command appropriate behavior directly. While some see the significance of the Endangered Species Act to be its commands to agencies to protect critical habitat, in fact one of its best attributes is more subtle: changing the incentives that agencies face so that they take action proactively, either out of concern about their impacts on sensitive species, or more often, concern about the possibility of being embroiled in a visible, political controversy.

Conflict over the owl case did have an impact on the outcome of the issues underlying owl conservation, and this attention to public land management was particularly appropriate for a resource whose direction and political foundation was forged largely in the 1950s and 1960s. It was time for rethinking the goals and methods of public forest management, and conflict over the owl issue did as much as anything else to force this rethinking. Individuals and organizations get stale when doing the same things they have been doing repeatedly, seemingly since the beginning of time. Corporate managers now understand that if you give a person a simple task to do repeatedly and endlessly, he or she will get less and less effective at it over time. Factory workers whose job is to screw in ten bolts per car for forty cars per day for 220 days per year for 25 years, not only screw in fewer and fewer bolts over time, but may also do other things that raise the interest level of their jobs, such as taking drugs and sabotaging the functioning of the cars they are producing.

Revitalization comes from new challenges and directions, as well as recognition, responsibility, and sharing in the success or failure of a set of activities. Revitalization on a societal level means iteratively redefining and recommitting to values that might be time-honored, but at least need periodic affirmation, and probably need regular upgrades to adapt to changing conditions. Such was the case with public forest management in the 1980s, and the owl issue did as much as any other issue or government policy to force a move toward revitalizing the policy processes and organizations involved with public forest lands.

Part of the reason to view this process of revitalization and renewal with optimism is that the core theories and operating principles of the Forest

Service are not wrong, they just got bogged down in the values of the 1950s and 1960s, and need to be updated in a real (not token) way for the 1990s. Many of the central themes established by Gifford Pinchot at the turn of the century—the greatest good for the greatest number for the longest period of time, science-based management, and multiple use— provide reasonable direction if their implementation is cognizant of the full range of values, participants, and information and expertise sources that exist today. For example, the greatest good for the greatest number for the longest period of time is not a bad goal statement. The problem in recent years is that its day-to-day interpretation was skewed far too heavily toward maximizing the yield of timber and other commodity outputs, rather than producing and protecting the full range of values inherent in our public estate, including those likely to be valued by future generations.

Other operating principles similarly have merit if they are modified for the times. Science-based management is still appropriate; the problem has been that the agency has not been true to its science, even though the scientific basis for managing terrestrial systems has developed considerably. Multiple use is also an appropriate concept if use is defined broadly to include land and wildlife preservation and the needs of future generations, and multiple uses are sought system-wide and not on every parcel of national forest land. Science-based, multiple value management, for the greatest good of the broadest set of publics, for current and future generations, is an effective and appropriate objective, and since it is in line with agency traditions, should be implementable.

Finally, expansion of the idea base underlying resource and environmental management can help in this revitalization process, by providing a whole host of creative possibilities for promoting and protecting the multiple values in public resources. New scientific concepts and methods, new technology for acquiring and manipulating information, new decisionmaking tools, and new ways of organizing the human activity involved in making and implementing collective choices provide resource and environmental managers with a lot of creative possibilities for trying out different ways to do their jobs.

The science base underlying resource and environmental management has changed considerably in the past two decades. Landscape-level ecology has risen as a serious field, with its focus on ecosystems and regional analysis of considerable relevance to land management decisions. Concepts of ecologically sensitive silviculture labeled the "New Forestry" provide forest managers with a broad set of tools with which to experiment and use to more effectively practice multiple resource management. New approaches to wildlife management allow a broadening of focus from a handful of game species and the management of species-level diversity, to a focus on ecosystem-level diversity, including interior-dwelling species and organisms at lower taxonomic levels.

The scientific understanding of landscape restoration has also increased, so that managers not only have the ability to manage intact systems better, but to restore the functioning of systems that have been trashed in the past. Approaches range from the management of exotic species that outcompete native ones to controlled burns to recreate the conditions favoring the growth of native prairie plant species. In the endangered species realm, new understanding of genetics and population ecology offers managers a set of approaches to restoring species through upgrading habitat, establishing corridors between habitat units, or captive breeding and subsequent reintroductions.

Some of these scientific advances have been made possible through changes in technology, and technological innovation also provides a new set of possibilities for creative resource management. The expansion of small-scale, diffuse information processing technology has created all kinds of possibilities. Better modes of acquiring and managing data about resource characteristics and conditions have come about through better remote sensing and geographic information system technology—both of which give managers enhanced means of viewing and understanding the landscape. The interconnection of scientists and managers through telecommunications technology, and the creation of global information-exchange networks, provides a marvelous set of tools for exchanging information and ideas. Other technological changes provide opportunities for the future, including a fundamental change in our ability to manipulate life through recombinant genetic techniques and an enhanced means of measuring environmental contaminants at extremely low levels. All technology can be misused, and some of our past problems came from the creation of new, poorly understood technologies. While we need to guard against a fundamental bias toward overconfidence in assessing new technologies, such new approaches will undoubtedly provide the raw materials for creative change, if they are carefully managed.

Technological change has also fostered the creation of new decisionmaking tools. Computer-based predictive analyses and simulations give decisionmakers new ways to think about alternative management directions. Population viability analysis and risk-assessment can provide indicators of the long-term efficacy of various management strategies arranged in a way that enables decisionmakers to make reasonable choices.

New ways of quantifying the benefits and liabilities of different actions, informed by nonmarket valuation techniques, can provide decisionmakers with a more robust calculus for evaluating alternatives.

Teleconferencing technology allows decisionmakers from around the globe to meet in the same time and space for interactive conversations and negotiations, without ever incurring travel costs. The power of FAX machines and conference calls was well-demonstrated in the negotiations among the environmental groups in responding to the 1989 Timber Summit, and the potential of electronic town halls was demonstrated in the 1992 U. S. presidential elections. The development of integrated data, video, and voice linkages across North America and throughout the world provides the opportunity for massive amounts of information to move rapidly through space in a way that is harder to control via political or military means. While information overload increasingly will be a problem, these changes create the possibility of all kinds of innovations in the ways that decisions are informed.

Finally, the development of innovations in the ways that human activity is organized to make and implement resource and environmental management strategies is a further reason to be optimistic that change can come about. New approaches to dispute resolution, public involvement, and planning create the possibility that collaborative settlements can be crafted that are well-informed and politically stable. New ways to lever administrative resources, such as through public–private partnerships, contracting out for services, and interagency working agreements, provide managers with a way out of the situation of constrained resources that currently face them. New visions of organizational management and leadership, including concepts of nonhierarchical organizational structures, teamwork and quality circles, can promote creativity and responsiveness at all levels of an organization. New types of policy instruments, including market-based incentive approaches and the use of nonmonetary, intrinsic motivation, have the potential to rise about the fairly simplistic, command-and-control approaches used to influence human behavior in the past.

Clearly the environmental and resource challenges that we face are enormous; and while there are reasons for optimism, there is an equal and opposite set of reasons for despair. But the fundamental question is: what choice do we have? A compelling and burgeoning set of problems will demand solutions, and innovation and change will occur simply because there is no other choice. Being proactive about fostering innovation, and evaluating and learning from it, can help us manage the process of change, mitigating its impacts on various segments of the population and potentially making it more efficient. We can stumble our way toward the future, or attempt to run toward it. The next two chapters provide a set of running shoes, by outlining a set of specific actions for "doing better."

12

Building More Effective Agencies and Decisionmaking Processes

Unlike owls, whose adaptability is defined by genetics and niche, organizations and policy processes, while constrained by traditions and turf, can learn from the past and improve, and there are a set of lessons from the spotted owl case that can enable our decisionmakers and decisionmaking processes to do better. It is possible to do better at anticipating, understanding, and responding to controversies like the spotted owl issue so that we avoid repeating what all would agree was not an optimal set of policy choices or organizational behaviors. Humans and human institutions are endowed with an amazing ability to learn, adapt, and change, and we need to fully exercise these capabilities for our own sake as well as for those of other lifeforms that necessarily depend on our benevolence.

This chapter begins a two-part description of the set of reforms needed to improve environmental and resource management in the 1990s and beyond. It focuses on specific ways to improve agencies and decisionmaking processes to avoid or better manage the problems described in previous chapters. The recommendations suggest the need to develop:

- New mechanisms to bridge the agency–nonagency boundary in order to build understanding and political concurrence;

- Altered approaches to organizational management, included updated notions of leadership;

- Improved means of gathering and analyzing information about resource problems, organizational possibilities, and political and social context; and

- Ways to promote a culture of creativity and risk-taking to generate more effective options for the future.

The discussion in this chapter focuses primarily on the U. S. Forest Service for several reasons: The spotted owl controversy was an issue that in part was a crisis of the FS's making, and it is appropriate to draw out of the case history a specific set of lessons for the FS. At the same time, the story of the FS is the story of many resource management agencies, at both state and federal levels, and reforms in its mission and organizational style can be viewed as models for other agencies operating within other political contexts. Finally, it is difficult to cast a specific prescription divorced from the particulars of context. Just as everyone is in favor of protecting endangered species until they turn up in their backyard, generic prescriptions have a sense of unreality about them. Testing whether they will work depends on how well they apply within specific organizational, political, and legal contexts.

While one might bemoan the problems of our decisionmaking processes and despair that change is unlikely or impossible, there are new approaches to be tried, and reasons to be optimistic that change is possible. New mechanisms for building understanding and political concurrence across the numerous elements of our society, expanded approaches to collecting and organizing information, and different modes of organizational management and leadership can all play a role in reestablishing the FS and other public resource agencies as models of wise land management and stewardship. It is possible to do better.

Bridging the Agency–Nonagency Boundary

Since American resource policy processes are fragmented politically and administratively, those who seek to create effective policy must build the political concurrence necessary to make policy changes. In addition, since adopted policies do not insure changes in behavior on-the-ground, it is necessary for proponents of policy change to work actively at sustaining the political concurrence underlying a policy direction. Concurrence means developing understanding, support, and ownership in certain directions. It represents a two-way street, where policymakers must understand and involve various interests in their decisionmaking, and at the same time educate and develop understanding for agency interests. For FS staff to be more effective than they have been in the past fifteen years in setting and implementing forest management policy, they must understand these realities of our policy process and work with them. In addition, as other boundary-spanning institutions have declined in American society, agencies like the FS have an important role to play in integrating values, interests, and information across a politically fragmented landscape.

At the upper levels of resource management agencies, developing political concurrence means working the political landscape in more extensive and creative ways than has been necessary in the past. Diversification is a reasonable long-term survival strategy in business and natural systems, and FS leaders should view diversifying their political base as an appropriate strategy for insuring organizational relevance in the future. The diversity of interests in forest management decisions is real, and implies both a challenging task for agency leadership, and an exciting set of possibilities because fragmented power allows the formation of new coalitions.

The range of interests with legitimate concerns and political power is large and includes traditional commodity interests such as timber, mining, and grazing; for-profit service providers, such as recreation concessionaires and equipment manufacturers; environmental groups, including those concerned with preservation, nongame wildlife conservation, hunting and fishing, wilderness recreation, and developed recreation; scientists, researchers, and professional associations; educators, particularly science teachers and university professors; associations of community action groups and local and state government; other federal agencies; and even social service groups, such as those concerned with alternative sentencing arrangements for punishing criminal behavior. Members of Congress and media organizations are interested in a similar range of concerns, and should be encouraged to participate in and support agency programs and priorities.

Methods for developing relationships should run the gamut from informal, ad hoc contacts to formal, fairly institutionalized approaches. Developing an advisory board with representatives from a broad set of these groups, and meeting annually or biannually, holding workshops for first-term members of Congress and/or Congressional leadership, Congressional staff, environmental reporters, academics, and representatives from

other countries, and promoting formal externally facilitated policy dialogues are all ways to develop the understanding and support necessary
to attain political concurrence.

At the ground level, developing political concurrence means modifying
top-down decisionmaking processes and employing a variety of approaches usually labeled as alternative dispute resolution.[1] These approaches include ad hoc problem-oriented groups, such as have been convened on some forests in response to forest plan appeals, as well as more
institutionalized approaches, including so-called "friends of the forest"
groups (collections of stakeholders who become integrally involved in
forest management decisions and their on-the-ground implementation).

The goals of these kinds of decisionmaking approaches are to foster a
two-way understanding of the real interests and constraints of stakeholders, including the FS, generate creative possibilities, build trust and relationships that allow this understanding and process of interactive creativity to emerge, and build ownership in resulting choices on the part of all
affected parties. While there has been a fair bit of agency rhetoric about
the benefits of alternative dispute resolution approaches, and some notable successes, much agency decisionmaking still operates from a public
involvement model that begins and ends with public concerns and comments, and the FS going to the mountain to invent solutions to these
concerns. This type of barbell model of public involvement—FS soliciting input on issues and concerns; FS analysis and decisionmaking; public
comment on FS proposed decisions—misses a lot of the positive benefit
of collaborative problem-solving approaches. It often does not generate
adequate understanding or creative solutions, and rarely promotes the
kind of intergroup relationships that will pay off in the long term.

While it is important for the agency to promote a process of collaborative problem-solving in forest policy decisionmaking, consensus is not
possible in all cases. One of the preconditions that enables alternative
dispute resolution to succeed is that no group must compromise a fundamental value or principle in order to settle.[2] In the spotted owl case, it
was hard to frame the dispute in such a way where this kind of compromise was not necessary. A second precondition for successful processes is
that the incentives facing the parties are such that they take the negotiating process seriously. In this case, fragmented policy processes provided
multiple opportunities for action, so it appeared that the major parties
never had the need to compromise because most thought they would
achieve their interests through other means.

Regardless of whether an all-encompassing settlement is likely, good process can yield the benefits of greater understanding, better relationships, and the possibility of cross-interest discussion that is constrained by the nature of our decisionmaking processes. While the FS's consultant concluded that the major parties to the owl dispute were too polarized for formal mediation to help out, my sense is that the agency still should have tried. Even if no settlement was possible, the scope of the subsequent policy debate would have been bounded by the attempt. Indeed, in the early 1980s, when there was more slack in the system because of the recession and buyback sales, the estimate of necessary owl habitat was lower, and the political environment was different, it is possible that a consensus-building approach could have succeeded.[3]

The resource base of the national forest system, and the image of the Forest Service as a professional organization, provide the agency with a marvelous set of resources with which to develop the relationships needed to promote collaborative problem-solving. Even though there are a lot of claims to the resources of the national forests, the system contains an enormous set of possible ways to meet those claims, given creativity and a solid understanding of the interests underlying those claims. In addition, the nature of the resource can be used to promote processes of collaborative problem-solving. For example, it is not hard to frame superordinate goals around national forest management that can encourage greater efforts at solving seemingly zero-sum problems. The ability to visually illustrate issues or concerns by walking or flying over the landscape is a potent and compelling tool that can be helpful in fostering collaborative decisionmaking. If nothing else, for policy-level issues, the opportunity to transport decisionmakers and interest group leaders to beautiful places in the American landscape, can, by getting them outside the Beltway or away from their everyday battlegrounds, foster new styles of interaction. It does not matter whether decisionmakers sign up for an interesting junket to the Tetons, the Cascades, or the Puerto Rican rainforest, the experience will affect them.

Clearly these are not costless relationships. At the forest level, building understanding and support may require some changes in organizational norms. Effectively involving representatives of interested groups may mean scheduling meetings in places and times that are inconvenient for FS employees, or compensating nongovernmental participants in some way, at least for the expenses of their involvement. Building relationships means changing the FS transfer policy, because one of the most frustrating thing for interest groups is to work at building a relationship with a FS employee, only to have that individual leave the region. Relationships are between individuals, not institutions, and the FS policy of moving staff members around breaks those relationships, even though it serves other legitimate ends. Since much that happens at the forest or district level does not get written down clearly, the loss of institutional memory resulting from the loss of staff members is detrimental to the ability to create long-term relationships and trust between forest management partners.

At both the field and policy levels, these kinds of relationships imply other costs. If the FS develops a better understanding of a group's true interests and it is clear to the affected groups that the agency truly understands their concerns, it may mean that the FS will have to respond to those interests in ways that are not consistent with agency traditions. In the spotted owl case, for example, the timber industry's concerns had as much to do with insuring certainty about future timber supplies as it did in insuring a specific future cut level. One of the primary concerns of the environmental groups was a lack of trust in the FS's ability and willingness to implement forest management policies in the direction in which the groups wanted. For both, the best solution might include the creation of additional dominant use areas defined by geographic boundary lines, including permanent old growth reserves and permanent timber production areas, even though these kinds of geographic set-asides are counter to traditional notions of multiple use and constrain agency discretion.

Developing relationships with other parties also tends to tie the involved agencies and groups to a collective set of interests. As described in Chapter 8, coalition-formation will both enable the FS to take action, and will bind it to the interests of the coalition. The best way to guard against being bound tightly, or bound to a set of interests wedded to the past, is to keep the power dynamics within the range of interests in flux. The opportunity that exists today is to forge crosscutting, dynamic alliances across the range of interests, not to bind the agency to a specific set of externally given directions that will become obsolete over time. Hence the agency should also keep looking ahead to involve new interests as they emerge.

The role of the FS in all this should be one of broker of political possibilities, advisor of technical realities, stakeholder with organizational and other interests of its own, and partner in managing a common property resource. The metaphor that is currently in vogue in the agency, where the FS is a business with the variety of societal interests and groups as its clients, is seriously deficient, for it implies a reactive, demand-satisfying role on the part of the agency. Nor is the image of the agency as a neutral, balance point for society an accurate or appropriate one, for it suggests that the agency has no interests of its own. In fact, the agency's responsibility is broader than that implied in a business–client model, it has organizational and substantive interests, and it should furthermore operate from a normative base, as outlined in Chapter 13.

While the business–client metaphor is problematic, the FS does have a legitimate and necessary role in building a dynamic and broad coalition of understanding and support for federal forest policy. This kind of dynamic coalition takes a lot of care and feeding and will no doubt require perpetual attention. It also means that the agency cannot count on specific members for support at all times. The environmental groups, for example, have been notoriously poor agency constituents, as they are better at opposing agency direction than supporting agency actions through testifying in appropriations battles and the like. While building this type of political concurrence at all levels of the agency takes a lot of effort and is inherently unstable, it is the only way for FS staff and leadership to truly develop an understanding of the diversity of interests present in public forest management, and it is their only chance at rebuilding the trust and support necessary to function effectively in the future.

There are other methods available to build the understanding, trust, and relationships necessary to achieve political concurrence. Using the national forests aggressively as an educational resource makes sense substantively and politically. As a place where science and natural system–human interaction can be demonstrated and seen, and as a focal point for collective decisions over common property resources, the national forests provide an unparalleled opportunity for physical, natural, and social science education relevant to all educational levels. Given the current concerns over educational preparedness and science education in secondary schools, connecting the resources of the national forest system to K–12 education is a natural. A similar concern at the undergraduate college level also provides opportunities for the FS. Given the current focus in graduate education on research, the size of the national forest landbase and the opportunity to engage in controlled experimentation suggest a set of important possibilities for better FS–university level ties.

Since the national forests and the FS are a visible presence in localities in at least one-third of the states, they provide real opportunities for promoting learning that has both local visibility and national and global implications. Indeed, one way out of the dependency relationships between local communities and the FS, where timber-based employment will inevitably decline over time, is to interest the children of these communities in a broader set of careers. The FS can promote such job imaging by encouraging educators to use national forest lands as a laboratory of physical, natural, and social processes. Since careers related to these processes are not limited to traditional forestry jobs, using national forest resources in this way may promote the development of careers that are not necessarily dependent on the FS.

But educational outreach should not be limited to those areas adjacent to national forest lands. The vast majority of American citizens, including a large number of young people, do not live in rural areas with economies based directly on natural resources. Nor are their values well aligned with traditional national forest activities or FS norms. To avoid establishing an active interchange of ideas and visions between the agency and the urbanized American population is not just to miss an opportunity, but probably to guarantee organizational irrelevance somewhere down the line.

Since education is a lifelong process, and part of the task in building political concurrence is in generating awareness of current issues in the broader public, the FS should work more extensively at creating opportunities and processes of public learning about the characteristics and processes evident on our federal lands and on the nature of the choices that face us as a society. Increased educational opportunities for forest visitors, heightened media contact at all levels of the agency, and expanded educational use of national forest lands by interest groups are all important ways to promote public learning. These approaches may include expanded use of national forest resources for interest group-led trips, and increased contact between FS employees and local residents through existing social networks such as those provided by churches, Rotary clubs, and the like. Providing more opportunities to local and national media to explore or showcase activities on the national forests represents another set of approaches. As issues become more global in scale, the national forest system can promote worldwide interchanges by encouraging foreign visitation, foreign student internships, and the exchange of foreign and U. S. expertise.

The political effects of increased educational outreach can be significant in the short term, and by influencing the attitudes and values of young people, will have important benefits for the FS in the future. Educational outreach can create shared images of federal lands and their management, can build ownership of national forest resources, and can provide another means of public participation in FS decisionmaking. These kinds of educational activities are a two-way street. They provide an additional window into public interests and values, while also providing information to the public. It would be wrong to view these kinds of approaches as a means to sell FS values and decisions to the public. Instead, they should be viewed as an approach to developing the two-way understanding necessary to direct and facilitate the process of building political concurrence.

Understanding and support can also be generated through more effective use of interagency networks. One of the clear lessons of the spotted owl controversy is that the FS should use interagency networks as a source of expertise and support for desired direction in ways that it did not in the evolution of the owl controversy. On one level, given the nature of future problems, better cross-agency relationships are essential. Just as landscape corridors provide a means for interchange between isolated wildlife populations, administrative corridors are necessary to deal with isolated agency jurisdictions, and problems like the impacts of global climate change and the loss of biological diversity due to habitat fragmentation.

On another level, interagency networks can provide an additional means of building political concurrence on desired policy direction. Science and professionalism provide a means of accessing policy supporters that cut across agency boundaries. While individuals develop allegiances to the organizations for which they work, they also relate to the values and norms of the professions that they practice. If the desired direction is consistent with sound professional or scientific norms (as will be argued below is necessary and appropriate), support can be generated across agency lines, even if turf issues dictate otherwise. Given the situation of technical pluralism that is played out in administrative appeals processes and courtrooms, the worse thing that can happen to an agency is to have its expertise challenged by a sister agency. Hence, making wise choices that are understood and supported by peer agency staff can help considerably in creating and implementing lasting management direction.

Approaches to building interagency networks range from ad hoc, problem-oriented task forces to the creation of formalized groups like the Interagency Grizzly Bear Committee. They can include participating in already existing regional institutions like the Oregon–Washington Interagency Wildlife Committee or fostering the creation of new entities, such as regional biodiversity councils. Professional associations such as the Society of American Foresters or the Wildlife Society can similarly be used to share problems and interests, and build understanding and support for policy direction.

Whatever form they take, interagency networks can provide a forum in which data negotiation can take place. One of the major problems in sorting out the technical information involved in the owl dispute was the range of assumptions underlying different groups' presentation of facts, theory, and ways to put the two together. When this kind of disagreement hits the policy arena, decisionmakers do not know what to do, except to throw up their hands and curse the scientists who appear not to be able to decide on anything but that they need more time and money to find the answers. Even if they do not yield definitive answers to technical questions, deliberate processes of analysis that involve experts across the spectrum of interests can at least clarify the uncertainties for policymakers, and often can bound the range of estimates and clarify competing theories.

Processes that promote this form of data negotiation are similar to those described above under the rubric of alternative dispute resolution, but they do not need to come to a settlement. In the best of worlds, they would involve not just experts from the involved government agencies, but experts from across the full spectrum of interests. Since the parties do not need to come to a single settlement point, they are not forced to make the hard choices involved in the trades usually required to settle a contentious dispute. As a result, it is easier for them to be involved because participation is less threatening. At the same time, their interests may remain fuzzy, an outcome that does not assist in resolving the conflict. The hope underlying such processes is that the common norms and principles of science can level the playing field somewhat, and encourage a productive dialogue about the technical dimensions that underlie most resource policy debates. At a minimum, these discussions should reduce and define the uncertainty facing decisionmakers, and assist them as they work toward resolution. Potentially it is also the first step toward building the understanding and relationships needed to promote a process of collaborative problem-solving.

A final way to build political concurrence and at the same time develop a broader set of expertise and values in the agency is to facilitate more interchanges of staff between the FS and other governmental and nongovernmental organizations. One theme that will be a part of effective organizational management in the future will be the ability to create permeable boundaries between interacting organizations, and permeability between units of an organization as well. The FS can be a leader in fostering these kinds of exchanges, including Intergovernmental Personnel Act-type arrangements between the FS and other federal agencies, state wildlife and natural resources agencies, and universities. In addition, exchanges between the FS and the forestry ministries of other countries could be productive. Finally, involving staff experts from nongovernmental interest groups in day-to-day management can also help develop the shared understanding and relationship needed to build political concurrence. These kinds of staff interchanges can range from two weeks or a month up to a year in duration. A summer program that in effect provides professional internships to University researchers or interest group staff members could be attractive to participants and effective for the agency. Some interest group representatives would view such a program as co-opting, and would opt out; but others would bring expertise, criticism, and energy to forest management.

In narrow terms, staff interchanges would not be a particularly efficient program, as anyone needs time to learn a new job. In these terms, presumably the most efficient staff deployment would keep staff members in the same jobs forever. At the same time, we know that productivity declines as job tasks get less challenging and interesting. In addition, since the environment of management always changes, stagnant staffing guarantees a mismatch between what the organization is doing and what it should be doing. The use of staff interchanges can add insights, energy, and diversity to an organization that should always seek these ends. In addition, the relationships that result will continue to benefit the agency long after the exchanges take place. Implementation of a staff interchange policy needs to be mindful of the desirability of maintaining stable relationships between forest units and outside groups, as described above. If implemented in a way that is cognizant of the potential for disruption at the local level, a short-term interchange program can yield some of the long-term benefits of the agency transfer policy without incurring its potential for breaking relationships and destroying institutional memory at the local level.

All of the above mechanisms for building the two-way understanding and relationships needed to develop political concurrence imply a different understanding about organizational control than is traditional in the FS. As walls are bridged, stakeholders are more completely involved in decisionmaking and ground-level management, interagency networks are built, and staff members are exchanged, traditional mechanisms of management control necessarily will have to change. Strong, fairly narrow norms of appropriate attitudes and behaviors will have to expand to include the diversity that opening agency boundaries will yield.

Altering the meaning of agency control is necessary for several reasons. Control tends to diminish creativity and innovation. You can put a wild animal in a cage and maintain some semblance of absolute control probably until its death, but in doing so, you lose the essence of the animal. In order to foster the creativity and mutually beneficial partnership arrangements needed to carry out effective forest policy in the 1990s and beyond, the FS needs to trade some degree of control over outcomes and events for ideas and innovation. For an agency that sought control over events through much of the spotted owl case as much as it pursued anything else, this process of giving up some degree of perceived control will be difficult. If nothing else, the model of a dynamic, permeable organization is messy, since everything does not fit into tight boxes on organization charts. In addition, some level of failure and inherent problems will occur. But given the situation that the FS faces, the risks are worth it.

The other reason to modify agency norms of control is that control is often more perception than reality. If an organization works hard at maintaining control of an issue through analysis and deliberate choice, and those choices do not endure, it is hard to argue that the organization is in fact in control of its destiny. The FS has worked hard and endured much, yet most forest management decisions do not hold up through endless appeals, and court and political battles. Agency leadership may view itself as firmly in control, but this view is probably more perceived than real. If it can shift some of its decisionmaking and management processes to work more productively toward political concurrence (building public, political, organizational, and interest group support for agency direction) it will be more likely to be in control of future direction.

Changing Organizational Management and Leadership Styles

Changes are also needed in agency norms and values, staff capabilities, relationships within the organization, and leadership styles. Some say you cannot teach an old dog new tricks, and there are those who argue that the FS cannot change in any serious way. While significant change is needed in agency behavior, many of the necessary changes are not inconsistent with traditional core values of the agency. Professionalism, an action orientation, and an image of serving human society while stewarding the landscape are traditions that are still appropriate today. But what they mean today is different, and they thus require new ways of operating.

Some of the norms that need modification have already been discussed, including an altered image of organizational control. Others similarly require updating for the times. The "Can Do" image of the agency needs to be updated from what has become the aggressive pursuit of unidimensional goals to a much more complex notion of action. It is wonderful to have an image as an agency that can get things done, as long as the ends that the agency is pursuing are appropriate and supported by the public and the political process. Unfortunately, Can Do has been translated to Must Do, at times in spite of scientific evidence and internal dissent.

Chapter 10 described the evolution of FS organizational management style from one of technically based, paternal benevolence before the second world war to a more quasi-industrial, military style after the war. A third organizational style is needed to meet today's challenges and political environment, and can be characterized as science-based and concurrence-seeking. It is a style that encourages a greater degree of questioning and discussion over organizational direction, is aware of a multiplicity of public values in forest systems, is supportive of a diversity of opinion within the agency, and that defines organizational product in a much more diverse way than simply cut levels. It is a style that rewards innovation and experimentation, and when the science base warrants caution, conveys that advice honestly to Congress and the Executive Office. This style is a more complex approach to organizational management that requires a lot more energy on the part of leadership than simply commanding behavior and carrying out orders. But the result will be that what is produced through management is richer and more appropriate to the times.

This updated organizational style needs to be more fully cognizant of uncertainty in decisionmaking, and recognize the cognitive bias toward overconfidence in our abilities to make comprehensive choices. One of the failures in the way that the spotted owl controversy was handled by the FS was that it was constantly forced into standard operating procedures that were ineffective at dealing with it. Managers need to recognize the possibility that existing SOPs will not effectively handle new situations.

How do they guard against the perverse effects of overconfidence? First, by constantly looking for mismatches between standard practice and current situations through introspection, and by encouraging and listening carefully to criticism of management choices from within and outside the agency. In addition, monitoring the effectiveness of decisions as they are implemented, framing contingencies that can be activated if future events do not turn out as anticipated, and preserving options in the choices that are made are all appropriate. While problematic, approaches to decision analysis that explicitly acknowledge uncertainty in decisionmaking can also help reveal how confident we are about what we think we know, and provide a baseline against which monitoring can identify whether we are right or not. Clearly decisionmakers need to act on incomplete information, and a lack of confidence can lead to analysis paralysis. But decisionmakers also need to keep in mind what cognitive psychologists tell us (and experiments and experience in cases like the spotted owl controversy demonstrate): that their decisionmaking processes have bias, and since this bias is somewhat systematic, its effects can be minimized by watching for it.

Another agency norm that needs examination is the Study It–Keep it Technical approach to agency decisionmaking. As described in the previous chapter, studying things endlessly will not generate faultless choices. Solutions to problems that are fundamentally human valuation questions cannot necessarily be divined from elaborate technical analysis, and to the extent that this reaction is a strategic response that aims at maintaining control over an issue, it delays the ultimate debate.

This can be worked on in several ways. Altering the process of choice, so that decisionmaking works off multiparty, collaborative problem-solving processes, or at least involves a process of multiparty data negotiation (both described above) are ways to avoid endless, technical analyses with which no one is satisfied. Providing ranges of outcomes, with honest assessments of the probability and magnitude of their impacts and the tradeoffs implied therein, gives decisionmakers a better basis for choice, and may get the agency out of the bind of endless cycles of analysis. Staff-level awareness can be built (for example, during midcareer shortcourses) that rarely is there one right answer to these kinds of issues. "Correct" answers are determined by some mix of technical, political, economic, and administrative realities, and if the FS has provided good quality information, advocated its interests well (without hiding these interests in a biased analysis), and done what it could to build political concurrence, it has met its obligations to society.

Other organizational norms need to be modified for the times: The metaphor of the agency as a "close family" needs updating to one of an "extended family," where individuals in other agencies and groups can be recognized as vitally concerned with the future of the national forests. Mechanisms to accomplish this transformation have already been outlined through the variety of outreach activities listed earlier. In addition, processes of in-house socialization that tend to promote isolation need to be modified. The FS is an important symbol for staff to identify with, but not to the exclusion of other members of the community whose concerns and knowledge make them interested in public forest management.

Finally, but very importantly, the set of norms that were described in Chapter 10 that overemphasized timber production as a component of planning, staffing, and decisionmaking need to be balanced as the agency shifts to a truly multiple value-based management scheme. Organizational means to promote such a shift include altered personnel evaluation systems, budgeting processes, and promotion criteria. Clearer messages from leadership that lend credence to this shift from within and outside the agency are also in order. In addition, some key promotions and retirements can be important symbols of a shift in emphasis.

Updating the value base and capabilities of FS staff involves changing personnel recruitment, selection, and training to better match the diversity of values in American society, and to provide state-of-the-art capabilities in forest science and democratic management. A couple of ideas have already been presented: enhancing the diversity of staff values and capabilities through staff exchanges resulting from more permeable agency boundaries, and revising agency transfer policy to promote better relationships at the ground level.

The agency could similarly benefit from changes in other staffing policies. The best way to incorporate a diverse set of values into an organization is to have the full range of values present in the organization's staff. Otherwise, underrepresented values tend to get caricatured and stereotyped in the decisionmaking process, as staff members try to articulate what those values are, yet do so from a cultural base that does not help them very much. At the entry level, the agency needs to recruit more heavily from segments of society and disciplines that have been underrepresented in recent FS staffing patterns. These groups include women and minorities (including native Americans) and urbanites whose values are increasingly important in determining national forest policy, but who have generally opted out of resource management careers. In terms of disciplinary background, staff members with nontraditional backgrounds in such areas as sociology, anthropology, rural development, or toxicology may be more qualified for some tasks needed in national forest management than individuals with traditional forestry training.

From the vantage point of an academic in the field, the best and the brightest of students who are already preselecting graduate careers in natural resources and environmental quality do not view FS employment as an exciting or intellectually challenging possibility, and this must be changed. At one time, FS employment was seen as a desirable goal, and a career in public land management as an attractive possibility. But the image that the agency has projected in the past fifteen years, and that was reinforced through the spotted owl issue, is that of a stodgy, conservative, boring organization whose leaders take stands that in the media's view are often on the wrong side of the issues. The Reagan administration's denigration of public service, the differential in salaries between public and private sectors, and personnel policies that make it difficult to cope with present-day family issues have further diminished the attractiveness of a FS career. This brain drain comes at a time when the agency needs quality staff more than ever, and must be dealt with to ensure both a high quality staffing base for the agency and an ability to adapt to current and future forest policy realities.

The FS should utilize a full set of recruitment devices to try to expand its staff capabilities and values. These include cooperative programs with universities, internships, and career-planning outreach to secondary schools. The educational outreach activities described above should also help young people to gain positive images of FS careers, as should educational materials at national forest recreation areas. The standard image of a FS employee, the robust white male, Smokey Bear hat in place, out cruising timber, needs updating for a potential skilled employment base that will not respond to that image.

The FS should also seek a diverse set of employees moving laterally into the agency at levels above entry-level. One of the main benefits of recruiting new workers primarily at low levels is that over the following five or ten years they can be socialized into agency norms, values, and ways of doing things. This makes sense for an agency seeking control and a predictable workforce, but as discussed above, tends to filter out individuals that can provide a necessary diversity of ideas and values. There are capable individuals working for state or other federal agencies, private industry, and nonprofit groups that might bring with them different skills or values to the 1990's FS. Recruiting capable individuals for higher levels of the organization (even as line officers) entails some risk, but given the situation the agency currently faces, is worth a try.

Clearly there is a need for more training programs for existing staff. Training via short courses, demonstration trips, and the like can enhance the capabilities of staff members while influencing their values and attitudes. Broadening the scientific knowledge of employees (especially that of higher level positions such as line officers) to include conservation biology, landscape ecology, and similar topics, as well as enhancing their understanding of political processes and economic development, and building their skills at interacting with other people, including dispute resolution and educational methods, are all necessary topics that can be addressed through a variety of training methods.

Midcareer training programs often influence staff attitudes more than specific individual capabilities. In this instance that reality is advantageous, since part of the problem seen in the spotted owl and other recent controversies is insensitivity to the full range of public and disciplinary perspectives and values. Particularly insidious is the effects of group socialization that fosters an "us-them" mentality. In the context of this research, it was amazing to speak with one group of stakeholders, and then another, and compare their images of each other. Most were fairly far off the mark, ascribing dastardly intent to haphazardly made choices, without ever testing these images by communicating with one another. Training programs that involve crosscutting allegiances and values can provide neutral ground on which productive discussions can be held and good relationships can be forged. These can include technical programs involving staff from different federal and state agencies, and courses on topics like negotiation that involve individuals from government agencies and nongovernmental organizations. The value differences that underlie organizational perspectives are often real, but a better understanding of these differences, real or perceived, will promote more effective interactions and decisionmaking down the line.

The agency also needs to work at establishing better relationships and understanding between subgroups within the FS family. This includes improving the relationships between researchers and managers, staff and line officers, and wildlife and timber staff members. The spotted owl controversy created a great deal of tension among these groups. As the agency gets beyond the owl issue and implements some of the other recommendations described in this book, some of this tension and misunderstanding will diminish. Rebuilding an image of shared purpose and recreating a positive external image of the agency will help. But the agency is large enough, with disparate enough subgroupings, that positive working relations between the groups should not be assumed. Again, midcareer cross-disciplinary training programs, ad hoc interdisciplinary workgroups, and other approaches that cut across these intraorganizational boundaries are all appropriate and important ways to build this shared understanding. Creating a six-month voluntary sabbatical program whereby every four years or so a staff member can work in a different area of the organization (a timber planner can assist with a fisheries enhancement project, a wildlife biologist can assist in timber sale prep work, or a district ranger can help carry out a research project, etc.) can help build cross-agency understanding at fairly low cost.

Many of the intraorganizational problems in the FS could be mitigated by better leadership throughout the agency, and leadership styles need updating to better deal with the issues of the 1990s and beyond. Chapter

10 described six roles played by formal agency leaders: articulating a vision and mobilizing support for it, facilitating staff aspirations, acting as an organizational change agent, translating and communicating between the agency and the outside world, modeling good behavior within the organization, and acting as a symbol of the agency to the outside world. It is easy to say that agency leaders should simply do better at carrying out these numerous roles and obligations of leadership, and that is part of the lessons from the owl case. Better scanning for new information and ideas as described below will also help agency leaders define a more robust vision for their agency.

Efforts to generate political concurrence will similarly force agency leadership to better understand the demands and needs of the political environment, and will acquaint policymakers with a clearer image of the aspirations and constraints of the agency as it implements its mandates. It is the job of agency leaders to constantly test the size and character of the political envelope, that is, the range of politically feasible possibilities, and keep pushing at its margins toward necessary change. It is also their obligation to be honest about the knowledge and beliefs of their staffs and themselves, and be true to the science-base that fundamentally is the reason for the agency's existence. This means having a working knowledge of the science base that currently underlies agency choices, and not getting isolated from organizational staff.

When issues and demands fly furiously at the top, there is a tendency for leaders to become isolated, partly because of their limited time and attention span, and partly because isolation is comfortable. We all seek to insulate ourselves from pain, particularly in times of high stress. Leaders tend to fall back on images, styles, and contacts with which they are comfortable, minimizing cognitive dissonance and the energy needed to learn new things. Unfortunately, the actions that leaders may be most comfortable with may well be inappropriate for the problems at hand. A leader's base of substantive knowledge and organizational needs comes largely from his or her experience before becoming a leader. Since they tend to generalize from their personal experiences, often their visions reflect images of the past more than images of the present or the future. One FS staff member described this dynamic in relation to the individuals that rise to the position of Chief in the agency: "The career paths of our Chief almost always guarantees that they will be great leaders of the agency, as it existed ten years in the past."

The best leaders rise above this tendency to carry outdated visions into leadership choices by working hard to offset this cognitive bias. Forging communication channels and relationships across the age and value sets of the agency, constantly working at updating their own knowledge of technical information and organizational aspirations, and finding some slack time and resources to continue to gather on-the-ground experience are all useful methods to increase leadership relevance. Since FS leadership is gathered from inside the organization, and the Chief is not a political appointee, FS leaders must work doubly hard at insuring that these biases do not inappropriately taint their vision for the agency.

One prescription for avoiding isolation is to require top leadership (Chief, Associate and Deputy Chiefs) to participate in periodic working sabbaticals that would place them on the ground in staff level positions. Perhaps four dispersed weeks per year per individual, two selected at random and two selected because of the high level of innovation ongoing would be an appropriate level of this activity. If a picture is worth a thousand words, then a week's worth of staff level work four times a year should be worth a forest plan-worth of images. Such a program would not only enhance leadership understanding, but would build staff appreciation for leadership. Having a Deputy Chief work on your district because what you are doing is seen as innovative would be a positive reinforcement that would promote on-the-ground innovation.

Creating Effective Information Flow

The Forest Service's legitimacy rests on the quality of its information and its ability to process that information in a way that is useful to society. Indeed, as issues become more information-intensive, agencies must expand their role as information-providers. If their information is of poor quality or is inadequate to inform policy choices, we would be better off relying on other groups or approaches to assist us in making collective choices. Policymakers need an honest assessment of current conditions, a reliable estimate of likely future conditions, a robust set of alternative choices, and an understanding of the tradeoffs involved in different choices. Yet the history of the spotted owl controversy suggests that they often get limited or biased information, based on information collection and analysis methods that are geared toward past problems, and that reflects the views of a limited set of informants defined as much by the old boy network as anything else. In addition, since information is power, much is not shared, and because information brings implications with it, it is often filtered and edited as it moves within and between organizations.

There are ways to improve information collection and analysis processes to avoid the problems evident in the spotted owl controversy. For example, individuals interviewed for this research were asked what they felt were the most important lessons for the FS from the history of the spotted owl issue, and many of them identified "Be Honest" to be a prime message they felt needed to be taken back to the agency. Since decisions are only as good as the information that informs them, if we do not have information that is as accurate as is available, we are wasting a lot of time. Many respondents felt that the FS and others had filtered information strategically, or interpreted information in an obviously biased manner. Others felt that agency officials had lied outright at times in the course of the debate. Since the only way that we can find creative solutions that meet as many concerns and interests as is possible is through analysis grounded in accurate information, and problem-solving that relies on some amount of trust in the good will of information-providers, misrepresentation limits the possibility for finding good policy solutions to pressing problems.

"Be Honest" is a simple message that is sometimes hard to implement. Clearly information represents power in a strategic interaction, and the negotiating process over policy choices often encourages misrepresentation. In addition, sometimes this misrepresentation is not fully conscious as the kinds of cognitive biases described in Chapter 10 influence the ways that officials hear and sort information provided to them. Nevertheless, the agency's obligation to the rest of society is to be as honest as possible. Besides, it is not difficult to test the validity of these statements, given the existence of parallel expertise outside of the agency. Where misrepresentation is seen to occur, the agency loses a lot of credibility. When faced with a question such as, "Are we currently overcutting areas in the national forests?", agency leadership could have replied honestly that "There are problems with cut levels in some areas. We serve the President and the Congress, and we can still cut at these levels if you so direct. But you should understand the tradeoffs that are implied by them, and the impacts that will occur as a result." It is the agency's responsibility and obligation by virtue of the role that the citizenry has bestowed on it to provide an honest assessment based on the best information and to indicate tradeoffs implicit in choices. It is not its obligation, or in anyone's long-term interest, to pull punches for political reasons.

There are other changes that can be made to improve the way that existing information is managed in the policymaking process. Some of the approaches to multiple party decisionmaking and data negotiation described in the previous section can be used to encourage the sharing of information across organizations, and the FS should work to collect and assess information from a broad set of sources, including those they have distrusted historically. In addition, agency officials need to take a second look at the form of information that they are producing for public consumption. Forest plans and other agency decision documents are covered by a veneer of technical complexity that makes it difficult for interested parties to access needed information. Within the agency, barriers among groups need to be modified to insure that information moves in a timely and productive manner. The interplay between researchers and managers is a good place to start.

Within the agency, practices that encourage the filtering of information as it moves upward need to be examined and changed. Obviously, top level leadership does not need, nor could it process, all the information that is generated at lower levels of the organization. At the same time, leaders need to know what is really happening in the decentralized offices and land units over which they exercise authority. Better sorting processes need to be instituted that filter information, but not to the point of hiding bad news. Those who constantly bring bad news are often a pain in the organizational butt, but their messages may be real and important for agency leaders to hear. Yes-men are great to have around, but only until the reality of a bad situation comes back to haunt you. At that point, yes-men become it-was-his-fault-men, and not very helpful to solving the problem at hand. A norm of shooting the messenger of bad news is clearly prevalent in the FS, as it is in many organizations, and needs to be worked on carefully to ensure that good quality information in adequate quantity moves to decisionmakers inside and outside the agency.

Expanding the information base that agency officials have available to them is also very important. Our understanding of how the world works and the problems facing managers will change over time, often in unanticipated ways, and mechanisms that collect and process information that help agency decisionmakers anticipate and respond to these changes are necessary, and probably well overdue. The FS needs to invest in ongoing monitoring and resource inventory in ways that have not been needed in the past. We need to know what the land base looks like, and what it is doing over time, in a manner that provides flexibility in combining this information in new and different ways, as problems arise and the needs of management change.

The FS should have an effective NFS-wide geographic information system that collects information based on biological, social, economic, and geographic characteristics that is at least one step removed from current problem interpretations. A set of information is needed that includes basic variables such as tree diameter, canopy closure, soil type, etc., in digital form, so that it can be sorted and combined in different ways to match different needs. The same sets of characteristic data can give us maps of commercially valuable timber, old growth areas, or spotted owl habitat, and can allow us to recombine and add variables at a later date to respond to inquiries about wholly unanticipated questions. The agency's inability to produce an accurate map of old growth or owl habitat even fairly late in the owl controversy was extraordinarily frustrating to decisionmakers. There is no good excuse today for not being able to provide this information: The technology has existed for a while, and the agency should be at the forefront of information processing techniques, grounded in adequate on-the-ground data collection and monitoring.

Monitoring and data storage techniques should be structured so that they are tracking environmental change and helping agency officials to forecast likely future states. Environmental conditions change over time, and even minor fluctuations can have significant effects on forest management. For example, global climate change that results in an increase in surface temperatures of only a few degrees Celsius can have significant effects on the ranges of numerous wildlife species, and the effectiveness of certain types of seed stock. These types of changes, along with others such as variation in precipitation amounts and pH, are important to track and assess, and just as important to be able to relate via inventory information to on-the-ground implications. Relationships with groups that are doing work on environmental change, such as NASA and the National Center for Atmospheric Research, should be forged. Collaborative research projects could be quite useful, and marketable, in this area.

Technical concepts, methods, and theories also change over time, and the agency should engage in ongoing scanning for new ideas. An active research program should utilize conferences and disciplinary and cross-disciplinary information networks to seek new ideas and approaches wherever they are developed. To do this, the agency should place as high a premium on interactions outside the agency as it does on fostering discussions within groupings in the agency. In an era of multiple value management, innovations may well emerge from a variety of disciplines, many of which have not been traditional concerns of forest managers. More interactions with land managers and resource policymakers from other countries, and their state-level counterparts within the U. S., may be another source of innovative ideas for American forest management. An active international forestry program and staff interchanges with state agencies and universities will help the FS staff tap into extra-agency sources of information and insight.

Not only will scientific fact and theories and management approaches change with time, but human values also will change, and the FS must do more in the way of value-scanning than it has done in the past twenty-five years. As described in Chapter 1, the spotted owl case represented a mismatch between FS management goals and styles and evolving public values in public land resources. The agency should have seen the handwriting on the wall sooner than it did, and it should have altered management styles faster than it did, when agency officials understood what had to be done.

A deliberate process of value-scanning is just as necessary as a systematic process of environmental monitoring and forecasting. Approaches to value-scanning include surveys that focus on American and agency values, an examination of global economic and sociologic trends, and periodic conferences that help the agency better assess the future and figure out how to respond to it. Forecasting approaches that largely rely on past economic trends to forecast future patterns, such as those contained in the RPA program, are only partly helpful in determining the future, as anyone who has engaged in energy demand forecasting over the past twenty years can tell you. The future will bring shifts in valuation and concerns that are only somewhat codified in past market choices, and we would be foolish not to look broadly for such changes.

One of the major failures of our policy processes is that they do not look very far downstream, and hence are not particularly proactive in avoiding upcoming problems or moving us to a desirable future state. Much of the promise of the FS comes in its ability to rise above these failings and incorporate a longer term sense of vision. Unfortunately, the spotted owl case suggests that the promise of vision has not been fully realized. The agency looked backwards, as much as forward, and lost much in the process.

Imaging alternative future states is something that our society is not very good at. Part of the reason for this is that Americans are very parochial and inward-looking as a society, and hence are not exposed to alternative models of what the future could hold. Part of this is because even in affluent America, most of us spend the vast majority of our time simply trying to cope with the day's events, and have little energy left for thinking broadly about the future. Planning this year's vacation is often as forward-thinking as we want to be. Part of this is because our institutions are generally committed to the status quo, and reinforce this direction through value and attitude-formation processes that are persuasive and subtle. Yet we know that the future will require changes. It is a luxury to believe that the current day's reality will always be the case, and it is clearly inappropriate to do so. Otherwise we would all have backyard bomb shelters, extensive collections of hula hoops, and spend our weekend days spraying DDT to control bugs.

The FS should experiment with different approaches to futuring, that is, mechanisms that image alternative future states. These activities can range from FS staff simply playing a more active role in local and regional planning activities up through supporting imaging workshops at the local level. Planning the future of a chunk of ground, as occurs in forest or unit plans, can serve as a trigger for these kinds of activities. Those who view government's role as a simply reactive one will say that imaging the future is not appropriate, and that the FS should simply respond to client demands. That image of government is legitimate; it just is not likely to be effective in a world of rapid technological and environmental change that demands shifts in human activities and values.

We would all be indebted to someone who encourages a more aggressive look at the future, and the FS, in partnership with foundations and other social institutions, can play a role in this process. Indeed, analyzing and imaging the future could be its most important and probably most legitimate role. If we simply wanted someone to prep timber sales and maximize the yield of wood fiber, deer, or visitor-days from national forest lands, we probably would be better off privatizing the public lands. The ability to make choices in our long-term collective interest is in fact the major promise of the public land system, an opportunity that only can be fulfilled through deliberate attention to the opportunities and values of the future.

Creating a Culture of Creativity and Risk-Taking

Better information can generate a more sophisticated understanding of current and future landscape conditions, and a more complete assessment of current and future values in public land resources, but understanding these things will not of itself cause changes in management practices. An organizational environment that fosters creativity and allows risk-taking needs to be developed to enable managers to try out new concepts and perspectives on the ground. Just as important, the agency needs to learn from these experiments and find ways to diffuse its new understandings throughout the organization.

Innovation represents an idea converted to action. Doing something different requires understanding possible ways to do things, and an environment that allows action. Some of the approaches described in the previous section will help FS staff members envision new ideas to try out on the ground. Developing an organizational environment that supports risk-taking is more difficult. It means concurring on and rewarding attempts to do things differently than standard operating procedure. It also means not penalizing well-intentioned failure. Both of these items translate into changes in staff performance evaluations, which tend to reward adherence to narrow goals carried out in time-honored ways.

Promoting risk-taking and innovation does not mean generating a total free-for-all. Organization goals must still be set and implementation monitored, but more latitude is needed to allow units to determine the nature of their goals and how they are to be carried out. Another approach is to reward units that cut costs in different ways by allowing them to use the savings for creative programs at the margins of their operations. Any such incentive program creates the possibility of encouraging perverse behavior, such as cutting necessary activities (or doing them poorly) in order to fund pet projects. Perverse incentives are always possible, but clearly a set of revised organizational incentives needs to be created and reinforced to encourage innovation.

Such innovation and risk-taking does not have to come from within the FS itself. Providing seed money and grants to nongovernmental parties, such as in the Challenge Cost-Share program, can help to promote innovation in forest management. Such programs should not be limited to those ideas brought forth by external parties as is often the case. Imaging new ideas and devising protocols to try them out can be a FS role, while implementation relies on nongovernmental sources.

The national forest system itself provides a marvelous set of resources for experimentation, as the forests within it represent an abundant set of landscapes for which controlled experiments can be devised and carried out. The national forest landbase provides diversity in types of natural systems and human demands, yet also provides great possibilities for replicating many key variables. Hence it is possible to devise experiments that control for many of the exogenous items that often frustrate researchers. Trying new ideas on a watershed-by-watershed, or district-by-district basis, allows us to test the utility of these potential innovations.

While encouraging innovation, risk-taking, and experimentation, it is important to keep in mind why we are doing this. The reason to try out new ways of acting is to do better throughout the system, and that means that experiments need to be monitored, successes and failures documented, and policy and management implications communicated broadly to agency and nonagency parties. We need to learn from experimentation so that we get better at matching forest management practices to current and future needs. Experiments need to have a deliberate process of evaluation built into them, and the lessons from attempted innovations must be communicated broadly within the agency. Newsletters, broader use of the DG message system, and enhanced use of video imagery to examine and convey organizational experiments are all in order.

One of the best things that the FS can do to assist the policy process is to be a source of tested concepts—of evaluated new ideas—that can then be considered for broader applicability. Innovation is sometimes useful as an end in itself, but that is not the end we seek here. Rather, more effective forest management and policy can result from expanded attempts at innovation and experimentation, and the sooner we get on with it, the more likely it is that solutions to issues like the spotted owl controversy will be available when we need them. Just like managing for a diversity of values will allow the agency to more easily adapt its management to the values of the future, a diversity of management approaches will help provide ways out of future crises. A monoculture of ideas is just as devastating to future survival as is a genetic monoculture.

Innovation and experimentation must relate to ongoing management activities in several ways. Trying out new ideas should not be the sole province of a research unit; rather, the process of innovation should be integral to the management of each national forest, and line officers should be evaluated on their efforts in these areas as well as their productivity in others. In addition, tested innovations must be given the opportunity to diffuse into normal management practice, and this means adopting courses of action that can change down the line. While it has the organizational and political problems described in Chapter 9, adaptive management is an important concept of forest management that must go hand-in-hand with an active program of experimentation and monitoring. If successful experiments cannot be adapted into management practice because options are precluded, the process of innovation becomes an intellectually appealing facade: a good deal for researchers, but a bad bargain for society.

Innovation and experimentation are not costless, since they require real dollars and staff time that could be employed in other ways. In addition, risk-taking is more likely to occur in an environment of some slack resources, otherwise failed risks cannot be absorbed. Traveling that deserted backroad that may get you to your destination in half the time is more appealing if you have a spare tire and an extra five gallons of gas. Slack resources make it easier for those who are willing to try something new to do so without drastic consequences if they fail.

Finding slack resources in a time of federal budget deficits is not easy. One way is to give budget flexibility at the local level, as described above, to encourage cost savings in one area that could be utilized in other areas. Another way is to create an innovation and experimentation endowment funded on a one-for-one match between Congressional appropriations and user fees. Even a small tax on use would yield a significant sum, whose interest could fund numerous demonstration projects on a competitive basis. A third way is to have FS units, perhaps in partnership with universities or nonprofit groups, actively competing for outside funding for innovative ideas, including support from private foundations and other government agencies like the National Science Foundation. This model has the added advantage of peer-reviewed evaluation of the concepts from outside the agency.

There are no doubt other ways to provide organizational slack, particularly as the need moves from money to staff or equipment. Interagency staff exchanges, using volunteer labor, soliciting equipment donations, or developing partnerships with private timber companies are all ways that the agency can find supplemental resources to try out new ideas. If the agency is at the forefront of forest science, and active in soliciting support for good ideas, funding probably will not be an overwhelming barrier.

Even though organizational slack helps foster a willingness to try out new ideas, the bottom line is that it is not the most critical variable. Some of the most innovative solutions to real problems are developed in times of significant resource constraints, such as in the aftermath of natural disasters where innovative ways are found to work around problems that would never have been conceived in a time of abundant resources. Humans are resilient and creative, and are most likely to be so when the incentives that they face encourage them. Changing the environment of day-to-day agency management is most likely as important a factor in promoting innovation and risk-taking, as is throwing money at a problem.

By employing a variety of new or modified approaches to political, organizational, and land management, the FS can regain an image as an effective federal agency and an innovative manager of public lands. By developing an organization with permeable boundaries; improved methods of monitoring, information processing, and value-scanning; and a style of organizational and external management that is grounded in science-based professionalism yet that recognizes the need to build political concurrence, the FS can be a model for other public land-managing agencies at the federal and state levels. Experimenting with innovative ideas and generating knowledge about their effectiveness are important activities that public agencies are uniquely qualified to carry out. The possibilities created by these new ideas and techniques can potentially help avoid further spotted owl controversies, or more likely, help us better respond to and manage them when they arise.

13

Building Better Policies

The spotted owl controversy points to the importance of the broader political and policy context in which resource management agencies and nongovernmental groups interact. There is obviously a need to update and reform a number of areas of federal natural resource and environmental policy. Public policies function by influencing human behavior in a variety of ways. By creating goals, they legitimize the actions of those who seek to move behavior in the direction of the goals. Through regulations, they seek to bind individuals and groups to specific directions, and through subsidies, grants, taxes, and technical assistance, they create incentives that encourage people to move in certain directions.

Public policies can also create opportunities for ongoing dialogue and cooperation, providing a basis for continued decisionmaking and motivation for parties with diverse interests to try to find compromises. The policy process also provides a means whereby compromises that are found can be enforced. Finally, by affecting patterns of power and influence, through the provision of information and opportunities for groups to act as watchdogs, policies can empower weaker parties in society. In doing so, they may increase the likelihood that certain long-term objectives are pursued by altering the strategic balance of power in implementation.

We need to harness all of these instruments and modes of public policy, and use them to overcome the problems evident in the owl case, and in natural resource and environmental management in general. The ambiguities and complexities evident in resource management today suggest the need to refocus and sharpen the objectives of natural resource policy, consistent with changing public values and expanding scientific knowledge. Since future issues will be complex, with hard choices required, expanded understanding and wisdom are required of citizens, scientists, and decisionmakers. Enhanced expertise and modes of public learning are also

357

needed to overcome the limited vision that is a structural reality of our political processes.

Other policy reforms are needed. Since resource policies, institutions, and the land base that they deal with will continue to be fragmented, better planning processes are needed that are capable of integrating diverse interests and long- and short-term objectives. In addition, stressed systems require a change in the incentives that promote inappropriate activities, and constrained fiscal resources require creative financing strategies. Finally, since we know that whatever reforms are put into place will bog down in day-to-day implementation and will become outdated over time, providing various parties with the opportunity to examine and challenge directions in a productive way is important to long-term effectiveness. This chapter will outline a set of policy reforms aimed at:

- Shifting and clarifying the objectives of resource policy;
- Changing incentives and providing opportunities; and
- Enhancing the capacity of individuals and institutions.

Shifting and Clarifying the Objectives of Resource Policy

One direction that should be pursued across many areas of resource and environmental policy is a stronger focus on ecosystem-level biotic health, with a focus on biological diversity as an indicator of the state of landscape ecology. As was described in Chapter 9, the Endangered Species Act's focus on species is a necessary but not sufficient approach to insuring the effective functioning of environmental systems. The ESA provides a clear bottom line that can be helpful in overcoming some of the problems described in the earlier chapters. We would not be where we are today without it—either in levels of conflict or understanding of the broader problem. It has been a successful program in raising public awareness, pressing for a clarification of societal values, and assisting some of the most critically sensitive organisms.

Those who condemn the act as a failure because it has not resulted in the delisting of many species, either do so strategically to weaken its impact, or do so honestly, but miss the point: The ESA is an important statement of societal values and an effective stop-gap measure. The regulatory program that it establishes helps bind us to important long-term objectives in the face of considerable incentives to favor short-term development. It has also provided an important mode of access into federal agency decisionmaking that has improved the quality of decisionmaking, and moved agency values in an appropriate direction. The ESA's reactive approach is one component of a much-needed, broader public program aimed at protecting biological diversity and insuring that environmental systems function effectively.

We need to use the full range of policy instruments to pursue these broader environmental objectives. These should include government-led efforts to generate information about the magnitude and location of areas most in need. Through national scale landscape ecosystem classification and gap analysis, organized in an accessible geographic information system, areas of high levels of biotic stress can be identified for high priority action of a variety of kinds. Government can also provide incentives to private landowners, and state and local governments, to act in ecosystem-conserving ways. Whether these incentives comprise technical assistance, tax breaks, or streamlined regulatory processes, the federal government can encourage private parties to act in appropriate ways. Expanding federal grants to states for creating biodiversity protection programs, using federal tax write-offs to reward individual and group conservation behavior, and establishing partnerships for cooperative management across an intermixed public and private landscape, are all viable mechanisms for inducing effective landscape restoration and protection practices.

The federal government should also use its fairly significant set of powers to influence behavior through regulation and by acting as an effective landowner. Numerous federal regulatory processes can be adapted to include a broader focus on biotic health. These include the Clean Water Act's section 404 dredge and fill permits, NPDES (national pollution discharge elimination system) permits, the Coastal Zone Management Act's planning and federal consistency provisions, natural resource damage assessment components of hazardous waste management laws, and impact statements required by NEPA. Changing the ESA at the margin to provide more focus on multispecies recovery plans, habitat conservation planning, and the faster consideration of candidate species would also move in the right direction. At the same time, changing the behavior of federal landowners would work directly on the problem and potentially set a more effective model for nongovernmental land managers. Raising the priority of biodiversity and ecosystem protection for such diverse landowners as the FWS, FS, and the departments of defense and energy, and increasing their capacity to pursue these objectives, could go a long way toward establishing core areas for protecting remnant ecosystems. Adding land and water resources to the federal portfolio through acquisition, purchase of development or water rights, claiming instream flow rights, and land swaps is also appropriate at the margins of existing areas.

The overall approach would be to establish a concerted effort to raise the priority of ecosystem-level protection through as many instruments of public policy as are available. Just as diversity insures evolutionary fitness on the part of a species, raising the priority of biotic health through a diverse set of policy instruments is more likely to move behavior in the right direction rather than having one law—the ESA—shoulder all the burden. Indeed, if at some point in the future, a political backlash develops that results in changes to the ESA, it is important that appropriate direction be well-integrated into other areas of law and practice. As the Reagan and Bush years suggest, it is also important that long-term conservation direction be able to weather shifts of the political winds. A diverse set of directives, management behaviors, personnel motivations, and legal structures is more likely to insure that the ship of state moves in an appropriate direction in the face of an ill political wind.

Another set of changes that would assist in aligning federal agencies more effectively with a mission of protecting biological systems would be to clarify the ends of resource management policy. The goals and methods established by law are often specified vaguely in order to secure passage (more specific goals would alienate too many potential supporters) and allow discretion in implementation. Indeed, the most durable law provides enough flexibility to allow implementers to adapt its means and message to changing times. For example, the U. S. Constitution has provided the direction for a governance process that has remained stable through more than 200 years of history because it provides broad objectives that can be interpreted as times change, but it also sets forth clear principles that bound our society.

Federal public lands policy also provides broad principles for agencies, such as multiple use and sustained yield, but by not prioritizing the objectives of resource management clearly enough, inordinate discretion is provided to agencies. Given the behavioral tendencies discussed above, the result of too much discretion has been to emphasize short-term, tangible goals (commodity production, large agency budgets) at the expense of longer term objectives such as conserving biological diversity. Multiple use was defined as maximizing commodity yields subject to minimum provisions for wildlife and other resources. One way to reverse the incentives to damage the resource base is to reverse this equation. Multiple use could be defined as maximizing provisions for wildlife and ecosystem health considerations, while allowing commodity and other consumptive uses only to the extent that they do not impede the primary objectives of long-term ecosystem health. In some respects, a scheme like this would make federal resource management similar to federal environmental regulation whereby human health-based standards set a ceiling on what is allowed.

The most effective public policies are those that define fairly clear ends, yet provide some flexibility in the means chosen to reach those ends, the capacity to agency staff to pursue the ends effectively, and the ability for nongovernmental watchdogs to press agencies when necessary. The ESA works quite effectively because it does most of these things. (Some would argue it is inappropriate because it is so effective.) Throughout resource and environmental policy, mechanisms are needed that establish appropriate, long-term performance standards, yet provide flexibility and foster creativity in meeting those standards. These include standards-seeking, incentive approaches to the regulation of pollution and providing more clarity in defining the ends of federal resource management policy, without necessarily prescribing the means to reach those ends. Both approaches require that the ends are monitored, and governmental and nongovernmental parties have the ability to intervene if the means of regulation and management are not moving in an appropriate direction.

One way to clarify public priorities for federal resource management would be to recast the public land systems in a way that elevates the priorities given to long-term objectives, without changing land ownership patterns in a significant way. I would not try to change ownership or management agencies, such as consolidating all biodiversity lands under a single agency, for two reasons: Such a change would require a huge amount of political support and would have to overcome serious implementation difficulties. In addition, diversity of ownership/management is more likely to yield effective implementation over time. It does not rely on the good graces of a single agency and set of political leaders. Rather, it puts the fate of public resources in multiple hands. Such diversity is more likely to yield an effective future, even if it guarantees some measure of failure and organizational sloppiness. Given a diverse ownership pattern, however, clear objectives and effective integrative forces are needed to overcome fragmentation of purpose and knowledge.

Similar to the approach taken by the Congressional designation of scenic rivers and wilderness, a three-tiered overlay could be placed over the landscape that would establish boundaries for the following three types of areas:

- *National Conservation Lands,* defined as land and water resources that are critical for the protection of native biological diversity, and the functioning of important ecosystems. These boundaries would be defined through ecosystem classification and gap analysis, and would include consideration of the designation of corridors providing for interchange of genetic material between isolated landscape units and for the adaptation of species to climatic changes. The primary objectives of these lands would be for landscape conservation and restoration, including research and experimentation to generate options for the future. These boundaries could be established around mixed ownerships, as in a greenline reserve, using federally owned lands as core areas and other ownerships as corridor areas.[1] Some land acquisition could go on in these areas, and land swaps aimed at consolidating large habitat units would be a priority. Direction for these units would be set by an interagency group of scientists and managers, with the input of an advisory group of other regional and national interests. Their actions would draw on an endowed conservation fund created by revamping the federal Land and Water Conservation Fund program (discussed below).

- *National Multiple Value Lands,* defined as the land and water resources that can be managed for sustained yield of a variety of goods and services, subject to overall performance standards including those insuring long-term sustainability and environmental quality.

- *National Commodity Lands,* perhaps managed by private sector operators under long-term leases with strict forest and range practice controls that include the ability of governmental and nongovernmental parties to monitor and adjudicate violations of those controls. Fragments of federal lands exist that are unlikely to serve as important contributors to environmental health objectives. The highest and best use of these lands may be in sustained yield commodity production. Regardless, any recasting of objectives of the public lands will require a political bargain to be made with commodity interests. There are areas of the landscape that could be wisely managed for intensive, sustained commodity production. Designating these areas for that use, with appropriate monitoring and environmental controls, would provide more certainty for commodity interests, and also would provide a political exchange so that more important lands can be set aside for conservation purposes.

Regardless of how it is done, it is important to bring about a set of changed priorities for land management in this country, so that long-term objectives have a chance when faced with the incentives provided by short-term development opportunities. At the same time, the response to the owl case suggests that all parties presently involved with the controversy would be happier with some certainty: the environmentalists, that certain objectives will be sought by agencies they do not trust; the timber industry, that the levels of commodities available from the federal lands are defined with greater certainty than has been possible in the past ten years, so that investment decisions can be made accordingly; and the agencies, that their decisions will endure. Surprisingly, all of these parties would be happier with boundaries that provide measurable direction and control. Recasting the land system in a way that increases the amount of area devoted to appropriate dominant use management is one way to move in this direction.

Changing Incentives and Providing Opportunities

While clarifying the objectives of resource policy will help, changing the pattern of incentives that encourage individuals and groups to overuse public resources, and providing more opportunities for regions to craft environmentally sound development strategies, would help to offset some of the problems evident in the owl case. A number of federal policies encourage the long-term degradation of public and private resources by establishing subsidies that promote overutilization; the stresses evident on natural systems would be alleviated considerably by reducing federal subsidies. Subsidizing the provision of goods and services by underpricing raw materials from public resources, or through direct payments to potential vendors, is an appropriate use of public power, when done in a manner that is compatible with a broad concept of the public interest and for a short time. Indeed, using land grants in the early 1800s as a means to promote internal development was appropriate, because such development was supported by much of the population, and land was the one resource available in abundance as a tool of public influence.

While subsidies can encourage certain behaviors in the short term, continued reliance on subsidies for marketed goods will lead to long-term impoverishment. If prices charged for marketed goods do not reflect their true cost, they will be overconsumed and hence the raw materials that comprise them will be overconsumed. With an unlimited resource base, overconsumption of natural resources would not be a problem, but since our portfolio of natural resources is finite, future citizens will suffer the costs of overconsumption. In addition, subsidies create a class of dependents whose lives depend on continued subsidies. It is rational and necessary for them to lobby for continuing the programs that have defined their existence. But the result is a continuing dependence on an unsustainable course of action, leading to festering controversies such as those evident in the spotted owl case.

Federal programs that provide subsidies to timber, agriculture, ranching, mining, and nuclear energy have served their earlier promotional purposes well, but they have also resulted in the degradation of natural systems, and the creation of a class of dependents. Timber interests have benefited from below-cost timber sales from national forest and BLM lands, agriculture by price supports, ranching from underpriced grazing allotments, mining from underpriced royalties, and nuclear energy by capping liability from accidents. While the American economy has benefited in the short term from these subsidies, over the long term, such practices have resulted in inefficiencies and the degradation of natural resources described in Chapter 11.

This argument is not intended to suggest that all publicly provided subsidies are wrong, nor is it intended to suggest that marketed goods that can be produced at a profit from public lands should be provided. Indeed, the provision of endangered species habitat most likely represents a subsidy to the species and its human supporters, and the harvest of valuable, old-growth timber from public lands in the Pacific Northwest probably is not below-cost. The critical questions here are what would happen without the subsidy, and who is its primary beneficiary? The dominant purposes of the public sector in resource management should be:

- To ensure that the use of national resources is carried out on a sustainable basis;

- To provide important public goods that would be undersupplied otherwise; and

- To protect and ensure the health of the natural and social legacies passed along to future generations, including the options present in a diverse land and cultural base.

When judged against these criteria, many subsidized programs currently in place would be diminished, and others might be enhanced. Habitat protection would rise in importance, and policies favoring the liquidation of old-growth reserves would decline.

Any shift in currently subsidized activities must be carried out in a manner sensitive to the fact that dependency relationships have developed, and the burden for eliminating them must necessarily be shared by the public sector. In the case of old-growth timber in the Pacific Northwest, such a weaning process includes job retraining, compensation, local economic development planning, and where possible, a phased reduction of timber cutting on public lands. Unfortunately, such a glide path to a more sustainable future would have been better implemented in the early 1980s, when there was more slack in the system. Nevertheless, the primary obligation of public sector land managers today is to shift resource management into a more effective set of practices for the future. There will be impacts from such a shift, as is evident in past economic translocations, current efforts to deal with the federal budget deficit, and the shift from massive public subsidies in eastern European economies. It is important to deal with such changes as humanely as possible, but it is not wise to do so by mortgaging the future.

Part of the reason that such booms and busts have typified resource-based economies is because of poor long-term planning at many levels of society. Better planning processes are needed to help overcome our bias toward short-term gains at the expense of long-term options. Chapter 7 described some of the historical reasons that Americans fear planning. In times of seemingly unlimited resources provided by an unending frontier, inadequate planning causes only localized problems. In addition, excessive planning can result in stagnation, diminished creativity and innovation, and analysis paralysis. As so many failed centrally planned economies have suggested, there is no reason to believe that we can either forecast future events well or keep less-benevolent forces (such as the use of planning institutions to maintain patterns of power or wealth) out of efforts to direct processes of social change. Yet inadequate planning in a world of finite resources leads to long-term social and natural system degradation. Hence, too much planning leads to impoverishment due to a lack of creativity and ideas; too little planning leads to impoverishment due to the excessive pursuit of short-term objectives. It is a dilemma that is inherent in our society, and must be dealt with via continued experimentation, learning, and adaptation.

Given the set of problems that we face, it is clear that more effective regional-scale planning institutions are needed. Such institutions can be forums whereby information can be exchanged and relationships developed among the diverse set of actors influencing the direction of a regional landscape. They can also be places where regional objectives can be defined, and alternative means for reaching those objectives examined, without necessarily prescribing means to reach ends in a way that stifles creativity. Regional planning groups can also work to manage the conflict inherent in a population with a diverse set of demands on a finite land base, and potentially assist disputing parties to find compromises where they are possible. They can also provide resources to enable well-motivated groups to move in an appropriate direction.

There are many ways that such regional discussion and direction can occur. In some places, breathing life into moribund regional planning institutions, such as river basin commissions and watershed associations, may be appropriate. In others, creating new entities such as biodiversity councils may be needed. In some places, professional groups such as the Society of American Foresters may play this role, collections of governmental agencies such as the Southeast Michigan Council of Governments may be effective, and associations of private sector entities such as Chambers of Commerce may function well. In others, visioning exercises done in an ad hoc way might be helpful. In areas where the federal government is a dominant landowner, federally initiated planning processes, such as the NFMA-created forest planning process or management planning in the Columbia River Gorge National Scenic Area, can play this role, as can federally funded community development personnel. Multiagency, problem-oriented task forces, such as those needed to deal with proposed changes in federal land management due to the needs of the owl, can also be important forces that promote region-scale consideration.

No matter what form the regional planning activity takes, its function is to help craft a regional future that includes a sustainable economy and a well-functioning environment. While in some places, a shift of power and authority from fragmented local interests to regional institutions might be desirable, an understanding of past regional governance attempts does not suggest that a significant reallocation of power is likely. Rather, the source of legitimacy of these institutions will come from the quality of their information, the effectiveness of the forums that they provide, and the strength of the ideas they develop. To the extent that such regional planning institutions are able to provide expertise and/or funding to motivate changed patterns of development or preservation activity at the local level, such activities would also increase that impact. Successful regional planning institutions have to provide something of value to the local parochial concerns that make them up. In some cases, this has been

money, but in others it has been authority, legitimacy, information, and relationships.

There are ways for the federal government to encourage regional-scale
planning and decisionmaking. One portion of a revamped federal funding
program (described below) could provide grants to these regional institutions aimed at promoting concepts of sustainability in development, and
the protection of important biotic resources. They could then either pass
them on to local-level units of government or nongovernmental groups,
or use them to develop expertise and/or data bases at the regional level.
Developing land-based information systems that are easily accessed by
individuals and groups across the region is another way to increase the
usefulness of such a regional institution to those groups who will determine its long-term effectiveness. Delegating specific federal powers such
as monitoring compliance with federal pollution control law is another
way to empower these groups, if it is done cautiously. Clarifying cumulative effects provisions of environmental impact assessment laws at the
state and national level might also force a broader consideration of future
directions. There is no one right way to encourage regions to consider the
long-term effects of their development trajectories on their ultimate ability to sustain such patterns economically and environmentally. At the
same time, it is unrefutable that such considerations are necessary. Using
federal law and inducements, information and technical assistance, and
the power of ideas are all mechanisms to promote such a consideration.

Enhancing the Capacity of Individuals and Institutions

To implement the above policy directions and to increase the capacity of
individuals, agencies, and decisionmaking institutions to improve, enhanced expertise, expanded knowledge, and creative financing mechanisms must be employed. In addition, enabling individuals and groups to
observe, evaluate, and criticize current behavior is important to maintaining long-term fitness. A technically complex world where interrelationships are more extensive and the impacts of actions appear at the
global-scale requires individuals and organizations to develop expertise
and understanding in order to be able to act on situations more effectively.
Clearly, administrative agencies can play a lead role as public educators,
and as places where state-of-the-art technical knowledge can be accumu-
lated, and the mechanisms described in Chapter 12 for updating the
knowledge and expertise base of the FS should be employed by many

other agencies. But the obligation to understand more completely, and make choices more effectively, rests at all levels of our society.

Decisionmaking institutions such as the Congress must find ways to update their members' understanding of complex policy issues, and develop their skills at asking appropriate questions and finding effective solutions. In the Congress, this means developing the technical understanding of staff members through participation in seminars, short courses, and the like. It also means maintaining resident sources of expertise, such as through the expanded use of Congressional fellowships that place agency and university experts on Congressional staffs. Members of Congress themselves must develop facility at understanding technically complex matters, and this need alone argues against arbitrary limitations on the number of years members should serve. Using hearings to facilitate debates on technical matters, rather than orchestrated presentations that are often ineffective at forcing interchange between witnesses, and relying more on externally facilitated policy dialogues, are two mechanisms for promoting better informed Congressional choices. A greater use of outside experts, including the Congressional Research Service, universities, and the National Academy of Sciences and National Research Council might also bring better knowledge into the decisionmaking process. Junkets to interesting places, such as numerous national forests, can be used as learning processes for individuals who generally become more and more isolated from the realities of the problems about which they are making decisions.

The courts also have not been a bastion of technical understanding, and the win–lose model prevalent in an adversarial judicial system tends to pursue process at a cost of understanding. While process is certainly important, there are ways to get around the inadequacies of the judiciary in understanding and ruling on technical questions. Using technically trained special masters as court-appointed mediators in technically complex situations is one way to expand the knowledge base of the judiciary, and promote consensus decisionmaking. Possibly creating new judicial structures, such as so-called science courts that adjudicate certain kinds of disputes, might also help.

The media has a potent role to play as an agent of public learning, and there are a variety of ways to enhance the effectiveness of media organizations. Updating the knowledge base of reporters, through learning sabbaticals at universities or government agencies, or hiring reporters with prior scientific knowledge, might improve the quality of the information conveyed to the public. While clearly both the nature of print and broadcast media and the need to compete in the marketplace encourage reporters to simplify and sensationalize the news, correspondents and editors should avoid pitching their information to the lowest common denominator of public understanding. The public has the capacity to understand information at a greater level of complexity than is often presented to them. Television in particular has a potent role in conveying knowledge and images, and it can go far beyond its current level of sophistication. For example, having a series of monthly hour-and-a-half specials on important policy issues, introduced by the President, and consisting of presentations by experts convened by the National Academy of Sciences, followed by interest group arguments and call-in question and answers, could help develop a shared understanding of problems and potential solutions. Linking these specials to curriculum materials in the public schools and community colleges and information in newspapers would help build more depth of understanding than is possible through a television presentation alone.

The fundamental obligation of citizens to express themselves through voting has been surpassed by an obligation to understand, for uninformed voting is perhaps worse than no voting at all. If the future will be controlled by those with skills and effective ideas, then an educated public is critically important, and there are ways for the media, agencies, corporations, existing social networks, public schools, and universities to assist in a public learning process. The roles of the media and the agencies have been discussed earlier. Corporations obviously have a major influence on the continuing education of their workforces. Every worker should have a continuing education strategy that includes updating skills, expanding understanding, and exploring issues relevant to his or her future effectiveness as worker and citizen. Social networks such as churches, Rotary Clubs, and neighborhood associations have historically played an important socialization and value-forming function, and they could do more today to update the knowledge and values of their members.

Formal and informal educational systems, while receiving much rhetorical support, must be seen as the most important institutions to our country's future. Public schools must reinvigorate their programs in science education and so-called social studies—an understanding of societies, and how diverse interests can be processed to make collective choices in an effective and equitable way. Schools are a vector into the minds and understanding of parents as well as students, and developing more community-level programs and outreach, and using more shared parent–child projects as a means of joint learning would expand the impact of formal educational institutions on the knowledge base of our society.

Universities help to form the skills and values of the future workforce, and training in natural resources and environmental management needs to be expanded to deal with the times. Resource managers and policy analysts must understand current science (including conservation biology and landscape-level ecology), ways to process information through analytic tools and information systems, and mechanisms for sharing that information and assisting decisionmaking at various levels of society. To do this, they need an understanding of human behavior at the individual and institutional levels, and skills at communicating and working with people. More than ever, natural resource management involves the management of humans—their conflicting wants and needs, and their impacts on resources. Some of the problems of the owl case could have been avoided if agency staff members had better human skills, and were more aware of the human factors in resource decisionmaking.

Colleges and universities also have a major obligation to the rest of society to share the wealth of their knowledge and their ability to act as a crucible for ideas. They must reward service to community and teaching as equally important to the success of academe as the production of academic research. They must also find ways to create more permeable boundaries: by offering programs to individuals out of "college age" through Elder Hostels, and the like, and by offering the benefits of their expertise to communities and government agencies. For example, graduate students provide a vast source of expertise capable of helping inform decisionmakers and citizens at many levels of society, yet they are greatly underused and overprotected from helping solve the problems in the "real world." There is a wealth of information, expertise, and educational resources in American society, including retirees and students, that is vastly underutilized. It would cost us little, yet we would gain much by mobilizing it.

In a time of constrained public sector fiscal resources, it is also necessary to utilize creative financing mechanisms to implement necessary policy changes. Expanding user fees is appropriate to support the management of public facilities and landscapes. While there are important equity questions involved in the institution of user fees on publicly owned resources, it is at least as fair to raise revenues directly targeted at users of public resources, as it is to raise them via income taxes that flow into the general treasury. User fees are also one means of controlling the demand for areas that are fragile or stressed by overuse.

A variety of targeted taxes could also be used to raise funds for resource management. Anglers and hunters have paid fees to support their habits for many years, and one result has been the validation of their demands on resource managers. It is appropriate to raise funds via excise taxes on goods commonly used on public lands, including ORVs, binoculars, and camping equipment. It is also appropriate to raise revenues via taxes on consumptive uses of nonrenewable resources—both as a means to discourage consumption, and to mitigate the impacts of such consumption. Auto and truck fuel taxes have been used for years in this way, as have motorboat fuel taxes and lease revenues from oil and gas development. Additional depletion taxes are appropriate, and could be placed on other energy resources, minerals, and water rights purchases. Land transfer fees as a proxy for taxes on land development are another way to raise funds for public resource management. Taxes on externalities such as pollution and waste are also appropriate, and probably preferable to many other kinds of taxes used to raise revenue to pay for government services. They have the dual benefits of raising revenues and discouraging behaviors that we want to discourage. There is a very compelling argument that so-called green taxes are preferable to taxes on labor (income) or investment (savings, capital gains) because these more common mechanisms to raise revenues discourage the activities we want to encourage (people working and investing/saving).[2]

If federal subsidies are reduced in public resource management, less utilization of resources may occur requiring less public sector employment, yet it may also raise revenues as the full costs of resource depletion are charged. It might also be appropriate to redirect some of the receipts generated from public resource management away from local interests and back into resource management. The pattern of automatically using a good sized percentage of timber receipts to pay for county services has created a network of dependent school and road systems. Certainly payments in lieu of taxes are appropriate in many areas where government is a major land owner; but a national public has a larger claim on the revenues accruing from public land management than they have been accorded in the past. Timber receipts are not an entitlement, and some additional portion of receipts should be recycled directly into management of the lands that created them.

No matter what means are used to raise revenues to finance effective land management, revenues from resource depletion should be earmarked for future land management and conservation activities (unless a drastic shift in taxation occurs toward green taxes that ultimately generate surpluses). Viewing use revenues as a way to subsidize other public demands is guaranteed to result in damaged ecosystems over time. Rather than using such revenues as a way to offset portions of the public debt on paper as is currently done, they should go directly into an off-budget fund earmarked for conservation purposes. A revamped Land and Water Conservation Fund should be created as a repository for depletion taxes, resource royalties, and use fees, and the interest on the fund should be used to finance new land conservation activities via a competitive grant process. Such a fund also could be used to accept voluntary, tax-deductible contributions of money or land from individuals or corporations. As has been the case in a number of states with income taxes, a check-off on individual income tax forms could be used as a way to raise additional revenues for the fund.

Sometimes important work can be accomplished with less public sector funding through public–private partnerships, utilizing the differing capabilities and skills of various public and private individuals and groups. Volunteers, retirees, social service groups, high school students, and others can assist public land managers in carrying out some of the functions required to manage the public estate. Indeed, if less commodity production occurs on public lands, and more stewardship and nonconsumptive uses occur, voluntary associations may be an important source of relatively free labor. Involvement of such groups has the additional benefits of forming public–private relationships and creating future constituents for public land agencies. These relationships clearly have a down side, as volunteers often are less reliable, need staff oversight, and often are not interested in providing the services in which the agencies are interested. Nevertheless, since one of the major recommendations discussed earlier was for land management agencies to create permeable boundaries, using voluntary groups in partnerships does in fact have multiple benefits.

Finally, there is a role for using public land exchanges as a means to achieve important public priorities, and for purchasing less than full rights to newly acquired properties. The current federal lands are a geographic patchwork set aside for a variety of reasons and sometimes by accident. There was no master plan in creating the current set of lands, and some are more valuable for public purposes than others. There is a role for carefully monitored land swaps to consolidate lands into core ecosystem and habitat areas, at the expense of lands at the periphery, or of less compelling value. And there is an important role for government to play in purchasing development rights or scenic easements on other properties. Both of these approaches can be used in a strategy to focus stewardship activities on large landscape blocks at a cost (both monetary and in public reaction) of less than that of full acquisition.

Beyond expertise and money, nonagency watchdogs are needed to insure that public resource management continues in an appropriate direction. Even under the best of circumstances, it is possible for programs to get off track. To monitor agency behavior, and keep implementation moving in the right direction, it is important to have opportunities for other parties to "fix" implementation. Both the public policy literature and research on environmental policy suggest the importance of so-called fixers in the implementation environment.[3] Fixers can be other agencies, elected officials, or nongovernmental organizations that observe implementation, note and publicize problems, and keep pressure on the primary implementing agents to do good work. While the fragmentation of political power inherent in an interest group society is problematic, we should note that the major reason that owl and old growth protection got onto agency and policy agendas was due to the efforts of nongovernmental interest groups.

Policies and policy processes can seed the implementation environment with opportunities for fixers to grow in several ways. Opportunities for judicial review can be provided to a wide variety of potential fixers by defining broadly the standing to sue. Reporting requirements that mandate that agencies identify and publicize what actions they are taking, including impact statements and annual progress reports, help potential fixers to understand administrative behavior, and lobby for change. Public participation requirements provide opportunities for other parties to understand what agencies are doing, and legitimize the input of potential fixers. Sunset provisions on administrative programs, and oversight hearings, require agencies to report on and justify their actions. Requiring agencies to organize information in a way that can be transported to other groups for independent analysis is more threatening to source agencies, but potentially very influential in affecting implementation.

Providing resources to nongovernmental groups so that they can participate in administrative decisionmaking is at times appropriate. Just as public funding of election campaigns has merits, public support for nongovernmental groups to be involved in collective choice decisions is appropriate in at least a limited way. Current efforts to amend the ESA to provide funding for parties to be involved in the habitat conservation planning process is an example of this approach. When combined with opportunities for carrying out data negotiation, it is in our collective interest that agency analysts be challenged to be the best they can be by encouraging the development of parallel sources of expertise. This approach can be problematic, and any potentially more significant use of it should be implemented cautiously, but one of the real barriers to nongovernmental groups being involved in agency decisionmaking is the limits of their resources, and any way we can deal with this is in our collective best interest.

Providing opportunities for fixers to function in the implementation environment does have the potential for bogging down decisionmaking, and earlier sections of this book have suggested that part of the problem with the inability of government to make binding decisions comes from nongovernmental groups having a great deal of access to decisionmaking processes. There are ways to streamline administrative and judicial review processes, so that they are not unending sources of indecision. The judiciary has the power to limit frivolous suits and combine lawsuits in efficient ways. Inserting more effective collaborative problem-solving and dispute resolution processes into administrative decisionmaking can yield better and more enduring decisions (even if the initial decision takes longer than a simple application of the administrative rulemaking process seemingly does).

The first priority for improving behavior should be to improve decisionmaking processes and organizational management strategies that cause the problems in implementation. Knowing what we know about ourselves, however, it is also necessary to provide opportunities for fixers to challenge decisions, and the means to listen to and act on their challenges. If decisionmaking processes are improved and effective means are provided for challenge, it is less likely that we will have to address the issue of how much judicial review should be provided. Better and more process-effective decisions will help solve the public policy impotence so evident in the owl case.

There are many other reforms indicated by the history of the spotted owl case, and many of these are generic to most areas of public policy. Clarifying and integrating the purposes of legislative authorization and appropriations committees, and improving the Congressional and agency budgeting processes are in order. Reforming campaign finance rules to limit the impact of PACs, and constraining the influence of single interest group lobbyists are needed. Finding more ways to incorporate a longer term focus into legislative institutions and their elected officials are also important. Creating incentives for innovation and risk-taking in many areas of government are needed as well. As noted in Chapter 7, some of the problems that gave rise to the owl case are problems generic to the

American decisionmaking process, and dealing with these problems would improve decisionmaking in many areas of social policy.

Public Resource Management as a Model of Excellence

New policy approaches, altered models of collective choice decisionmaking, new patterns of political coalition-formation, new methods of information collection and analysis, new ways of managing forest systems, and new ways of administering and leading resource agencies are all there for the taking. While hard choices will continue to prevail in federal resource management, we are not doomed to repeat the events of the spotted owl controversy. We can learn from history and adapt our decisionmaking processes to better deal with the political realities and issue characteristics that most likely will prevail in the future.

Not the least of our needs are experiments in collective choice decisionmaking. The public lands can become a laboratory for experimentation in how we make choices in a democratic society in technically complex times. The unique promise of public lands managed for multiple values is in the process of choice that they imply: Can we collectively find and implement solutions to differences of opinion about what is valued and what the future holds and do so in a way that is enduring and scientifically legitimate? If we simply seek the production of any one use or commodity from public lands, then we would probably be better off with managers geared toward producing those uses most efficiently, which might mean private sector management of priced goods such as timber, and nonprofit sector management of nonpriced goods, like wilderness.

As a storehouse of multiple values and processes, and an obvious place where social and natural resource objectives come together, the national forests in particular provide a great opportunity to test whether our governmental system is sound. Can a diverse society effectively and humanely solve its differences, and do so in a way that is both supportive of diversity and environmentally sound? The spotted owl case points to a lot of problems in our public decisionmaking processes, but there are clearly ways to improve. Our decisionmaking processes are experiments from which we can learn and attempt change. We may not succeed, but we have no choice but to try.

Part of the reason that the spotted owl issue persisted as long as it did is that decisionmaking was out of step with the values and needs of the times, and the normative base of the national forest system needs clarification for the times we live in. Stewardship—caring for the land—

implies a long-term perspective and a sense of humility that have evaded
FS decisions in recent years. While serving a set of clients is a popular
current FS metaphor, it is seriously flawed, as it is too reactive to current
demands, assumes we know more than we do, and is heavily oriented
toward maintaining the status quo. Clearly the agency needs to build and
seek political concurrence in its future decisions and operations. But that
does not obviate the need to operate from a normative base appropriate
to the long-term future of the national forests as a component of our society. There is a role for the agency beyond simply being the neutral balance
point of multiple demands.

If we look at what museums store and value about the past, it is the
quality of ideas and forces of inspiration as revealed in a society's culture,
science, technology, and lifestyle, and the nature of relationships among
individuals, societies, and societies and the environment in which they
survive. They also serve as testament to the hubris of the past: the crazy
ideas and the under- and over-estimated risks of various actions. We look
back now and are amazed that anyone could have sprayed DDT liberally
on children without worrying about it, and are bemused that anyone
could have thought that "duck and cover" would protect you from the
harmful effects of nuclear fallout.

What does this mean for public resource management? It means serving as a source and testing ground for those new ideas and human interrelationships, being cognizant of the likelihood that what we think we
know today will prove erroneous in the future, and constantly seeking an
understanding of the needs of the future. *What future generations will
value from the decisions of today's managers is the knowledge that they generate, approaches that they identify and test, and just as important, the options
that they maintain* so that future generations can take full advantage of the
new insights and possibilities that will develop with time.[4]

In this light, the role of the resource agencies as stewards of the public
lands is to promote and evaluate new ideas and represent the interests of
future generations of humans and nonhuman lifeforms. Maintenance of
economic infrastructure is an important function for societies, and the
national forests have a role in producing commodities to support the lifestyles of current generations of humans. But it is not the highest and best
use of much of the national forests as a system of lands, values, choices
and options for the future.

There is no doubt that the federal land management agencies have begun a process of change along these lines, and change is occurring at
different rates at various places in the agencies. For example, most of the

recommendations that have been outlined above are consistent with and supportive of recent FS directions. Nevertheless, "new perspectives," the "new forestry," "ecosystem management," and "landscape ecology" and whatever other slogans arise in an agency fond of such labels, must bear fruit in order for the agency to regain an image of trustworthiness and legitimacy, an image that has been tarnished by the events of the spotted owl and other recent controversies. The window of possibilities is open now, and change must not be of a token or simply rhetorical nature. As Toynbee has said, "The piano tuner has been in the house for a long time now, and the hour has come for a sonata to be played or even a concerto." The public lands are a rich symphony of sights, sounds, ideas, and possibilities, a priceless American feature. The U. S. Forest Service and its peer agencies can be a seedbed of innovation, ideas, experiments, and options. While we can bemoan the failings of our agencies and decisionmaking processes as revealed in the spotted owl case, the response to the controversy can be a springboard to this exciting future.

Notes

Introduction

1. The cover picture of a roosting spotted owl was headlined, "Who Gives a Hoot? The timber industry says that saving this spotted owl will cost 30,000 jobs. It isn't that simple." 135 *Time* cover, June 25, 1990. See also Ted Gup, "Cover Story: Owl vs. Man," 135 *Time* 56–65, June 25, 1990.

2. In 1988, the Forest Service estimated that 44% of Oregon's economy and 28% of Washington's economy were either directly or indirectly dependent on national forest timber. [USDA-Forest Service, *Final Supplement to the Environmental Impact Statement for an Amendment to the Pacific Northwest Regional Guide, Volume 1, Spotted Owl Guidelines* (Portland, OR: Pacific Northwest Regional Office, 1988), Summary-27.] Their figures included jobs in log harvesting, transportation, and milling, as well as jobs in restaurants and stores that supported the timber workers. The Forest Service figures were disputed by environmental groups.

The 1993 Supplemental Environmental Impact Statement (SEIS) analysis (that chose FEMAT Option 9 as the Clinton administration's preferred alternative) indicated that roughly 5% of the region's jobs were directly employed in timber harvesting and processing, down from 10% in the early 1970s. While the change was partly a result of declines in timber employment, it reflected to a greater extent rising employment levels associated with economic diversification in the region. Total employment rose from 1.6 million in the early 1970s to 2.7 million in the late 1980s. While the overall economy became less dependent on timber over this time period, many areas outside the Portland and Seattle metropolitan areas were still heavily timber-based in the early 1990s. [USDA-Forest Service and USDI-Bureau of Land Management, *Draft Supplemental Environmental Impact Statement on Management of Habitat for Late-Successional and Old-Growth Forest Related Species Within the Range of the Northern Spotted Owl* (Portland, OR: U. S. Government Printing Office, July 1993), 3 & 4–119.]

3. 11.6 million acres was the amount of land proposed by the Fish and Wildlife Service in May 1991 as critical habitat for the northern spotted owl. 56 *Federal Register* 20821, May 6, 1991.

4. A zero sum situation is one in which one group can increase their benefits only at the cost of those of another group: what one wins another loses. Dividing

up a pie between individuals represents a zero sum situation where all the server can do is to split what is on the table between those who will consume the pie. A zero sum, win–lose situation is contrasted with a win–win situation where the size of what is on the table can be expanded, or the pie can be split in a new way in which all parties have their needs satisfied.

5. While almost no one requested confidentiality, I have not specified the sources of some of the quotations where I thought individuals were vulnerable to retribution.

Chapter 1

1. In Pinchot's biography, he is quoted as saying that "Forestry is tree farming . . . To grow trees as a crop is Forestry. Trees may be grown as a crop just as corn may be grown as a crop. The farmer gets crop after crop of corn, oats, wheat, cotton, tobacco, and hay from his farm. The forester gets crop after crop of logs, cordwood, shingles, poles, or railroad ties from his forest. . . ." Gifford Pinchot, *Breaking New Ground* (New York: Harcourt Brace, 1947), 31.

2. Marion Clawson, *The Federal Lands Revisited* (Washington, D.C.: Resources for the Future, 1983), 72.

3. Clawson, *The Federal Lands Revisited,* 74–76.

4. Clawson, *The Federal Lands Revisited,* 75.

5. There have been a variety of ways described to view the organizational objectives of the Forest Service. One critical analysis suggests that the agency is primarily a budget-maximizer. See, for example, Randal O'Toole, *Reforming the Forest Service* (Covelo, CA: Island Press, 1988).

6. Not everyone viewed timber management and outdoor recreation as compatible activities, nor was the FS necessarily trusted to protect outdoor recreation resources. The longstanding battle between the conservationists, allied with the FS, and the preservationists, supportive of the NPS, reflected distrust of the FS. In the preservationists' view, an agency oriented toward timber harvest would be unable to protect scenic landscapes. Indeed, Ben Twight has argued that had FS leaders been somewhat more accommodating to the interests of aesthetic-oriented recreationists, the NPS and several national parks might never have been created. Ben W. Twight, *Organizational Values and Political Power: The Forest Service Versus the Olympic National Park* (University Park, PA: Pennsylvania State University Press, 1983), 107.

7. U. S. Outdoor Recreation Resources Review Commission, *Outdoor Recreation for America* (Washington, D.C.: U. S. Government Printing Office, 1962).

8. Charles F. Wilkinson and H. Michael Anderson, "Land and Resource Planning in the National Forests" 64 *Oregon Law Review* 344, 1985.

9. See, for example, Julia Wondolleck, *Public Lands Conflict and Resolution* (New York: Plenum, 1988), 119–152.

10. Herbert Kaufman, *The Forest Ranger* (Baltimore, MD: Johns Hopkins, 1960).

11. These concerns about excessive demands being placed on national forest resources were part of the impetus for passage of the Multiple-Use Sustained-Yield Act of 1960.

12. Hana U. Lane *et al.,* eds., *The World Almanac and Book of Facts 1982* (New York: Newspaper Enterprise, 1981), 204.

13. Opinion Research Corporation data cited in J. Clarence Davies and Barbara S. Davies, *The Politics of Pollution,* 2nd edition (Indianapolis: Pegasus-Bobbs Merrill, 1975), 82.

14. Opinion Research Corporation data cited in Davies and Davies, *The Politics of Pollution,* 82.

15. P. L. 91–190, January 1, 1970, 83 Stat. 852–856, 42 U.S.C. 4321–4347.

16. P. L. 93–205, December 28, 1973, 87 Stat. 884–903, 16 U.S.C. 1531–1543.

17. 16 U.S.C. 1533. The listing process is generally referred to as the Section 4 process.

18. 16 U.S.C. 1536. The interagency consultation process is generally referred to as the Section 7 consultation process.

19. Steven L. Yaffee, *Prohibitive Policy: Implementing the Federal Endangered Species Act* (Cambridge, MA: MIT Press, 1982), 38.

20. Yaffee, *Prohibitive Policy* 57.

21. Interview with Eric Forsman, July 27, 1989.

22. Memorandum from Howard M. Wight, Leader, Oregon Cooperative Wildlife Research Unit to Dr. Thomas S. Baskett, Chief, Division of Wildlife Research, "Timber Harvest—as it affects spotted owl habitat", July 31, 1972.

23. Memorandum from Wight, July 31, 1972.

24. Memorandum from Wight, July 31, 1972.

25. Memorandum from Spencer H. Smith, Director, BSFW to John R. McGuire, Chief, USFS, August 18, 1972.

26. Letter from John R. McGuire, Chief, USFS, to Spencer H. Smith, Director, BSFW, September 26, 1972.

27. Letter from Eric Forsman, OSU to Robert McQuown, District Ranger, Klamath Ranger District, Winema NF, May 2, 1973.

28. Letter from Robert McQuown, District Ranger, Klamath Ranger District, Winema NF to Eric Forsman, May 9, 1973.

29. Leon W. Murphy, "Strix and his mate lucky birds?", *Oregon Wildlife,* June 1978.

30. Interview with Eric Forsman, July 27, 1989.

31. Letter from Floyd W. Collins, City of Corvallis Public Works Department to Eric D. Forsman, OSU, April 18, 1973.

32. Letter from Collins, April 18, 1973.

33. Letter from Eric Forsman, OSU, Dept of Fisheries and Wildlife to Floyd Collins, City of Corvallis, May 2, 1973.

34. Interview with Eric Forsman, July 27, 1989.

35. Oregon Endangered Species Task Force (OESTF), Minutes, First Meeting, June 29, 1973, 1.

36. Interview with Eric Forsman, July 27, 1989.

37. Interview with Charlie Bruce, August 3, 1989.

38. Interview with Chuck Meslow, August 3, 1989.

39. Interview with Chuck Meslow, August 3, 1989.

40. OESTF Minutes, June 29, 1973, 1.

41. OESTF Minutes, June 29, 1973, 1.

42. OESTF Minutes, June 29, 1973, 1.

43. OESTF Minutes, June 29, 1973, 2.

44. OESTF Minutes, June 29, 1973, 2.

45. Interview with Eric Forsman, July 27, 1989.

46. Interview with Chuck Meslow, August 3, 1989.

47. Bill Nietro, "Chronology of Events Related to the Spotted Owl Issue", (Portland OR: BLM Oregon State Office, unpublished report), 2–3.

48. Contained in Manual 1603 Supplemental Guidance, and cited in Bill Nietro, "Chronology of Events Related to the Spotted Owl Issue", (Portland OR: BLM Oregon State Office, unpublished report), 1–2.

49. Interview with Eric Forsman, July 27, 1989.

50. OESTF Minutes, August 22, 1973, 1.

51. Interview with Charlie Bruce, August 3, 1989.

52. Interview with Chuck Meslow, August 3, 1989.

53. Interview with Chuck Meslow, August 3, 1989.

54. OESTF Minutes, August 22, 1973, 2.

55. OESTF Minutes, March 12, 1974, 1.

56. OESTF Minutes, September 25, 1974, 1.

57. OESTF Minutes, September 25, 1974, 3.

58. OESTF Minutes, November 17, 1976, 1–2.

59. Eric Forsman, "A Preliminary Investigation of the Spotted Owl in Oregon", (Corvallis, OR: OSU Department of Fisheries and Wildlife, June 1976), unpublished MS thesis.

60. Forsman, "A Preliminary Investigation," unnumbered abstract.

61. Forsman, "A Preliminary Investigation," 115–118.

62. Forsman, "A Preliminary Investigation," 113.

63. Forsman, "A Preliminary Investigation," 118.

64. Eric D. Forsman, E. Charles Meslow, and Monica J. Strub, "Spotted Owl Abundance in Second-Growth Versus Old-Growth Forests in Western Oregon", Oregon Cooperative Wildlife Research Unit, pre-publication draft forwarded to the FS-R6, BLM and ODF&W on January 7, 1977, 1.

65. Interview with Eric Forsman, July 27, 1989.

66. Interview with Charlie Bruce, August 3, 1989.

67. Sikes Act, P. L. 93–452, s 201a and 201b, 88 Stat. 1371. The Sikes Act

extension was passed as an outgrowth of the 1970 Public Land Law Review Commission whose major recommendation was that public land management agencies should move to a dominant use management scheme in which protection of rare and endangered species should always take precedence. Needless to say, their major recommendations were not enacted, and the Sikes Act extension was called "totally unnecessary" by the USDA-Forest Service since wildlife conservation was already a statutory objective provided by the Multiple Use Sustained Yield Act of 1960. Nevertheless, the Sikes Act's requirement that federal agencies assist the states in managing state-listed species was interpreted as a mandatory, and not a discretionary, duty. See Michael Bean, *The Evolution of National Wildlife Law, Report to the CEQ* (Washington, D.C.: U. S. Government Printing Office, 1977), 155–156.

68. Memorandum of Understanding between the BLM and the Oregon Wildlife Commission, updating the March 20, 1961 Memorandum, and cited in Nietro, "Chronology of Events Related to the Spotted Owl Issue", 3.

69. Interview with Bill Nietro, August 1, 1989.

70. P. L. 94–579, s 702(b), 90 Stat. 2743, 43 USCA 1701 et seq.

71. OESTF Minutes, November 11, 1976, 1.

72. Mayo W. Call, BLM Avian Biologist, "Report on Status of Spotted Owl in Oregon," November 16–19, 1976, unpublished report, 3.

73. OESTF Minutes, November 11, 1976, 1.

74. Interestingly, this approach of establishing concentrated habitat conservation areas was quite similar to the approach adopted much later by the Interagency Committee of Scientists in 1990. See Jack Ward Thomas *et al., A Conservation Strategy for the Northern Spotted Owl,* report of the Interagency Scientific Committee to Address the Conservation of the Northern Spotted Owl, Portland, Oregon, April 2, 1990.

75. Mayo W. Call, BLM Avian Biologist, "Report on Status of Spotted Owl in Oregon," November 16–19, 1976, unpublished report, 4.

76. M. W. Call, "Report on Status of Spotted Owl in Oregon," 5–6.

77. Memorandum from Regional Office, FS Region 6, to Oregon Forest Supervisors, "Tentative Fish and Wildlife Program Objectives Through FY 1990", October 8, 1976.

78. Memorandum from Regional Office, FS Region 6, October 8, 1976.

79. Memorandum from Leon Murphy, Director of Fish & Wildlife, Region 6 to Oregon Forest Supervisors, "Status of Spotted Owl", November 24, 1976.

80. Memorandum from Leon Murphy, Director of Fish & Wildlife, Region 6, to Oregon Forest Supervisors, "Status of Northern Spotted Owls on Oregon Forests", October 14, 1976.

81. Memorandum from Murphy, November 24, 1976.

82. OESTF Minutes, December 13, 1976, 1.

83. OESTF Minutes, December 13, 1976, 2.

84. OESTF Minutes, October 17, 1977, 2.

85. Robert Maben, Oregon Department of Fisheries & Wildlife, "Enclave parameters", October 13, 1977, unpublished report.

86. OESTF Minutes, October 17, 1977, Appendix A, "Management Area Parameters".

Chapter 2

1. Letter from Murl Storms, BLM Oregon State Director, to Bob Stein, Chairman, OESTF, January 11, 1978; Letter from R. E. Worthington, Regional Forester, Region 6, to Robert Stein, Chairman, OESTF, February 21, 1978.

2. Memorandum from BLM Oregon State Director, to District Managers, Western Oregon, "Land Use Guide - Spotted Owls", February 16, 1978.

3. OESTF Minutes, February 23, 1978, 2. The original letter is from, Letter from J. E. Schroeder to Robert Stein, February 3, 1978, quoted in Eric Forsman & E. Charles Meslow, "The Spotted Owl", in Roger L. DiSilvestro, ed., *Audubon Wildlife Report 1986,* (New York: National Audubon Society, 1986), 754.

4. Katherine Barton, "Wildlife on BLM Lands", in Amos Eno and Roger DiSilvestro, *Audubon Wildlife Report, 1985,* (New York: National Audubon Society, 1985), 348.

5. USDOI-BLM, *Final Environmental Impact Statement on Timber Management in Western Oregon* (Portland, OR: Oregon State Office, March 5, 1975). Page I-23 states that "Current plans call for the harvest of the old-growth timber on lands available for timber production over the next twenty year period."

6. Interview with Bill Nietro, August 1, 1989.

7. Section 6, National Forest Management Act of 1976, October 22, 1976, P. L. 94–588, 90 Stat. 2949.

8. For example, in a May 1977 memo to forest supervisors that indicated support for the OESTF's interim guidelines, the regional forester noted that "we are all aware of the desirability of diversified vegetation in our overall management of the National Forests. This concept is now legislatively directed in the National Forest Management Act. Plant diversity, without question, includes extended rotation 'old-growth' components. The direction to be resolved is not 'if,' but 'how much' and 'where.' Memorandum from Acting Regional Forester Donald Morton to Westside Forest Supervisors, "Westside Habitat for Indicator Species Spotted Owl and Associated Species," May 17, 1977, 1977, 1.

9. "Interim Position Statement, Siuslaw National Forest Older Forest Communities," appended to: Memorandum from Deputy Regional Forester, James Torrence, to Forest Supervisors, "Older Forest Communities," December 19, 1977.

10. Glenn Patrick Juday, "Old Growth Forests: A Necessary Element of Multiple Use and Sustained Yield National Forest Management," 8 *Environmental Law* 497–522, 1978.

11. 122 *Congressional Record* S17274, September 30, 1976.

12. Memorandum from John E. Lowe, Assistant Forest Supervisor, Willamette NF to Regional Forester, R-6, "Spotted Owl Management", October 18, 1979.

13. Memorandum from Deputy Regional Forester James Torrence to Oregon Forest Supervisors, "Spotted Owl Management Plan", December 19, 1977.

14. Memorandum from F. Dale Robertson, Forest Supervisor, Mt. Hood NF to Regional Forester, "Spotted Owl Management Plan", February 6, 1978.

15. Letter from Storms, January 11, 1978. [See Note #1, this section.]

16. Letter from Worthington, February 21, 1978. [See Note #1, this section.]

17. See, for example, memorandum from Acting Regional Forester Donald Morton to Forest Supervisors, "Westside Habitat for Indicator Species Spotted Owl and Associated Species", May 17, 1977, 1.

18. Interview with Eric Forsman, July 27, 1989.

19. Siuslaw Older Communities Policy statement, Siuslaw National Forest, unpublished document.

20. John H. Beuter, K. Norman Johnson, and H. Lynne Scheurman, *Timber for Oregon's Tomorrow: An Analysis of Reasonably Possible Occurrences* (Corvallis, OR: OSU School of Forestry, Forest Research Laboratory, Research Bulletin 19, January 1976), 43, 53.

21. Quoted in Jim Kadera, "New regional forester eyes bigger harvests," *The Sunday Oregonian,* July 10, 1977, B7.

22. See, for example, BLM Oregon State Office Instruction Memo 78-116, July 3, 1978 in which several BLM district offices requested relief from their own pair allocations.

23. Most of the material on early wilderness activities in Oregon is taken from the excellent history of the wilderness movement and the National Forests by Dennis M. Roth, *The Wilderness Movement and the National Forests: 1980–1984* (Washington, D.C.: U.S. Government Printing Office, August 1988), USDA FS History Series FS-410.

24. Roth, *The Wilderness Movement and the National Forests,* 25.

25. Comments of Robert Wazeka, Sierra Club, quoted in Roth, *The Wilderness Movement and the National Forests,* 25.

26. The FS had of course dealt with wildlife-oriented conservation groups for many years, but the organizations that the FS staff saw as legitimately interested in national forest management were very different from the Oregon Wilderness Coalition or OSPIRG. For example, on July 7, 1978, Dale Jones, Director of Wildlife Management sent a copy of Leon Murphy's article about owl protection to a set of groups he felt represented the interested nongovernmental conservation groups. These groups were: National Audubon Society, National Wildlife Federation, Wildlife Management Institute, The Wildlife Society, the Izaac Walton League, and the International Association of Fish and Wildlife Agencies. All of these groups can be seen at this time as fairly old line conservation groups. OWC and OSPIRG were a very different breed.

27. Quoted in Roth, *The Wilderness Movement and the National Forests,* 26.

28. Roth, *The Wilderness Movement and the National Forests,* 26.

29. Cameron LaFollette, "Saving all the Pieces, Old Growth Forest in Oregon", (Portland, OR: Oregon Student Public Interest Research Group, 1979), unpublished report.

30. Cameron LaFollette, "Spotted owl management plan: a reply", *OSPIRG Impact,* June 1978, 5.

31. LaFollette, "Spotted owl management plan: a reply", 5.

32. Cameron LaFollette, "What's all this fuss over spotted owls?" *Earthwatch Oregon,* Feb/March 1979, 22.

33. LaFollette, "What's all this fuss over spotted owls?" 22. "Forest Service gets criticism from Cascade holistic group", *Eugene Register-Guard,* November 3, 1979.

34. Cameron LaFollette, *Analysis of the Oregon Interagency Spotted Owl Management Plan,* (Eugene, OR: Cascade Holistic Economic Consultants, Forestry Research Paper #6, September 1979).

35. Oregon Wilderness Coalition, Cascade Holistic Economic Consultants; Land, Air, and Water; and Lane County Audubon Society, Appellants; National Wildlife Federation; Oregon Wildlife Federation; Portland Audubon Society, Intervenor-Appellants, "Statement of Reasons Before the Chief, USFS; In the Matter of Region VI Spotted Owl Habitat Guidelines", February 11, 1980.

36. National Wildlife Federation; Land, Air, and Water; Lane County Audubon Society, Oregon Wildlife Federation; Portland Audubon Society, Appellants, "Statement of Reasons Before the Interior Board of Land Appeals, USDOI; In the Matter of BLM Oregon State Office Spotted Owl Habitat Guidelines." IBLA 80–370, February 25, 1980.

37. Letter from James F. Torrence for Regional Forester R. E. Worthington to Terrence L. Thatcher, Pacific Northwest Resources Clinic, March 20, 1980, 2–3.

38. Letter from Torrence, March 20, 1980, 3.

39. 40 *CFR* 1508.23.

40. 40 *CFR* 1500.6(d)(1).

41. Letter from Torrence, March 20, 1980, 7.

42. Decision of the Chief of the Forest Service on an Appeal of a Decision of the Regional Forester, Region 6 Spotted Owl Habitat Guidelines, August 11, 1980, 3.

43. Decision of the Chief of the Forest Service, August 11, 1980, 4.

44. Letter from Joseph R. Blum, FWS Area Manager, Olympia WA, to Tom Campion, Seattle Audubon Society, March 25, 1981, 2.

45. Roth, *The Wilderness Movement and the National Forests,* 27.

46. Umpqua Wilderness Defenders, "Request for Administrative Review" and "Notice of Appeal", September 7, 1978 and November 6, 1978.

47. Michael D. Axline, University of Oregon Law School; Land, Air and Water, "Statement of Reasons for Appeal of French Creek Leave and Annie Timber Sales, Willamette National Forest," December 14, 1979.

48. Howard R. Postovit, Zoology Department, North Dakota State University,

"A Survey of the Spotted Owl in Northwestern Washington," Forest Industry Resource and Environment Program, NFPA, 1979, unpublished report.

49. Letter from Dr. Robert Vincent, Philomath, OR to Jim Geisinger, Executive Director, Douglas Timber Operators, Roseburg OR, January 26, 1979, 3.

50. Oregon-Washington Interagency Wildlife Committee, "Proposed Revision of the Oregon Interagency Spotted Owl Management Plan, Draft Revision," February 26, 1981.

51. Letter from Robert Stein, Chairman OWIWC to Regional Forester, Oregon State Director, State Forester, Oregon Dept of Forestry, March 6, 1981.

52. Bill Nietro, "Chronology of Events Related to the Spotted Owl Issue", (Portland OR: BLM Oregon State Office, unpublished report).

53. Forsman and Meslow, "The Spotted Owl", 750. [See Note #3, this section].

54. Barton, "Wildlife on Bureau of Land Management Lands," 374.

55. Barton, "Wildlife on Bureau of Land Management Lands," 374.

56. Forsman and Meslow, "The Spotted Owl", 750.

57. Quoted in Harry Hoogesteger, "Loggers resent protection of owls," *The Coos Bay World,* Tuesday, June 16, 1981, 1–2.

58. Quoted in Hoogesteger, "Loggers resent protection of owls," 1–2.

59. All data are from R. Dennis Hayward, Field Forester, "What You Always Needed to Know About the Northern Spotted Owl—But Didn't Know to Ask: An Industry White Paper", (Eugene OR: North West Timber Association, May 1981).

60. Quoted in Hoogesteger, "Loggers resent protection of owls," 1–2.

61. Quoted in Hoogesteger, "Loggers resent protection of owls," 1–2.

62. BLM Oregon State Office Instruction Memorandum No. 81–365, change 1, May 12, 1981.

63. Memorandum from U. S. Department of the Interior, Solicitor's Office, to BLM Director, "Legal Adequacy of the Bureau's Proposed Multiple Use Management Policy for the O&C Lands in Western Oregon", September 8, 1981.

64. Memorandum from BLM Director Robert Burford to Oregon State Office Director, "Criteria for Application of O&C Forest Policy", July 15, 1982.

65. USDOI-BLM, *South Coast-Curry Timber Management Plan, Proposed Decision,* (Portland, OR: BLM Oregon State Office, September 23, 1982).

66. "BLM-ODFW Agreement for Spotted Owl Habitat Management on BLM Lands in Western Oregon", signed by John Donaldson, Director, ODFW and William Leavell, Oregon State Director, BLM, September 26, 1983.

67. See the discussion of the BLM's response to the Coos Bay controversy in Jay Heinrichs, "The Winged Snail Darter", *Journal of Forestry* 214, April 1983.

68. Forsman and Meslow, "The Spotted Owl", 751.

Chapter 3

1. Interview with Eric Forsman, July 27, 1989.

2. For example, the 1980 report of the U. S. Council on Environmental Quality, the last annual report issued by the Carter Administration's CEQ, contained a

chapter on the need to act to protect biological diversity, defining diversity as consisting of two parts: genetic diversity, which represents the genetic variability among individuals in a species; and ecological diversity or species richness, which represents the number of species in a community of organisms [*The Eleventh Annual Report* (Washington, D.C.: U. S. Government Printing Office, 1980), 32]. The 1985 CEQ Report added a third component of diversity, habitat or natural diversity, including the variety and number of natural habitats and ecosystems [*The 1985 Annual Report* (Washington, D.C.: U. S. Government Printing Office, 1985), 273].

3. U. S. Office of Technology Assessment, *Technologies to Maintain Biological Diversity* (Washington, D.C.: U. S. Government Printing Office, 1987), 3. One of the more influential scientific articles was Reed Noss, "A Regional Landscape Approach to Maintain Diversity," 33 *BioScience* 11:700–705, December 1983.

4. See, for example, J. F. Franklin *et al.*, *Ecological Characteristics of Old-Growth Douglas-Fir Forests,* General Technical Report PNW-118 (Portland OR: Pacific Northwest Forest and Range Experiment Station, 1981). The researchers had pushed for a concentrated program of research on old growth forests, and their management, and the FS created an Old Growth Research, Development, and Application (RD&A) program in 1982, but it did not produce much of great use in owl management. Funding was slim, and studies generally focused on characterizing the overall old growth community, rather than on species-specific work. It took the researchers several years of working on community studies before they were confident enough to focus on species that were most associated with the old growth system. One of the more useful outgrowths of the old growth research in the first half of the 1980s was an interim definition of old growth published by the FS in July 1986. The Old-Growth Definition Task Group established "interim definitions of old-growth forests . . . to guide efforts in land-management planning until comprehensive definitions based on research that is currently underway can be formulated." USDA-FS, *Interim Definitions for Old-Growth Douglas-Fir and Mixed-Conifer Forests in the Pacific Northwest and California* (Portland, OR: Pacific Northwest Research Station, Research Note PNW-447, July 1986).

5. For basic material covering the topic of conservation biology, see, for example, M. E. Soulé and B. A. Wilcox, eds., *Conservation Biology: An Evolutionary-Ecological Perspective* (Sunderland, MA: Sinauer Associates, 1980). Work related to island biogeographic theory includes T. E. Lovejoy *et al.*, "Edge and Other Effects of Isolation on Amazon Forest Fragments", in Michael Soulé, ed., *Conservation Biology: The Science of Scarcity and Diversity* (Sunderland, MA: Sinauer, 1986), 257–285. The original work on island biogeography was documented in R. H. MacArthur and E. O. Wilson, *The Theory of Island Biogeography* (Princeton, NJ: Princeton University Press, 1967). Research on the effects of forest fragmentation includes L. D. Harris, C. Maser, and A. W. McKee, "Patterns of old growth harvest and implications for Cascades wildlife," *Transactions, 47th North American Wildlife and Natural Resources Conference* 374–392, 1982.

6. 36 *CFR* 219.3(g)

7. 36 *CFR* 219.12(g)

8. 36 *CFR* 219.12(g)(6)

9. Mark L. Shaffer, "Minimum Population Sizes for Species Conservation" 31 *BioScience* 2:132, February 1981.

10. Hal Salwasser, Thomas Hanley, and Richard Pederson, "Fisheries, Wildlife, and Sensitive Plants in National Forest NFMA Planning: A Perspective on Management Indicator Species, Viable Populations, and Diversity," (San Francisco, CA: USFS Region 5, August 1980), draft manuscript.

11. "Spotted Owl Management Proposal for Region 6, USFS", review draft, circa late fall 1980.

12. Memorandum from Kirk Horn, Wildlife Biologist to the files, "Region 6 Spotted Owl Management—Viable Population Procedures", November 18, 1980.

13. "Spotted Owl Management Proposal for Region 6, USFS", review draft, circa late fall 1980.

14. See, for example, O. Franklin and M. Soulé, *Conservation and Evolution* (Cambridge: Cambridge University Press, 1981).

15. Memorandum from Leon Murphy, Director of Fish & Wildlife, Region 6 to Meeting Participants, "Spotted Owl Coordination Meeting—Summary", January 5, 1981.

16. "Spotted Owl Management Proposal for Region 6, USFS", January 8, 1981, 17.

17. Letter from E. B. Chamberlain, Jr., FWS Assistant Regional Director, Federal Assistance to R. E. Worthington, Regional Forester, December 1, 1980.

18. Memorandum to Forest Supervisors, Westside Forests, from James Torrence, Deputy Regional Forester, "Range-Fish-Wildlife Training Session", November 5, 1981.

19. Memorandum from Jeff Sirmon, Regional Forester to Forest Supervisor, Gifford Pinchot NF, "Spotted Owl Habitat", March 11, 1982.

20. Memorandum from Sirmon, March 11, 1982.

21. See, for example, Dennis Hayward, "Will the Spotted Owl Become Our Billion Dollar Bird?", *The Log*, July 1982, 14–17.

22. Letter from R. E. Worthington, Regional Forester to Senator Bob Packwood, August 27, 1981.

23. Letter from John B. Crowell, USDA Assistant Secretary, to R. Max Peterson, Chief, FS, "Spotted Owls", July 14, 1981.

24. The Draft Plan's research statement clearly indicated the rising interest in old growth protection and management as a credible scientific and policy issue: "The issue of the spotted owl and maintenance of old-growth is the present focus of attention. It is, in our opinion, only a symptom of a much larger and more complex issue. That is the question of old-growth as a special or unique habitat for certain species of wildlife." USDA-FS, "Appendix I draft," *Draft Pacific Northwest Regional Plan* (Portland OR: Pacific Northwest Region, April 1981).

25. Wilkinson and Anderson, "Land and Resource Planning in the National Forests, 299. [See Note #8, Chapter 1.]

26. Quoted in Wilkinson and Anderson, "Land and Resource Planning in the National Forests," 299.

27. Interview with Hal Salwasser, June 12, 1989.

28. Memorandum from J. B. Hilmon, Associate Deputy Chief to Regional Foresters, "Wildlife and Fish Viable Populations in Forest Planning," February 24, 1982.

29. Interview with Hal Salwasser, June 12, 1989.

30. 36 *CFR* 219.12 (1984).

31. USDA-FS, "Minimum Management Requirements Report", July 25, 1985 draft, 11.

32. USDA-FS, "Minimum Management Requirements Report", 11.

33. In meetings of forest biologists in August, 1982, it was decided that viability requirements should not be specified for all species, but rather those "genuinely threatened with a loss of viability." In specifying these requirements, one "concern that arose was that in defining the MMR species and their requirements for viability, we not forget that the object here was to define *minimum* levels, not desired levels. This was recognized throughout the development of the direction. Frequent attention was drawn to only do what was minimally necessary. Often, where available information was less than that desired to make a decision, standards were set at a level just below a level that the biologists were comfortable with." USDA-FS, "Minimum Management Requirements Report", 12.

34. Memorandum from Regional Forester, Region 6, to Forest Supervisors, "Regional Guidelines for Incorporating Minimum Management Requirements in Forest Planning," February 9, 1983.

35. Memorandum from Regional Forester, Region 6, February 9, 1983, 4,25.

36. Memorandum from Regional Forester, Region 6, to Forest Supervisors, "Clarification of Wildlife Minimum Management Requirements Direction", April 16, 1984.

37. Minimum Management Requirements Report", 16.

38. Memorandum from Dennis Hayward, North West Timber Association, to Region 6 Technical Committee, "The Viable Population Syndrome", October 1, 1982.

39. Subcommittee on Minimum Legals, "Region 6 Technical Committee Analysis of Minimum Legal Requirements for Modeling Benchmarks and Alternatives", (Portland OR: Forest Industry Region Six Planning Technical Committee, 1982), 5–6.

40. Letter from Dennis Hayward, North West Timber Association, to Charles Hartgraves, Director, Land Management Planning, FS, March 15, 1983, 4–5.

41. Subcommittee on Minimum Legals, "Region 6 Technical Committee Analysis", 6.

42. Letter from Hayward, March 15, 1983, 6.

43. Interview with Andy Stahl, July 24, 1989.

44. Roth, *The Wilderness Movement and the National Forests,* 36.

45. A good summary of the SAF task force report is found in Jay Heinrichs, "Old Growth Comes of Age", *Journal of Forestry* 776–779, December 1983.

46. Karl Bergsvik *et al., Report of the SAF Task Force on Scheduling the Harvest of Old-Growth Timber,* (Bethesda, MD: Society of American Foresters, 1983), 11,24–25.

47. Interview with Andy Stahl, July 24, 1989.

48. IBLA 84–241, February 26, 1985.

49. Interview with Andy Stahl, July 24, 1989.

50. National Wildlife Federation, Oregon Wildlife Federation, Lane County Audubon Society, and Oregon Natural Resources Council, "Reply of Appellants to Chief's Responsive Statement, Appeal of Region VI Regional Guide and EIS", January 7, 1985, 1–2.

51. FS Chief Max Peterson, "Appeal of Region VI Regional Guide, Responsive Statement", December 12, 1984, 1–2.

52. Peterson, "Appeal of Region VI Regional Guide, Responsive Statement", 3–4.

53. National Wildlife Federation *et al.,* "Reply of Appellants to Chief's Responsive Statement," 3.

54. National Wildlife Federation, Oregon Wildlife Federation, Lane County Audubon Society, Oregon Natural Resources Council, "Joint Statement of Reasons and Petition for Preparation of a Joint Agency Environmental Impact Statement, Appeal of Region VI Regional Guide and EIS", October 18, 1984, 4–5.

55. National Wildlife Federation *et al.,* "Joint Statement of Reasons," 5.

56. Peterson, "Appeal of Region VI Regional Guide, Responsive Statement", 5–6.

57. Eric Forsman and E. Charles Meslow, "Old-Growth Forest Retention for Spotted Owls—How Much Do They Need?" in Ralph Gutierrez and Andrew Carey, *Ecology and Management of the Spotted Owl in the Pacific Northwest* (Portland OR: USDA-FS, Pacific Northwest Experiment Station, September 1985), 58–59.

58. George Barrowclough and Sadie Coats, "The Demography and Population Genetics of Owls, with Special Reference to the Conservation of the Spotted Owl (*Strix occidentalis*)" in Ralph Gutierrez and Andrew Carey, *Ecology and Management of the Spotted Owl in the Pacific Northwest* (Portland OR: USDA-FS, Pacific Northwest Experiment Station, September 1985), 74–85.

59. Peterson, "Appeal of Region VI Regional Guide, Responsive Statement", 6.

60. National Wildlife Federation, Oregon Wildlife Federation, Lane County Audubon Society, and Oregon Natural Resources Council, "Reply of Appellants to Chief's Responsive Statement, Appeal of Region VI Regional Guide and EIS", January 7, 1985, 11.

61. National Wildlife Federation, Oregon Wildlife Federation, Lane County

Audubon Society, Oregon Natural Resources Council, "Joint Statement of Reasons and Petition for Preparation of a Joint Agency Environmental Impact Statement, Appeal of Region VI Regional Guide and EIS", October 18, 1984, 9.

62. Peterson, "Appeal of Region VI Regional Guide, Responsive Statement", 8.

63. Peterson, "Appeal of Region VI Regional Guide, Responsive Statement", 9–10.

64. Peterson, "Appeal of Region VI Regional Guide, Responsive Statement", 10–11.

65. National Wildlife Federation, Oregon Wildlife Federation, Lane County Audubon Society, and Oregon Natural Resources Council, "Reply of Appellants to Chief's Responsive Statement, Appeal of Region VI Regional Guide and EIS", January 7, 1985, 15.

66. Letter from Andy Stahl, NWF, to Regional Forester, January 3, 1985.

67. Letter from Richard T. Bailey, Director of Planning and Special Projects, Industrial Forestry Association to Mr. John B. Crowell, Jr., USDA Assistant Secretary for Natural Resources and Environment, January 7, 1985, 2–3.

68. Letter from Bailey, January 7, 1985, 3.

69. Letter from Bailey, January 7, 1985, 8.

70. Letter from Bailey, January 7, 1985, 5–6.

71. Letter from Bailey, January 7, 1985, 10.

72. Letter from Douglas W. MacCleery, USDA Deputy Assistant Secretary for Natural Resources and Environment, to R. Max Peterson, Chief, FS, "USDA Decision on Review of Administrative Decision by Chief of the Forest Service related to the Administrative Appeal of the R-6 Regional Guide and EIS", March 8, 1985, 13.

73. Letter from MacCleery, March 8, 1985, 11–12.

74. Letter from MacCleery, March 8, 1985, 12.

75. While the details of the remand decision as to what the SEIS was to contain reflected the clandestine input of Hal Salwasser, that does not explain why MacCleery chose to have the FS undertake the analysis, an action that organizationally had as many risks as opportunities.

Chapter 4

1. Interview with Dick Holthausen, July 19, 1989.

2. Interview with Bruce Marcot, July 21, 1989.

3. Interview with Dick Holthausen, July 19, 1989.

4. Memorandum from Jeff M. Sirmon, Regional Forester to Chief, "Road Map for Responding to the Remand of the R-6 Regional Guide," May 9, 1985.

5. Interview with Tom Ortman, October 5, 1989.

6. Interview with Dick Holthausen, July 19, 1989.

7. Interview with Bruce Marcot, July 21, 1989.

8. Memorandum from Gary E. Cargill, Associate Deputy Chief to Regional Forester, Region 6, April 8, 1985.

9. Interview with Kathy Johnson, July 17, 1989.

10. Interview with Larry Fellows, July 19, 1989.

11. USFS, SEIS ID Team, "Regional Guide Supplement Issues and Concerns", drafts, July 9, 1985, August 20, 1985; and USDA-Forest Service, *Draft Supplement to the Final Environmental Impact Statement for an Amendment to the Pacific Northwest Regional Guide, Volume 1, Spotted Owl Guidelines* (Portland OR: Pacific Northwest Regional Office, July 1986), 1–8.

12. Draft of working agreement between FS Region 6 and Northwest Executive Consultants, Inc., June 17, 1985.

13. Gerald Oncken, Northwest Executive Consultants, Inc., "Report to Forest Service regarding interview findings with timber industry and conservation groups," August 18, 1985.

14. Interview with Kathy Johnson, July 17, 1989.

15. Interview with Kathy Johnson, July 17, 1989.

16. The alternatives are summarized from USDA-FS, "Proposed Spotted Owl SEIS Alternatives", September 11, 1985, Spotted Owl SEIS ID Team draft.

17. Interview with Bruce Marcot, July 21, 1989.

18. Interview with Tom Ortman, October 5, 1989.

19. Interview with Bruce Marcot, July 21, 1989.

20. Interview with Bruce Marcot, July 21, 1989.

21. USDA-FS, *Draft Supplement to the Final Environmental Impact Statement for an Amendment to the Pacific Northwest Regional Guide, Volume 2, Appendices* (Portland OR: Pacific Northwest Regional Office, July 1986), B-1.

22. USDA-FS, *Draft Supplement to the Final Environmental Impact Statement, Volume 2*, B-45–46.

23. Summarized from "Interdisciplinary Team Draft Evaluation Criteria for the SEIS", unpublished FS working paper, September 11, 1985.

24. Hal Salwasser, "My notes from R-6/5 Meetings in September", ca. September 23, 1985.

25. Interview with Bruce Marcot, July 21, 1989.

26. Interview with Tom Ortman, October 5, 1989.

27. Interview with Bruce Marcot, July 21, 1989.

28. Interview with Tom Ortman, October 5, 1989.

29. The ranges on these numbers reflect the fact that Alternative F provided for 1000 acres of designated habitat, with an additional 1200 acres as potential habitat. The low figure in these ranges assume only 1000 acres of habitat are provided per SOHA, and the higher figure assumes 2200 acres per SOHA.

30. USDA-FS, *Draft Supplement to the Final Environmental Impact Statement, Volume 1*, S-15.

31. USDA-FS, *Draft Supplement to the Final Environmental Impact Statement, Volume 1*, S-2.

32. USDA-FS, *Draft Supplement to the Final Environmental Impact Statement, Volume 1,* S-2.

33. Letter from Russell Lande, University of Chicago, to Andy Stahl, NWF, Portland, June 12, 1985.

34. Letter from Terrence L. Thatcher, NWF to Douglas MacCleery, June 14, 1985.

35. Letter from Douglas MacCleery, Deputy Assistant Secretary for Natural Resources and Environment, USDA to Terrence L. Thatcher, NWF, July 1, 1985.

36. Interview with Andy Stahl, July 24, 1989.

37. Interview with Amos Eno, May 24, 1989.

38. Summarized from William R. Dawson *et al., Report of the Advisory Panel on the Spotted Owl,* Audubon Conservation Report No. 7, (New York: National Audubon Society, 1986), 7.

39. Comments of the Northwest Forest Resources Council, Portland OR, on the Draft SEIS, November 17, 1986.

40. USDA-FS, "Draft: Public Response Report for the Draft SEIS", unpublished draft report, January 26, 1987, 2–3.

41. USDA-FS, "Draft: Public Response Report for the Draft SEIS", 4.

42. For example, in the Final SEIS, the costs of the different seven alternatives in terms of lands suitable for timber production were: A-0 acres; C-71,800 acres; D-184,400 acres; F-347,700 acres; G-461,700 acres; M-977,800 acres; L-2,561,200 acres. USDA-FS, *Final Supplement to the Environmental Impact Statement, Volume 1,* Summary-39.

43. Interview with Dick Holthausen, July 19, 1989.

44. Interview with Bruce Marcot, July 21, 1989.

45. Interview with Dick Holthausen, July 19, 1989.

46. Interview with Grant Gunderson, July 31, 1989.

47. Interview with Grant Gunderson, July 31, 1989.

48. Interview with Bruce Marcot, July 21, 1989.

49. Interview with Dick Holthausen, July 19, 1989.

50. Interview with Tom Ortman, October 5, 1989.

51. USDA-FS, *Final Supplement to the Environmental Impact Statement, Volume 1,* 1–3.

52. USDA-FS, *Final Supplement to the Environmental Impact Statement, Volume 1,* Summary-23.

53. Interview with Andy Stahl, July 24, 1989.

54. 52 *Federal Register* 34396–34397, July 23, 1987.

55. Interview with Andy Stahl, July 24, 1989.

56. Memo from Mark Shaffer to Jay Gore, Regional Office, Region 1, November 11, 1987.

57. Memorandum from FWS Regional Director, Region 1, to Director, U. S. FWS, "Twelve-Month Petition Finding—Northern Spotted Owl", December 17, 1987, 2.

58. Memorandum from FWS Director to Regional Director, Region 1, "Interagency Agreement with Forest Service on Spotted Owl", December, 4, 1987.

59. Harriet Allen, "What's Happening to Spotted Owls in Washington?" Remarks presented at the January 15, 1988 Washington Wildlife Commission meeting, 8.

60. Spotted Owl Subcommittee, Oregon–Washington Interagency Wildlife Committee, "Draft Interagency Management Guidelines for the Northern Spotted Owl in Washington, Oregon, and California," March 28, 1988, 1.

61. Notes written on top of "Draft Interagency Management Guidelines for the Northern Spotted Owl in Washington, Oregon, and California", prepared by Harriet Allen, Charles Bruce, and Eric Forsman, October 1987.

62. This order was modified the next month, so that the BLM was prohibited from selling old growth stands within 2.1 miles of known owl nest sites.

63. Interview with Dick Holthausen, July 19, 1989.

64. Interview with Bruce Marcot, July 21, 1989.

65. *Peninsula Daily News,* March 1, 1989.

Chapter 5

1. 54 *Federal Register* 26666, June 23, 1989.

2. 54 *Federal Register* 26666, June 23, 1989.

3. U. S. General Accounting Office, *Endangered Species: Spotted Owl Petition Evaluation Beset by Problems* (Gaithersburg, MD: U. S. General Accounting Office, February 1989), Report #GAO/RCED-89–79.

4. See, for example, U. S. Congress, Committees on Agriculture and Interior and Insular Affairs, *Management of Old-Growth Forests of the Pacific Northwest,* Joint Hearings, No. 101–35, June 20, 1989.

5. Oregon Congressional delegation, "Outline of Short-Term Solution to Timber Supply/Old Growth Problem", unpublished one-page document presented at the Timber Summit, June 24, 1989, Salem, Oregon.

6. The average timber harvest level from Oregon and Washington national forests for the 5-year period from 1985 to 1989 was 4.097 bbf per year, roughly equal to the levels outlined in the Hatfield–AuCoin proposal. Harvest levels in 1987 and 1988 were at an all-time high of 4.428 bbf in 1987 and 4.382 bbf in 1988. Data taken from USDA-FS, *Final Environmental Impact Statement on Management for the Northern Spotted Owl in the National Forests, Volume 1* (Portland OR: USDA-FS, January 1992), Chapters 3 and 4–191.

7. Letter from Rick Brown, Oregon Ancient Forest Alliance, *et al.,* to Senator Mark Hatfield *et al.,* June 27, 1989, 2.

8. Ancient Forest Alliance, "Old Growth/Timber Supply Interim Solution," July 14, 1989.

9. Memorandum from Brock Evans, National Audubon Society, to Audubon

Northwest Leaders, "Ancient Forest Campaign: Sequence of Events Leading Up to Senate Appropriations Bill Floor Vote, July 26th," August 16, 1989, 2.

10. Letter from F. Dale Robertson, FS Chief, to Senator Mark Hatfield, July 17, 1989.

11. Memorandum from Evans, "Ancient Forest Campaign," 2.

12. Letter from Robertson, July 17, 1989.

13. U. S. Congress, House of Representatives, "Making Appropriations for the Department of the Interior and Related Agencies for the Fiscal Year Ending September 30, 1990, and for Other Purposes," *Conference Report* No. 101–264, October 2, 1989, 87.

14. *Seattle Audubon Society v. Robertson,* 914 F. 2d 1311 (9th Cir. 1990).

15. Kathie Durbin, "Court frees 16 timber sales," *The Portland Oregonian,* March 26, 1992, A1.

16. F. Dale Robertson, USFS, John F. Turner, USFWS, Cy Jamison, BLM, James M. Ridenour, NPS, "A Charter for an Interagency Scientific Committee to Address the Conservation of the Northern Spotted Owl," October 1990.

17. Thomas *et al., A Conservation Strategy for the Northern Spotted Owl,* 1.

18. Thomas *et al., A Conservation Strategy for the Northern Spotted Owl,* 2.

19. 9 *Land Letter* 12:4, April 20, 1990.

20. 9 *Land Letter* 12:4–5, April 20, 1990.

21. 9 *Land Letter* 12:5, April 20, 1990.

22. Thomas *et al., A Conservation Strategy for the Northern Spotted Owl,* 9.

23. Thomas *et al., A Conservation Strategy for the Northern Spotted Owl,* 5.

24. "Ancient Forest Protection Act of 1990", HR 4492, 101–2, April 4, 1990.

25. Quoted in John Lancaster, "Northern Spotted Owl Is 'Threatened'", *The Washington Post,* June 23, 1990, A1.

26. Quoted in John Lancaster, "Lujan: Endangered Species Act 'Too Tough,' Needs Changes", *The Washington Post,* May 12, 1990, A1.

27. Quoted in Andrew Rosenthal, "President Skirts 2 Ecology Issues", *The New York Times,* May 22, 1990, A10.

28. Quoted in Lancaster, "Northern Spotted Owl is 'Threatened,'" A14.

29. Condensed from "Yeutter and Lujan Annouce (sic) Five-Point Plan to Preserve Owl and Protect Jobs", USDA News release 831–90, June 26, 1990.

30. See, for example, Gerald Seib, "Bush Delays Oil, Gas Drilling for Decade In Large Areas Off California and Florida," *Wall Street Journal,* June 27, 1990, A14.

31. Keith Ervin, "Sketchy administration plan may still bring new round of environmentalist lawsuits," *The Seattle Times/Seattle Post-Intelligencer,* September 23, 1990, B1.

32. Quoted in Alyson Pytte, "Bush's Modest Proposal," *Congressional Quarterly,* September 29, 1990, 3105.

33. Quoted in Pytte, "Bush's Modest Proposal," 3105.

34. Quoted in Pytte, "Bush's Modest Proposal," 3105.

35. "Bush tries to duck an environmental choice," *The Minneapolis Star Tribune,* September 16, 1990.

36. 55 *Federal Register* 40413, September 28, 1990.

37. Rob Taylor, "A Plan to Boost Logging", *Seattle Post-Intelligencer,* October 6, 1990, A1.

38. Taylor, "A Plan to Boost Logging," A1.

39. Hearing transcript at 384, reprinted in U. S. District Court, Western District of Washington, "Memorandum Decision and Injunction," Seattle Audubon Society *et al.* v. John L. Evans *et al.,* No. C89–160WD, May 23, 1991.

40. 55 *Federal Register* 26114, June 26, 1990.

41. U. S. District Court, Western District of Washington, "Order Granting Plaintiffs' Motion for Summary Judgment and Motion to Compel Designation of Critical Habitat," Northern Spotted Owl *et al.* v. Manual Lujan *et al.,* No. C88–573Z, February 26, 1991.

42. 56 *Federal Register* 20821, May 6, 1991.

43. 56 *Federal Register* 20821, May 6, 1991.

44. Indeed, the final critical habitat designation was revised downward to 6.9 million acres when it was published on January 15, 1992. U. S. Department of the Interior, Fish and Wildlife Service, *Recovery Plan for the Northern Spotted Owl—DRAFT* (Washington, D.C.: U. S. Government Printing Office, April 1992), 4.

45. Estimates by University of Washington professor Bruce Lippke, cited in Bill Dietrich, "Both sides say owl-habitat case is critical," *The Seattle Times/Seattle Post-Intelligencer,* May 5, 1991, B-5.

46. U. S. District Court, Western District of Washington, "Memorandum Decision and Injunction," Seattle Audubon Society *et al.* v. John L. Evans *et al.,* No. C89–160WD, May 23, 1991, 33–34, 20.

47. U. S. District Court, Western District of Washington, "Memorandum Decision and Injunction," 34.

48. Quoted in Jim Simon, "Gardner blasts U.S. on owl issue," *The Seattle Times,* May 8, 1991, G-1.

49. Quoted in Christopher Hanson, "Lawmakers condemn Bush for delays on spotted owl solution," *Seattle Post-Intelligencer,* May 30, 1991, A3.

50. Quoted in Hanson, "Lawmakers condemn Bush for delays on spotted owl solution," A3.

51. Quoted in Hanson, "Lawmakers condemn Bush for delays on spotted owl solution," A3.

52. Quoted in Clyde Weiss, "Dicks warn feds to 'close ranks' on owl plan", *Aberdeen WA Daily World,* March 6, 1991, A-9.

53. For example, a Congressionally convened panel of scientists unveiled a set of 14 owl-protection options in July 1991, and indicated that nothing less than the ISC report recommendations would adequately protect owls and other old growth-associated wildlife. The so-called "Gang of Four" (John Gordon, Dean of the Yale Forestry School; FS biologist and University of Washington professor Jerry Franklin; FS biologist Jack Ward Thomas; and Norm Johnson, professor of forest management at Oregon State University) was asked by two Congressional

subcommittees to frame a series of options to resolve the controversy, and report on the various economic impacts. The most sweeping option identified by the group would result in an annual harvest level of less than 700 million board feet at a cost of more than 60,000 direct and indirect jobs. See, for example, Christopher Hanson and Rob Taylor, "Panel reveals spotted-owl options," *Seattle Post-Intelligencer,* July 25, 1991, A1.

54. Quoted in "Spotted owl recovery team has first meeting," *Amberdeen WA Daily World,* March 5, 1991, A-5.

55. Quoted in "Spotted owl recovery team has first meeting," A-5.

56. *The Washington Times,* April 24, 1991.

57. See, for example, *Seattle Weekly,* May 29, 1991.

58. See, for example, Sharon LaFraniere and Bill McAllister, "Hatfield Inquiries Grow," *The Washington Post,* May 29, 1991, A1.

59. M. Lynne Corn, *Spotted Owls and Old Growth Forests,* CRS Issue Brief Order Code IB90094 (Washington, D.C.: Congressional Research Service, The Library of Congress, January 27, 1992), CRS-9.

60. Quoted in Tom Kenworthy, "'God Squad' To Ponder Spotted Owl", *The Washington Post,* October 21, 1991, A17.

61. Quoted in Kenworthy, "'God Squad' To Ponder Spotted Owl", A17.

62. See, for example, Tom Kenworthy, "Owl Supporters, Interior At Loggerheads Again," *The Washington Post,* February 13, 1992.

63. Quoted in Jeff Barnard, "Train wreck theory gaining steam in owl fight," *The Daily Astorian,* March 27, 1992.

64. Quoted in Kenworthy, "Owl Supporters, Interior At Loggerheads Again."

65. The FWS published a draft recovery plan in April 1992, which would establish 196 designated conservation areas (DCAs) to provide approximately 7.5 million acres of federal lands as the primary habitat for the owl. The draft plan estimated impacts at 18,900 timber jobs and 13,200 jobs in related sectors. Lost or reduced wages were estimated at $1.4 billion over a twenty-year period. The draft plan recognized that "The conservation of northern spotted owls is a difficult public policy issue. It is important to achieve recovery in a way that is appropriate under the Endangered Species Act, yet also managerially and economically efficient." USDI-FWS, *Recovery Plan for the Northern Spotted Owl—DRAFT,* ix, xiii.

66. The FS produced a final environmental impact statement on spotted owl management in January 1992. USDA-FS, *Final Environmental Impact Statement on Management for the Northern Spotted Owl.* Its preferred alternative, and proposed action, was to use the Interagency Scientific Committee report (the Jack Ward Thomas committee) as the basis for management of owl habitat in the Pacific Northwest. The environmental groups appealed the agency's decision and were rewarded with a decision by Judge Dwyer in May that the agency had not included information produced subsequent to the 1990 ISC report, and that the EIS failed to take into account the impact that an owl plan would have on 32 other old-growth-dependent species. Dwyer extended the injunction on sales he first

initiated in May 1991, and gave the FS until September 1993 to revise its plan. See "Feds appeal logging ban," *Seattle Post-Intelligencer,* August 27, 1992, B3.

67. See, for example, Michael Weisskopf and David Maraniss, "When Irresistable Force Met Arkansas' Timber Industry: Did Bill Clinton Sell Out or Was He Bowing to the Inevitable?", *The Washington Post National Weekly Edition,* June 29–July 5, 1992, 11–12.

68. See, for example, Secretary Babbitt's comments to the Environmental Grantmakers Association reprinted in *The Washington Post,* February 26, 1993.

69. Quoted in Jeffrey St. Clair, "Ancient Forests in the Balance," 13 *Forest Watch* 5, April/May 1993.

70. Quoted in Tom Kenworthy, "Clinton criticized for shift on public lands," *The Washington Post,* April 2, 1993, A8.

71. Jeff Mapes, "Clinton vows NW forest summit; Bush eyes changes in species act," *The Portland Oregonian,* August 27, 1992.

72. The "conference" was downgraded rhetorically from a "summit" because President Clinton was to meet with Japanese leaders in Canada the next day for an economic summit, and administration officials did not want to confuse the two in the public's mind.

73. Quoted in "'Forest Summit': Everyone's Wary," *The Seattle Times,* November 27, 1992, D6.

74. Quoted in "'Forest Summit': Everyone's Wary," D6.

75. Quoted in "'Forest Summit': Everyone's Wary," D6.

76. "Improving U. S. Economy Cited as Driving Up Lumber Prices," *The Washington Post,* March 20, 1993.

77. See, for example, Jim Pissot, "Timber Troubles: The Spotted Owl is Not the Cause of the Northwest Forest Crisis," *The Washington Post,* April 2, 1993.

78. Quoted in Tom Kenworthy and Ann Devroy, "Clinton Pledges 'Balanced' Solution to Forest Policy Crisis in 2 Months," *The Washington Post,* April 3, 1993.

79. Closing remarks of President Bill Clinton at the Northwest Forest Conference, April 2, 1993, as reprinted in 13 *Forest Watch* 30, April/May 1993.

80. Office of the Press Secretary, The White House, "Media Advisory: Inter-Agency Groups to Craft Northwest Forest Policy Options," April 14, 1993.

81. Memorandum from the Forest Conference Executive Committee to the Forest Conference Inter-Agency Working Groups, "Statement of Mission," May 7, 1993.

82. See Note # 53, this section.

83. Forest Ecosystem Management Assessment Team, *Forest Ecosystem Management: An Ecological, Economic, and Social Assessment* (Portland, OR: U. S. Government Printing Office, July 1993), I-2.

84. The 50–11–40 rule called for maintaining 50% of forested land within each quarter township (9 square miles) in forested condition, with stands of trees averaging at least 11 inches diameter at breast height and with a stand canopy closure of at least 40%.

85. All data are summarized from Forest Ecosystem Management Assessment Team, *Forest Ecosystem Management: An Ecological, Economic, and Social Assessment*. Good tables of this information are presented in USDA-Forest Service and USDI-Bureau of Land Management, *Draft Supplemental Environmental Impact Statement on Management of Habitat for Late-Successional and Old-Growth Forest Related Species Within the Range of the Northern Spotted Owl* (Portland, OR: U. S. Government Printing Office, July 1993), S-6, S-18.

86. Quoted in Kathie Durbin, "Forest Protection Takes New Turn," *The Portland Oregonian*, June 10, 1993, A1.

87. Forest Ecosystem Management Assessment Team, *Forest Ecosystem Management*, III-29.

88. Kathie Durbin, "Timber Industry Accused," *The Portland Oregonian*, June 5, 1993, B1.

89. Durbin, "Forest Protection Takes New Turn;" Tom Kenworthy, "Logging is Top Goal in Policy Memo," *The Washington Post*, June 18, 1993, A1.

90. Kenworthy, "Logging is Top Goal in Policy Memo."

91. Tom Kenworthy, "Foley Casts Doubts on Clinton's Northwest Timber Plan," *The Washington Post*, June 29, 1993.

92. Quoted in Paul Koberstein, "Clinton vs. Foley: House speaker is furious at plan to protect Northwest forests," *High Country News*, July 26, 1993, 8.

93. Quoted in Kenworthy, "Logging is Top Goal in Policy Memo."

94. Sierra Club, "Urgent Action Alert: Your Calls Are Needed to Secure the Fate of Our Ancient Forests," June 15, 1993.

95. Betsy Marston, "The Clinton Forest Plan in Brief," *High Country News*, July 26, 1993, 8.

96.USDA-Forest Service and USDI-Bureau of Land Management, *Draft Supplemental Environmental Impact Statement on Management of Habitat for Late-Successional and Old-Growth Forest Related Species Within the Range of the Northern Spotted Owl*.

97. Quoted in Tom Kenworthy, "Clinton to Slash Logging," *The Washington Post*, July 2, 1993, A1.

98. Quoted in Kenworthy, "Clinton to Slash Logging."

99. Quoted in Kenworthy, "Clinton to Slash Logging."

100. See, for example, Kenworthy, "Foley Casts Doubts on Clinton's Northwest Timber Plan."

101. Tom Kenworthy, "Interior, Allies at Loggerheads Over Timber," *The Washington Post*, September 23, 1993.

Chapter 6

1. While no one really knows the full extent of suitable owl habitat existing before the 1800s, an estimate of 17.5 million acres in 1800 is commonly used. Recent estimates of suitable habitat range from two to seven million acres, with

reduction continuing at a rate of one to two percent per year. Population size estimates range from two to four thousand owl pairs, with declines in the same range as the habitat losses. Recent estimates of home range size indicate that a median home range of 3000 to 5000 acres is typical of most areas, while the median amounts of old growth within a home range varies from 600 to 4600 acres. Thomas *et al., A Conservation Strategy for the Northern Spotted Owl,* 65–67, 207, 212.

2. Comments by wildlife biologist Chuck Meslow at the April 2, 1993 forest conference indicated that the recovery team for the northern spotted owl considered some 482 species of plants and animals as associated with old growth in the Pacific Northwest. For an earlier assessment, see Andrew B. Carey, "Wildlife Associated with Old-Growth Forests in the Pacific Northwest," 9 *Natural Areas Journal* 3:151–162, July 1989. Carey's assessment reported close association or dependence for twelve species, including the Vaux's swift, the Olympic salamander, and the northern flying squirrel.

3. Dating back to its origins in the corrupt and inept General Land Office, the BLM's expertise, resources, and priorities for its lands were always suspect, and an easy target for criticism. Proponents of nongame wildlife conservation for years had questioned the commitment of the Fish and Wildlife Service to doing anything other than growing ducks and eliminating livestock predators.

4. One response from the 1980s FWS was to create a scheme whereby fees were paid by developers into a mitigation or research fund allowing development to take place while in theory at least adding to the potential for protecting the affected endangered species. The so-called Windy Gap approach to carrying out the FWS statutory mandates was highly controversial, raising fundamental questions about the commitment of the federal agencies to protection, and the appropriateness of different kinds of strategies for dealing with the expanding set of conflicts between endangered species and human-generated development.

5. For example, new proposed diversions of water for development and other purposes from the Upper Colorado and North Platte river systems generated jeopardy opinions from the FWS in the early 1980s because of their impacts on endangered fish and bird species. It seemed like development would have to come to a standstill in these river basins. Issues over spotted owl and salmon protection in the Northwest appeared even more problematic.

6. Dan Wyant, "Timber shortage plagues industry," *Eugene Register-Guard,* February 9, 1989, 1C.

7. For example, some forty percent of the employment in Skamania County, Washington was directly in timber-related industries. Approximately 27 percent of Douglas County, Oregon's 1987 workforce was in timber jobs. See Kathie Durbin, "Fear of job losses unites NW timber towns", *Portland Oregonian,* April 30, 1989.

8. Jeffrey Pressman and Aaron Wildavsky, *Implementation* (Berkeley, CA: University of California Press, 1973).

9. Yaffee, *Prohibitive Policy*, 32–57. [See Note #19, Chapter 1.]

10. Note that the issue of below cost timber sales is controversial, and the FS argues that the approach taken to evaluating returns on investment is misleading. The data presented in the text are taken from Richard E. Rice, *National Forests: Policies for the Future, Volume 5, The Uncounted Costs of Logging* (Washington, D.C.: The Wilderness Society, 1989), 2; Appendix.

11. In 1972, Forsman knew of 37 owl nest sites. In 1990, after millions of dollars and countless human-hours had been spent on inventory work, the Inter-agency Scientific Committee reported that, "Inventory and monitoring of owls have been occurring, piecemeal throughout the range since the early 1970s. Not until the mid-1980s, however, have these efforts been extensive enough to begin providing reasonably good information about the distribution and abundance of owls throughout their range. These results indicate about 2000 pairs located during the last 5 years, representing some unknown fraction of the true number of pairs. Because a census of the total population is not available, we have no statistically reliable population estimate. Recent claims of actual counts of some 6000 birds in 1989 are not out of line with other information from monitoring and inventory efforts." Thomas *et al., A Conservation Strategy for the Northern Spotted Owl*, 20.

12. The second growth stands that supported owls tended to exhibit old growth characteristics resulting from "sloppy" forestry a number of years ago. See, for example, Thomas *et al., A Conservation Strategy for the Northern Spotted Owl*, 19–20.

13. Operational definitions of home range expanded from 300 acres per pair in the early 1970s to 1000 acres per pair in the early 1980s, as radiotelemetry work began to give researchers an image of owl movement across large areas of space. As the level of research effort increased, and methods of tracking individual owls improved, a more complicated picture emerged, with home range sizes varying according to the productivity of the habitat. Generally, as you move north in the owl's range, it takes more acreage to produce equivalent survival conditions. Operationally, that reality led to a variable notion of home range size. For example, the Forest Service's 1986 Draft SEIS employed figures of 1000 acres in the southern part of the owl's range up to 2700 acres for the Olympic Peninsula in Washington. USDA-FS, *Final Supplement to the Environmental Impact Statement, Volume 1,* Summary-21. The 1990 Thomas report identified home range sizes ranging from 1000 acres in the south to more than 30,000 acres in the north, with medians ranging from 1411 to 9930 acres. Thomas *et al., A Conservation Strategy for the Northern Spotted Owl*, 211.

14. The interim definition was published in July 1986 by the Old-Growth Definition Task Force. USDA-FS, *Interim Definitions for Old Growth Douglas Fir and Mixed Conifers.* In his 1989 testimony at the House of Representatives hearing on old growth Forests, Chief Dale Robertson noted that "There is a lot of debate and disagreement about the appropriate definition of old growth . . . A nationally

consistent definition is essential to achieving a common reference point in discussing old growth." A draft definition authored by Tom Spies and Jerry Franklin was attached to the testimony. F. Dale Robertson, "Statement before the Committees on Interior and Insular Affairs and Agriculture", Joint Hearing Concerning the Management of Old Growth Forests in the Pacific Northwest, June 22, 1989.

15. George T. Frampton, Jr., "Testimony before the Joint Hearing of the U. S. House of Representatives, Interior Subcommittee on National Parks and Public Lands and the Agriculture Subcommittee on Forests, Family Farms, and Energy, on Old Growth Forests in the Pacific Northwest", June 20, 1989, 3.

16. Robertson, "Statement before the Interior and Agriculture Subcommittees", June 22, 1989, 5.

17. V. Alaric Sample and Dennis C. LeMaster, *Assessing the Employment Impacts of Proposed Measures to Protect the Northern Spotted Owl* (Washington, D.C.: American Forestry Association, Forest Policy Center, 1992), i. This report provides an excellent overview of the varying economic analyses prepared in response to the report of the Interagency Scientific Committee (the Jack Ward Thomas report) and the proposal of the FWS to designate critical habitat for the owl. Four sets of studies are examined by the AFA report: analyses by the FS, the American Forest Resources Alliance, The Wilderness Society, and the Scientific Panel on Late-successional Forest Ecosystems (the Gang of Four), a group of four scientists commissioned by the U. S. House of Representatives Committee on Agriculture in 1991.

18. Sample and LeMaster, *Assessing the Employment Impacts*, 31.

19. For example, the FS only considered the effects of changes on national forest lands, while the AFRA depicted changes in levels on all ownerships—a difference that inflated the employment impact numbers. Multipliers are used to extrapolate from the loss of a job in an industry such as loggers to the total job loss incurred since others are employed in firms that service the industry and its workers, such as restaurants that feed the loggers. Obviously the choice of a multiplier can have dramatic effects on economic projections.

20. For example, the FS estimated that the final plans would reduce timber harvesting on the national forests by 635 MMBF per year, while AFRA projected a much greater impact, 2,142 MMBF per year. See Sample and LeMaster, *Assessing the Employment Impacts*, 32.

21. See, for example, David A. Lax and James K. Sebenius, "The Negotiator's Dilemma: Creating and Claiming Value," *The Manager as Negotiator: Bargaining for Cooperation and Competitive Gain* (New York: The Free Press, 1986), 29–45.

Chapter 7

1. It could be argued, however, that the diversity requirements contained in NFMA required the agency to take action despite the owl's unlisted status,

particularly since the owl was listed on both the Oregon and Washington state lists of sensitive species.

2. The metaphor of a game has been used to describe institutional behavior before. See, for example, Eugene Bardach, *The Implementation Game: What Happens After A Bill Becomes A Law* (Cambridge, MA: MIT Press), 1977.

3. USDA-FS, *Final Supplement to the Environmental Impact Statement Volume 1,* IV-46.

4. Sharon LaFraniere and Bill McAllister, "Hatfield Inquiries Grow," *The Washington Post,* May 29, 1991, A1.

Chapter 8

1. See, for example, John Lancaster, "Bush Environmental Choice Appears Defeated", *The Washington Post,* November 17, 1989, A18, and 8 *Land Letter* 26:1, October 1, 1989.

2. Indeed, since the media rides on the horse of economic prosperity, as the economy deteriorates, so do the resources available to the media. Coverage of some issues, including science and environment, tend to lose out in the battle over what beats are covered. Hence, science beats have declined over the past five years. For example, in 1988, 95 U. S. newspapers had science sections. In 1992, 51 of the 95 sections had been dropped due to declining advertising. Of the remaining 44 sections, 21 had been reduced in size, and many had focused their attention on health and fitness topics, rather than the broader topic of science. Fred Jerome, "Bad News for Science News," *The New York Times,* September 27, 1992.

3. Andy's Stahl's comments are taken from a partial transcript of the proceedings entitled, "Old-Growth's Last Stand?", (Eugene OR: Sixth Annual Western Public Interest Law Conference, University of Oregon School of Law, March 5, 1988), 15–16.

4. Quoted in Elliot Diringer, "Loggers Sorry They Spotted Owls", *San Francisco Chronicle,* May 20, 1991, A11.

5. The original Ancient Forest Alliance included 6 national groups, 10 regional groups, 23 local chapters of Sierra Club or Audubon, and 17 other groups. As of October, 1989, the Alliance listed 83 affiliated organizations.

6. USDA-FS, *Final Supplement to the Environmental Impact Statement Volume 2,* G1–3.

7. Letter from Lynn Burns, Winston, Oregon, to Regional Forester, Pacific Northwest Region, November 14, 1986.

8. Interview with Michael McCloskey, February 22, 1989.

Chapter 9

1. Act of June 12, 1960, P. L. 86–517, sec 4(a), 74 Stat. 215, 16 USC 528–531.

2. Witness the response of environmental groups to the forest management

strategy adopted by the Clinton administration in July 1993. While a significant amount of land was to be protected from timber sales, a portion was still open to salvage operations and thinning. Calling these provisions a huge loophole, environmental groups did not trust the FS to implement them in an environmentally sound manner. See, for example, Gwen Ifill, "Clinton Backs a $1 Billion Plan to Spare Trees and Aid Loggers," *The New York Times,* July 1, 1993, A1.

3. Act of June 12, 1960, PL 86–517, sec 4(a), 74 Stat. 215, 16 USC 528–531.

4. Memo from Allan Lampi, Director, Land Management Planning, USFS, to Forest Supervisors and Directors, Region 6, "Minimum Management Requirements", September 24, 1982.

5. For an excellent discussion of the problems of forest planning, see Wondolleck, *Public Lands Conflict and Resolution.*

6. Letter from Sydney Herbert, Conservation Chair, Lane County Audubon Society to Michael Kerrick, Forest Supervisor, Willamette National Forest, October 12, 1981.

7. Interview with Andy Kerr, July 20, 1989.

8. A former representative from New Mexico, a state where BLM lands and cattle are abundant, Lujan also indicated that he viewed BLM lands as "a place with a lot of grass for cows," and saw nothing wrong with grazing in national parks. Cass Peterson, "Questioning Lujan's Interior Motives," *The Washington Post,* March 22, 1989, A17.

9. Katherine Barton and Whit Fosburgh, "The U. S. Forest Service," in DiSilvestro, *Audubon Wildlife Report 1986.*

10. Peterson, "Questioning Lujan's Interior Motives," A17.

11. "Will Owls Have a Foe at Forest Service?" *The Washington Post,* March 30, 1989.

12. Interview with Dr. Robert M. Wolcott, EPA Representative to the Endangered Species Committee, November 19, 1992.

13. Interview with Wolcott, November 19, 1992.

14. Originally, this metaphor was written as the "chrysalis" of expertise, but my daughter Anna reminded me that butterflies emerge from a chrysalis, not moths, and butterfly seemed too pretty a term to associate with the emergence of experts from smoke-filled rooms.

Chapter 10

1. For a description of the Progressive Conservation Movement, see Samuel P. Hays, *Conservation and the Gospel of Efficiency: The Progressive Conservation Movement 1890–1920* (New York: Atheneum, 1975).

2. For a discussion of the evolution of the expertise paradigm, see Wondolleck, *Public Lands Conflict and Resolution.*

3. Other recent surveys of FS employees have come to the same conclusion: loyalty to the organization is seen by employees as a behavior that is rewarded highly by the agency. See, for example, James J. Kennedy and Thomas M. Quigley,

"How Entry-Level Employees, Forest Supervisors, Regional Foresters and Chiefs View Forest Service Values and the Reward System," The Sunbird Conference, Second Meeting of Forest Supervisors and Chiefs, Tucson, Arizona, November 13–16, 1989.

4. For example, there is a literature on police forces as public bureaucracies, and it is understood that the sergeants who oversee divisions of street-level cops are reluctant to pass information critical of their units upwards to their leadership, in part because the sergeants' performance is evaluated by how well their street units are doing. See, for example, James Q. Wilson, *Varieties of Police Behavior* (Cambridge, MA: Harvard University Press, 1968).

5. These conclusions about the problematic response of the FS to a changing environment are consistent with other case studies of FS behavior. For example, Twight's classic study of the creation of the Olympic National Park concluded that the FS's "tenacious commitment to its value orientation" prevented it from responding to new demands, and as a result, agency leaders lost an opportunity to expand political support. See Twight, *The Forest Service Versus the Olympic National Park,* 116.

6. See, for example, Thomas E. Cronin, *The State of the Presidency* (Boston, MA: Little, Brown, 1975), 85–116.

7. President Jimmy Carter is probably the best recent example of a leader who was ineffective at building political concurrence for his policy objectives, items that he believed were good ideas and hence would be self-supporting.

8. For an excellent discussion of the effects of cognitive bias on decisionmakers judgment, see Max Bazerman, *Judgment in Managerial Decision Making* (New York: John Wiley, 1986).

9. For example, in research in summer camps in the 1950s, investigators had rival cabins go on a jelly bean hunt, after which they showed cabin members pictures of a partially filled jelly bean jar. The researchers told half the boys that their cabin had collected the beans, and half that the opposing cabin had collected them, and both were asked to estimate how many jelly beans were in the jar. What they found was that the boys overestimated the value of their own group's contributions and underestimated the value of the opposing group's contributions.

Chapter 11

1. Cited in Graham Hueber, "Americans Report High Levels of Environmental Concern, Activity," *The Gallup Poll Monthly,* No. 307, April 1991, 6.

2. All Gallup citations are from Hueber, "Americans Report High Levels of Environmental Concern, Activity," 6–8.

3. Richard L. Berke, "Oratory of Environmentalism Becomes the Sound of Politics," *The New York Times,* April 17, 1990, A1+.

4. Memorandum from Greenberg/Lake, The Analysis Group and The Tarrance

Group to The Nature Conservancy and The National Audubon Society, "Bipartisan poll results", January 8, 1992.

5. The results of the *Weekly Reader* survey were reported in Carol Zimmerman, "What You Should Do, Mr. President," *Parade Magazine,* January 17, 1993, 8+.

6. Much of what is driving the age structure of the United States is the existence of the Baby-Boom generation, defined as individuals born in the fifteen year period after World War II (1945–1960). The Boomers constitute roughly a third of the American population. As they age, the average age of the American population will increase, and the demands of the Boomers will rise in prominence. In 1960, some 35.7% of the American population was age 17 or under; by 2010, it is estimated that only 22.2% will be under 17. At the same time, only 29.2% of the American population was age 45 or older in 1960. Estimates indicate this age class will make up some 41.7% of the population in 2010. In the year 2000, the age classes dominated by the Boomers and their children will represent close to half the American population. (Population percentages for 1960 are from U. S. Bureau of the Census, "Table No. 13. Total Population, by Age and Sex: 1960 to 1989," *Statistical Abstracts of the United States: 1991* (Washington, D.C.: U. S. Government Printing Office, 1991), 13. Estimated population percentages in 2000 and 2010 are from U. S. Bureau of the Census, "Table No. 18. Projections of the Total Population by Age, Sex, and Race: 1995 to 2010," *Statistical Abstracts of the United States: 1991,* 16.)

7. See, for example, Clara S. Schuster, *The Process of Human Development,* 3rd edition (Philadelphia: Lippincott, 1992), 540; or Morris Massey, *The People Puzzle* (Boulder, CO: Morris Massey Associates, 1975).

8. Memorandum from Greenberg/Lake and The Tarrance Group, "Bi-partisan poll results," 3.

9. See, for example, the most important national problem as perceived by the American public, 1935–1991 as reported in George Gallup, Jr., *The Gallup Poll: Public Opinion 1991* (Wilmington, DE: Scholarly Resources, Inc., 1992), 197; George Gallup, Jr., *The Gallup Poll: Public Opinion 1990* (Wilmington, DE: Scholarly Resources, Inc., 1991), 84; George H. Gallup, *The Gallup Poll: Public Opinion 1979* (Wilmington, DE: Scholarly Resources, Inc., 1980), 94; George H. Gallup, *The Gallup Poll: Public Opinion 1978* (Wilmington, DE: Scholarly Resources, Inc., 1979), 159–160.

10. Reported in Mark Uehling, "All-American Apathy," *American Demographics,* November 1992, 32.

11. Frances Fox Piven and Richard A. Cloward, *Why Americans Don't Vote* (New York: Pantheon Books, 1988), 161.

12. Bill Whalen, "Nation: More Voters on Rolls May Not Mean More at Polls," *Insight,* August 21, 1989, 18.

13. James Ridgeway, "The Moving Target: Pain and Suffrage," *Village Voice,* August 9, 1988, 16–17.

14. Reported in Uehling, "All-American Apathy," 33.

15. Cited by Senator Edward Kennedy, "National Student/Parent Mock Elections," *Congressional Record,* daily edition, March 20, 1990, S2704.

16. As reported in "Notes and Comment: U. S. Voters Don't," *Allied Industrial Worker,* March 1988, 2.

17. By combining information from several survey questions, the Harris Poll's Alienation Index attempts to track the public's sense of powerlessness and disenchantment with the people running the nation's major institutions. When the measure was first constructed in 1966, 29% of respondents felt alienated. It reached an all-time high of 62% in 1983, then improved during the 1980s economic expansion. It has risen rapidly in the past few years, increasing from 54% in 1988 to 61% in 1990. Information on the Harris Poll Alienation Index is taken from Uehling, "All-American Apathy," 31.

18. The information on confidence in institutions is taken from George Gallup, Jr., "Confidence in Institutions—Trend," *The Gallup Report,* Report No. 279, December 1988, 30, and George Gallup, Jr., "Confidence in Institutions—Trend," *The Gallup Report,* Report No. 288, September 1989, 21. Respondents were asked, "I am going to read you a list of institutions in American society. Please tell me how much confidence you, yourself, have in each one—a great deal, quite a lot, some, or very little." What is reported in the text is those individuals indicating "a great deal" or "quite a lot."

19. For definitions of option and existence values, see, Gardner Brown, "Valuation of Genetic Resources," in Gordon Orians *et al.,* eds., *The Preservation and Valuation of Genetic Resources* (Seattle: University of Washington Press, 1990), 214–215. Also, for option values see Burton Weisbrod, "Collective Consumption Services of Individual-Consumption Goods," 78 *Quarterly Journal of Economics* 471–477 (1964). For existence value see John Krutilla, "Conservation Reconsidered," 57 *American Economic Review* 777–786 (1967).

20. Timothy Egan, "National Parks: An Endangered Species," *The New York Times,* May 27, 1991, 7.

21. See Timothy Egan, "Forest Service Abusing Role, Dissidents Say," *The New York Times,* March 4, 1990, 27.

22. See, for example, Brad Knickerbocker, "Everything's Not OK at the Western Cattle-Ranching Corral," *Christian Science Monitor,* February 28, 1991, 10; and U. S. General Accounting Office, *Rangeland Management: Forest Service Not Performing Needed Monitoring of Grazing Allotments* (Washington, D.C.: U. S. Government Printing Office, May 16, 1991), Report No. RCED-91-148.

23. Egan, "National Parks: An Endangered Species," 7.

24. Michael deCourcy Hinds, "Cries of Poverty in Cradle of Liberty," *The New York Times,* October 14, 1990, p. I-22.

25. Egan, "National Parks: An Endangered Species," 7.

26. Memorandum from the Region One Forest Supervisors to Chief Dale Robertson, "An Open Letter to the Chief from the Region One Forest Supervisors," November 1989, 1–2. Portions of this letter were reprinted in the national

media early the next year. See Egan, "Forest Service Abusing Role, Dissidents Say," 27.

27. "SUNBIRD: Feedback to the Chief," unpublished electronic message, 1989.

28. Indications of problems within the national forest system also came from the regional forester level—some of the highest line officers within the agency. In a well-publicized 1991 flap that highlighted the politicization of agency decisionmaking, Region 1 Regional Forester John Mumma, the only wildlife biologist to attain the regional forester rank in FS history, was removed from his position allegedly for refusing to meet timber quotas that in his view would violate federal environmental laws, even after he repeatedly made his case to FS Chief Dale Robertson. Scott Sonner, "'Tip of iceberg' in park scandal: Two officials cite White House pressure to disregard environmental laws," *The Denver Post,* September 25, 1991, 1A. While the Chief disputed Mumma's explanation of the situation, he acknowledged in Congressional testimony that logging levels might have to be revised downward to meet environmental concerns. Scott Sonner, "Chief denies pressure to log national forests," *The Denver Post,* September 27, 1991.

29. Steering Committee of the 75th Anniversary Symposium, *National Parks for the 21st Century: The Vail Agenda,* Washington, D. C., April 1992, 9.

30. Steering Committee of the 75th Anniversary Symposium, *National Parks for the 21st Century,* 12.

31. See, for example, "FWS Confirms Harmful Refuge Uses," *Land Letter,* February 20, 1991, 5; and "FWS Director Defends Refuge Management," *Land Letter,* April 1, 1991, 6–7.

32. See, for example, Wendy Smith Lee, "The National Wildlife Refuge System," in DiSilvestro, *Audubon Wildlife Report 1986,* 440–447. Concentrations of selenium up to 4,200 parts per billion were found, while the Environmental Protection Agency recognizes 10 ppb as a safe level for drinking water.

33. Information on Kesterson's current status was taken from a talk by Dr. Ted LaRoe, U. S. Fish and Wildlife Service, at the School of Natural Resources and Environment, The University of Michigan, December 7, 1992.

34. Lee, "The National Wildlife Refuge System," 441.

35. Lee, "The National Wildlife Refuge System," 446.

36. Data on population declines of mallards, salmonids, striped bass, lake trout, oysters, nongame birds, and wetland and prairie grasslands are taken from a talk by Dr. Ted LaRoe, U. S. Fish and Wildlife Service, at the School of Natural Resources and Environment, The University of Michigan, December 7, 1992.

37. U. S. Office of Management and Budget, "Historical Tables," *The Budget of the United States Government: Fiscal Year 1993—Supplement* (Washington, D.C.: U. S. Government Printing Office, 1992), Part 5, 60–61, 65–66.

38. The increase represents a growth rate of approximately 2.7% per year, considerably less than the inflation rate, which averaged roughly 4.5% per year over this period. Ibbotson Associates, *SBBI, Stocks, Bonds, Bills, and Inflation, 1992 Yearbook* (Chicago IL: Ibbotson Associates, 1992).

39. Cynthia Lenhart, "Federal Fish and Wildlife Program Budgets," in William J. Chandler, ed., *Audubon Wildlife Report, 1988/89* (San Diego, CA: Academic Press, 1988), 767–768. For example, from 1981 to 1987, while the FS's budget increased from $2.113 to $2.324 billion, its value in constant dollars declined 14%. The NPS's budget declined approximately 6% in real terms, while the BLM's evidenced roughly a 5% decrease in purchasing power over the 1981–1987 period. Katherine Barton, "Federal Fish and Wildlife Agency Budgets," in Roger L. DiSilvestro, *Audubon Wildlife Report, 1987,* (Orlando FL: Academic Press, 1987), 333, 340, 346.

40. Barton, "Federal Fish and Wildlife Agency Budgets," 74, 324.

41. Barton, "Federal Fish and Wildlife Agency Budgets," 337.

42. Barton, "Federal Fish and Wildlife Agency Budgets," 344.

43. U. S. General Accounting Office, *Wildlife Protection: Enforcement of Federal Laws Could Be Strengthened* (Washington, D.C.: U. S. Government Printing Office, 1991), Report No. RCED-91-44.

44. Michael Mantell, Phyllis Myers, and Robert B. Reed, "The Land and Water Conservation Fund: Past Experience, Future Directions," in William J. Chandler, *Audubon Wildlife Report, 1988/89* (San Diego, CA: Academic Press, 1988), 257–9.

45. Barton, "Federal Fish and Wildlife Agency Budgets," 349.

46. U. S. Bureau of the Census, "Table No. 516. Federal Civilian Employment and Annual Payroll, By Branch: 1970 to 1990," *Statistical Abstracts of the United States: 1992,* 112th ed., (Washington, D.C.: U. S. Government Printing Office, 1992), 331.

47. U. S. Bureau of the Census, "Table No. 516," 331.

48. 1985 data are from Barton and Fosburgh, "The U. S. Forest Service," 33; 1988 data are from Cynthia Lenhart, "Federal Fish and Wildlife Program Budgets", in William J. Chandler, ed., *Audubon Wildlife Report, 1989/90* (San Diego, CA: Academic Press, Inc., 1989), 553.

49. Lenhart, "Federal Fish and Wildlife Program Budgets", *Audubon Wildlife Report, 1989/90,* 558.

50. U. S. General Accounting Office, *Rangeland Management: BLM Efforts to Prevent Unauthorized Livestock Grazing Needs Strengthening* (Washington, D.C.: U. S. Government Printing Office, 1991), Report No. RCED-91-17.

51. "State Parks Show Effects of Reductions in Budgets," *The New York Times,* July 8, 1990, section 12 , 16.

52. Total funding for the department was $278.6 million in 1986, rising in absolute terms to $301.9 million in 1991. Putting these figures in constant (1991) dollars, however, reveals a decline from $347.7 million in 1986 to $301.9 million in 1991—a thirteen percent drop in purchasing power from the 1986 levels. Raw data were from a personal communication with Dennis R. Adams, Chief, Office of Budget and Federal Aid, Michigan Department of Natural Resources, December 1992.

53. Lenhart, "Federal Fish and Wildlife Program Budgets", *Audubon Wildlife Report, 1989/90,* 558.

54. Indeed, one of the early endangered species program managers noted that at the level of staffing prevalent in the 1970s, the job of listing alone would take some 6,000 years to complete. See discussion in Yaffee, *Prohibitive Policy,* 71–72.

55. Barton, "Federal Fish and Wildlife Agency Budgets," 330.

56. U. S. Department of Interior, Fish and Wildlife Service, "Box Score: Listings and Recovery Plans," 17 *Endangered Species Technical Bulletin* 16, March-August 1992.

57. U. S. Office of Management and Budget, *The Budget of the United States Government, FY1993, Supplement* (Washington, D.C.: U. S. Government Printing Office, 1992), Part Five, 14, 97.

58. See, for example, John M. Berry, "The Economy: The Recovery is Now Squarely in Clinton's Court: The rising public expectations may be tough to satisfy," *The Washington Post National Weekly Edition,* January 11–17, 1993, p. 19.

59. See, for example, Jeffrey M. Berry, *The Interest Group Society,* 2nd edition (HarperCollins Publishers, 1989), pp. 18–19.

60. Quoted in Berry, *The Interest Group Society,* 17.

61. In 1974, 608 PACs were recognized by the Federal Election Commission, growing to more than 4700 in 1990. In 1976, PACs gave $12.5 million to candidates running for federal office; in 1990, that amount grew to $159 million. In 1976, PACs were the source of only one-quarter of the campaign revenues for winners of seats in the U. S. House of Representatives; in 1990, PACs represented almost half of the campaign revenues used by House winners. Larry Makinson, *Open Secrets: The Encyclopedia of Congressional Money and Politics,* 2nd edition (Washington, D.C.: Congressional Quarterly, 1992), 6, 10.

62. Personal communication with Michael McCloskey, Chairman, Sierra Club, November 6, 1992.

63. Quoted in Michael Oreskes, "American Politics Loses Way As Polls Displace Leadership," *The New York Times,* March 18, 1990, 22.

64. Quoted in Oreskes, 22.

65. As the Boomers age and live longer, they will continue to demand their share of the public sector pie, including the costly programs of social security, federal retirement, and assisted health care.

66. While previous waves of empowerment were largely white male-dominated, the ascension to power of other social groups, while laudatory and legitimate, will tend to increase the fragmentation of the political landscape.

67. Term limitations will tend to promote fragmentation and diminish institutional capacity, as more individuals become involved in decisionmaking arenas, as institutional memory declines, and as the stability provided by pre-existing relationships is lost. Ironically, what term limits would ensure is that interest group lobbyists and bureaucrats would be the most stable, and knowledgeable, components of the policymaking apparatus, elements that are the most interest-specific of all.

68. The movement includes regional or national groups such as the Wilderness Impact Research Foundation, the National Council for Environmental Balance,

and the Center for the Defense of Free Enterprise, and more localized groups like the Idaho-based Blue Ribbon Coalition and the Oregon-based Public Lands Coalition. See, for example, Margaret L. Knox, "The Wise Use Guys," 2 *Buzzworm: The Environmental Journal* 6:30–36, November/December 1990.

69. See, for example, Sharon Begley and Patricia King, "The War Among the Greens: Grass-roots environmentalists charge 'sellout'," *Newsweek,* May 4, 1992.

70. For example, in December, 1990, the Council ordered EPA to drop its proposed ban on the incineration of lead batteries, the source of sixty percent of the lead in U. S. garbage, and the likely contributor to some 400,000 babies born each year with high blood levels of lead. The council also killed a regulation that would have required cities with solid waste incinerators to recycle a quarter of their trash, and proposed more than 100 changes in Clean Air Act regulations. See Paul Rauber, "Industry's Friend in High Places," 76 *Sierra* 5:42–43, September/October 1991; and Tim Beardsley, "Executive Fix: Is the Competitiveness Council Overstepping Its Bounds?" 266 *Scientific American* 3:106, March 1992.

71. Material on the Competitiveness Council's proposed revision of the wetlands manual is taken from "Testing of Wetlands Manual Shows Major Losses," 10 *Land Letter* 32:1–2, December 1, 1991.

72. Witness the defeats of President Jimmy Carter in 1980, who had argued that national sacrifice was needed to overcome the country's addiction to foreign oil, and Walter Mondale in 1984 who told voters during the presidential campaign that he would raise taxes if elected.

73. The FS was acting under a consent decree that evolved out of a lawsuit in California (Region 5). The decree committed them to increasing significantly the percentage of women in their workforce, and the agency extended this mandate as a goal for the entire national forest system.

Chapter 12

1. See, for example, Wondolleck, *Public Lands Conflict and Resolution.*

2. Philip J. Harter, "Dispute Resolution and Administrative Law: The History, Needs, and Future of a Complex Relationship," 29 *Villanova Law Review* 1393, 1984.

3. There were a couple of attempts to institute collaborative decisionmaking in the owl case, none of which succeeded. One of these was created in the 1989 appropriations bill that provided a one year compromise (Section 318). It mandated the creation of multiparty working groups to advise the FS on timber sales on a forest-by-forest basis. By and large, these working groups did their job, though they could hardly be called good models of dispute resolution. Their task was prescribed and constrained from the start, the number of members was prescribed by law, and the participants were preselected by the FS. Organizations that were not represented viewed the working groups with extreme suspicion.

Representatives of some of the mainstream environmental groups involved throughout the case claimed not to know some of the "environmental representatives" selected by the forest supervisors to participate in the working groups.

Chapter 13

1. A designation process as is described in the text would not be easy, but there are ways to make it work. An independent commission, whose recommendations could be vetoed only in their entirety by a two-thirds vote of the Congress, could provide the right kind of vision and political cover needed to carry out what would be perceived as a major change, with associated sociopolitical impacts. Such a process has been used successfully to identify and designate military bases for closure.

2. Lecture by Robert Repetto, World Resources Institute at the School of Natural Resources and Environment, The University of Michigan, February 10, 1993.

3. The concept of a fixer was described by Bardach in *The Implementation Game* as a benevolent high official who kept pressing for effective implementation. I expanded on the fixer concept in *Prohibitive Policy* to include nongovernmental organizations as well. Indeed, at a recent conference on endangered species management held in Ann Arbor, Michigan on January 8–9, 1993, most of the participants involved in specific cases of endangered species recovery noted that the primary reason that an action was achieved was due to the involvement of a nongovernmental fixer—in most cases, an environmental group that kept pressing the federal agencies for action.

4. While history tells us that ideas have always been a significant agent of change, changes in global political power and economic structure may make them more so for the near-term future. The dispersed political power inherent in the post-Cold War world make dominance by force and ideological allegiance less likely than it has been for some time. At the same time, an interconnected economic system where raw materials, labor, and capital can move fairly rapidly across space makes national borders and the economic wealth within them less relevant and controlling. It is possible to go too far with this argument, but here is where it leads: The direction of future societies, and the power accruing to different elements within, will be based more on the efficacy and viability of the ideas that they generate, not the least of which are environmentally sound ways to promote sustainable development. The United States has provided a model of governance and economic well-being that is viewed by many in the rest of the world as something to strive for. Innovative ways to respond to human needs while simultaneously being effective at managing natural resources and safeguarding environmental quality can be a portion of our national legacy. Using public natural resources as a testing ground for such innovations can help make this possible.

Glossary of Acronyms

ASQ	Allowable Sale Quantity
BLM	U.S. Bureau of Land Management
BUMP	Biological Unit Management Plan
CDF&G	California Department of Fish and Game
CHA	Critical Habitat Area
CHEC	Cascade Holistic Economic Consultants
DSEIS	Draft Supplemental Environmental Impact Statement
EIS	Environmental Impact Statement
ESA	Endangered Species Act (1973 and amendments)
FEMAT	Forest Ecosystem Management Assessment Team
FLPMA	Federal Land Policy and Management Act (1976)
FS (or USFS)	U.S. Forest Service
FSEIS	Final Supplemental Environmental Impact Statement
FWCA	Fish and Wildlife Coordination Act
FWS	U.S. Fish and Wildlife Service
GAO	U.S. General Accounting Office
HCA	Habitat Conservation Area
HMP	Habitat Management Plan
IDT	Interdisciplinary Team
ISC	Interagency Scientific Committee (the Jack Ward Thomas committee)
MIS	Management Indicator Species
MMR	Minimum Management Requirement
MVP	Minimum Viable Population
NEPA	National Environmental Policy Act (1969)
NFMA	National Forest Management Act (1976)
NFPA	National Forest Products Association
NPS	U.S. National Park Service
NWF	National Wildlife Federation
O&C	Oregon and California Railroad lands (managed by BLM)
ODFW	Oregon Department of Fish and Wildlife
ODLCD	Oregon Department of Land Conservation and Development
OESTF	Oregon Endangered Species Task Force
ONRC	Oregon Natural Resources Council

OSPIRG	Oregon Student Public Interest Research Group
OSU	Oregon State University
OWC	Oregon Wilderness Coalition; also Oregon Wildlife Commission
OWIWC	Oregon-Washington Interagency Wildlife Committee (grew out of OESTF)
PAC	Political Action Committee
RPA	Resources Planning Act (1974)
SCLDF	Sierra Club Legal Defense Fund
SEIS	Supplemental Environmental Impact Statement
SOHA	Spotted Owl Habitat Area
SOMA	Spotted Owl Management Area
SOMP	Spotted Owl Management Plan
UWD	Umpqua Wilderness Defenders
WDOW	Washington State Department of Wildlife

Index

Accountability, diminished, 190–91
Actors in American decisionmaking processes, 207–33
  elected officials, 212–14
  individual personalities, 207–10
  interest groups, *see* Interest groups
  media, 210–12
  science and scientists, 231–33
Adaptive management approach, 82, 103, 145–46, 148, 149, 253–55, 354
Administratively withdrawn areas, 145
Advocate science, 51–52, 171
Allen, Harriet, 110
Allowable Sale Quantity (ASQ), 95, 103, 242
  timber summit of 1989 and, 117–21
Alternative dispute resolution, 224, 327, 332–33, 342, 344
  data negotiation, 338, 342
  SEIS analysis and, 89–90, 273–74
American decisionmaking processes, xix–xx, 184–206
  actors in, *see* Actors in American decisionmaking processes
  challenging administrative choices, 195–96
  competitive perspective, 197–98
  creative solutions and, 196–99
  crisis management, 199, 201–202
  fragmented policy processes, *see* Policy processes, fragmented
  ideal process, 184–85
  incremental decisionmaking, 201
  planned change, 200–201
  politics and, 198–99, 241–43
  process- and not outcome-orientation, 195
  quality of information and, 202–206
    see also Information
  resource management and, 195
  short-range perspectives, 199–201
  stability of, 199–200
  zero sum preconception, 196

American Farm Bureau Association, 313
American Forest Resource Alliance (AFRA), 125, 174–75, 313
American Forestry Association, 174
American Museum of Natural History, 77, 98
American Ornithologists' Union, 99
Ancient Forest Alliance, 119–20, 222, 223
Ancient Forest Protection Act, 126
"Ancient Forests: Rage Over Trees," 226, 231
Apollo XI, 10
Appalachian Mountain Club, 200
Associated Press, 114
Association of Forest Service Employees for Environmental Ethics (AFSEE), 322
Atkins, Chester, 209
AuCoin, Les, 105, 118, 119, 230
  timber summit and, 118–21
Audubon Blue Ribbon Panel of scientists, 98–99, 102, 108

Babbitt, Bruce, 140
Baby-Boomers, 289–90, 323
Bailey, Betsy, 88
Barrowclough, George, 77, 98
Beasley, Lamar, 111
Beuter, John, 43, 244
Biological diversity, 38, 144, 162, 250
  evolution of concept of, 58–60
  as objective of resource policy, 358–60
Biological Unit Management Plans (BUMPs), 32
Blomquist, Jim, 139
Blum, Joseph, 50
Bogy II timber sale, 110
"Bottom line," 181
Boundary-spanning institutions, decline of, 311–12
Bruce, Charles, 20, 25, 29
Budgets, decline in agency, 305–309

ABOUT THE AUTHOR

Steven Lewis Yaffee is a faculty member in the School of Natural Resources and Environment at The University of Michigan where he teaches courses in natural resource policy and administration, negotiation skills, American environmental history, and biodiversity and public policy. His research focuses on understanding and improving public decisionmaking processes as they influence the management of natural resources, and exploring the behavior of administrative agencies and interest groups as they are involved in implementing public policies. He has worked for more than fifteen years on federal endangered species policy, and is the author of *Prohibitive Policy: Implementing the Federal Endangered Species Act* (Cambridge, MA: MIT Press, 1982).

Dr. Yaffee received his Ph.D. in 1979 from the Massachusetts Institute of Technology in environmental policy and planning and has earlier degrees in natural resources. He has taught at MIT and the Kennedy School of Government at Harvard and has been a researcher at the Oak Ridge National Laboratory and The Conservation Foundation/World Wildlife Fund.